Stafford Library
Columbia College
1001 Rogers Street
Columbia, Missouri 65216

WITHDRAWN

Arithmetic for Teachers

With Applications and Topics from Geometry

GARY R. JENSEN

American Mathematical Society
Providence, Rhode Island

Partially supported by a grant from the Howard Hughes Medical Institute.

2000 *Mathematics Subject Classification.* Primary 97–01, 97B50, 97D50, 00A06.

For additional information and updates on this book, visit
www.ams.org/bookpages/foa

The St. Louis Post-Dispatch article quoted on pp. 361–362
is reprinted with permission of the Associated Press.

Cover photograph courtesy of Rick Antuono.

Library of Congress Cataloging-in-Publication Data
Jensen, Gary R., 1941–
  Arithmetic for teachers : with applications and topics from geometry / Gary R. Jensen.
    p. cm.
  Includes bibliographical references and index.
  ISBN 0-8218-3418-5 (alk. paper)
    1. Arithmetic—Study and teaching (Elementary)—United States. 2. Geometry—Study and teaching (Elementary)—United States. I. Title.

QA135.6.J46 2003
372.7′2—dc22   2003060517

  Copying and reprinting. Individual readers of this publication, and nonprofit libraries acting for them, are permitted to make fair use of the material, such as to copy a chapter for use in teaching or research. Permission is granted to quote brief passages from this publication in reviews, provided the customary acknowledgment of the source is given.
  Republication, systematic copying, or multiple reproduction of any material in this publication is permitted only under license from the American Mathematical Society. Requests for such permission should be addressed to the Acquisitions Department, American Mathematical Society, 201 Charles Street, Providence, Rhode Island 02904-2294, USA. Requests can also be made by e-mail to reprint-permission@ams.org.

  © 2003 by the American Mathematical Society. All rights reserved.
  The American Mathematical Society retains all rights
  except those granted to the United States Government.
  Printed in the United States of America.
  ∞ The paper used in this book is acid-free and falls within the guidelines
  established to ensure permanence and durability.
  Visit the AMS home page at http://www.ams.org/
  10 9 8 7 6 5 4 3 2 1   08 07 06 05 04 03

Contents

Preface	vii
Chapter 1. Counting	1
§1.1. Sets	1
§1.1 Exercises	11
§1.2. Whole Numbers	13
§1.2 Exercises	22
§1.3. Measurement	24
§1.3 Exercises	37
Chapter 2. Whole Number Arithmetic	39
§2.1. Addition	39
§2.1 Exercises	45
§2.2. Subtraction	47
§2.2 Exercises	56
§2.3. Multiplication	59
§2.3 Exercises	73
§2.4. Division	76
§2.4 Exercises	88
§2.5. Dividing the Whole	90
§2.5 Exercises	98
Chapter 3. Whole Number Computation	101
§3.1. Numeration	102

§3.1 Exercises	111
§3.2. Computing Sums	111
§3.2 Exercises	117
§3.3. Computing Differences	118
§3.3 Exercises	124
§3.4. Computing Products	125
§3.4 Exercises	133
§3.5. Computing Quotients	135
§3.5 Exercises	150
Chapter 4. Number Theory	153
§4.1. Factors and Primes	154
§4.1 Exercises	165
§4.2. Euclid's Algorithm	168
§4.2 Exercises	176
§4.3. Unique Prime Factorization	177
§4.3 Exercises	186
Chapter 5. Rational Numbers	189
§5.1. Representation by Fractions	190
§5.1 Exercises	196
§5.2. Comparing Fractions	196
§5.2 Exercises	203
§5.3. Addition of Rational Numbers	205
§5.3 Exercises	210
§5.4. Subtraction of Rational Numbers	211
§5.4 Exercises	218
§5.5. Multiplication of Rational Numbers	219
§5.5 Exercises	227
§5.6. Division of Rational Numbers	229
§5.6 Exercises	238
Review Exercises	240
§5.7. Ratio and Proportion	241
§5.7 Exercises	253
Chapter 6. Decimals	257
§6.1. Decimal Expansions	257

§6.1 Exercises	265
§6.2. Arithmetic of Decimals	265
§6.2 Exercises	272
§6.3. Repeating Decimals	273
§6.3 Exercises	280
§6.4. Arithmetic of Infinite Decimals	280
§6.4 Exercises	293
§6.5. Archimedes's Estimate of π	294
§6.5 Exercises	300
§6.6. Percentage	300
§6.6 Exercises	308
§6.7. Exponents and Scientific Notation	309
§6.7 Exercises	313
Chapter 7. Integers	315
§7.1. Negative Numbers	315
§7.1 Exercises	318
§7.2. Integer Addition	318
§7.2 Exercises	325
§7.3. Integer Subtraction	326
§7.3 Exercises	332
§7.4. Multiplication of Integers	332
§7.4 Exercises	340
§7.5. Division of Integers	341
§7.5 Exercises	344
§7.6. Negative Rational Numbers	345
§7.6 Exercises	345
Chapter 8. Clock Arithmetic	347
§8.1. Clock Addition, Subtraction, and Multiplication	348
§8.1 Exercises	353
§8.2. Clock Division	354
§8.2 Exercises	359
Chapter 9. RSA Encryption	361
§9.1. A Preliminary Example	362
§9.2. Decoding	364

§9.2 Exercises	368
§9.3. Nonprime Number of Hours	368
§9.3 Exercises	371
§9.4. Raising Numbers to Large Powers in $\mathbf{Z}_p$	372
§9.4 Exercises	374
§9.5. Signatures	374
§9.5 Exercises	376
Bibliography	377
Index	379

Preface

This is a textbook for a college course in mathematics for prospective elementary school teachers. From the mathematics curriculum of grades K-6, it covers the foundations of arithmetic, a selection of topics from geometry, and a wide range of applications. A method of visualizing a problem through the use of line segment diagrams is stressed throughout the text. Being a math textbook for prospective teachers, this book always emphasizes explanations: Of the concepts, of how to solve a problem and of how the concepts relate to the solution of a problem.

Most topics are presented in greater depth than is usual for an elementary school classroom. The elementary school teacher will need this greater depth of knowledge while explaining concepts, computation procedures, and how to solve problems. Depth of knowledge also helps the teacher to see how different topics relate to each other. It can also give insight into where children might have difficulties.

Students from the United States have ranked poorly on international mathematics tests. Singapore and China have ranked at or near the top. It is possible that this difference in ranking is due in part to a difference in mathematical preparation and expertise of the teachers. Liping Ma's 1999 book *Knowing and Teaching Elementary Mathematics* [**Ma99**] gives evidence that elementary school teachers in China have a deeper level of mathematics preparation than their counterparts in the United States. Indeed, she provides evidence that excellent teaching of mathematics at the elementary school level requires that the teacher be an expert in school mathematics. As she put it, the goal is for teachers to achieve Profound Understanding of Fundamental Mathematics, or PUFM for short. To achieve PUFM is to

> Know how and know why.

She also makes the case that prospective teachers can become experts in school arithmetic if they are taught the foundations.

Preliminary versions of this book have worked well for students in my mathematics course for elementary education majors at Washington University. Many of these students have had no mathematics beyond high school algebra. Although some students bring to the course a long standing fear of mathematics, they all soon discover that they can understand explanations which start at the beginning and are given in everyday English. They begin to enjoy learning the foundations of arithmetic. They have known arithmetic since childhood, but usually not well enough to understand why the basic operations work, how you would go about confirming that they work, or how you would explain to others, especially to school children, why they work. For these students, the key to learning this material is getting all the facts, expressed at the right level, with all the details, and at a pace slow enough to allow proper absorption. This book attempts to provide this key.

The presentation of material in this book is in the old-fashioned style of definition, theorem, proof used in Euclid's *Elements*. The number line is the focus of every development. The first step towards understanding a concept is to have a clear definition of it. For example, addition of whole numbers must be defined in some specific way. Having chosen the definition, we then prove as theorems the equivalence of alternative definitions. With this foundation, the students understand that different ways of thinking about addition, as the context dictates, are all equivalent formulations of the same concept. Addition is not context-dependent. A single, correct definition gives us firm ground to fall back to at times of puzzlement or doubt. It also gives us a base from which to start the argument of a proof.

The word theorem is seldom seen in elementary mathematics books. No apology for its use should be required, as mathematics is about theorems. If some property is true, then some other property is true. That is a theorem. Any teacher, any educated person, should understand the meaning of theorem, as well as the variations on its statement contained in the converse and contrapositive. There is no merit in banishing these concepts from all but advanced mathematics courses. School arithmetic comprises a body of theorems, and school teachers, as well as students, are capable of understanding them, provided that the theorems are presented at the correct level and in sufficient detail.

The mention of proofs scares mathematics students at all levels. But a proof is nothing but an explanation for why a theorem is true, and everybody wants to know that. Students' antipathy to proofs comes from their experience of proofs which don't explain anything to them. Formal, symbol-laden arguments, based on imprecise definitions, do not explain anything. Believing that we learn inductively, from special cases and examples to the general case, I have presented most proofs in the form of a special case that exhibits the general argument. For example, one does not prove that $2 + 3 = 3 + 2$ by showing that each side equals 5, but by explaining that each side is the number of elements in the union of disjoint sets, one containing 2 elements and the other containing 3 elements, and such a union is independent of order. If that argument is understood for this special case, it is understood for all cases.

A proof is also an exercise in problem solving. The problem is, "why is this theorem true?" It frequently happens that understanding why the theorem is true is the key to understanding how the theorem is applied to problems.

Problem solving is given a position of highest priority in today's elementary school mathematics curriculum. It is emphasized here, too, always within the context of applying the mathematics concepts being taught. A realistic goal in problem solving is to learn how to apply these concepts to a wide range of problems. As the concepts are mastered and the ability to apply them is learned, a mathematical way of thinking about problems will develop.

I have made special effort to avoid the use of formulas in the solution of problems, as these lead too easily to rote solution, without the need for any thought or understanding. The emphasis is on a visualization achieved through the use of line segment diagrams. These diagrams are not easy to construct, as they depend on a complete understanding of what the problem is and of how the arithmetic and geometric concepts enter into its solution. The diagrams become an excellent means of explaining the solution of a problem. In many cases no additional words are needed. The diagram explains it all.

Each section of the text is designed to be the lesson for a one and a half hour class. Oral exercises occur frequently in each section as a means to keep the student involved with the material. Students should try to answer these oral exercises while they read the text, and these exercises should be discussed in class.

The written exercises at the end of each section are of several types. Some provide activity for the student to enter into the details of the proofs and of the concepts. Some are exercises which might be seen in an elementary school classroom. Others are exercises meant to challenge the prospective teacher's grasp of the concepts and ability to apply these concepts to solve problems. Students are urged to make line segment diagrams whenever possible.

In my experience, prospective elementary school teachers need a great deal of instruction on how to write the solutions to many of the written exercises. They have had very little experience in writing proofs, or of explaining what they are doing in a mathematical problem. Students will not begin this course possessing the ability to write this kind of explanation. Learning this ability is one of the goals of the course.

For this text, students will need paper (often graph paper), pencil, straight edge, protractor, and compass. Calculators should be avoided until Chapters 8 and 9, where calculators that do long division (with remainder) and factoring would be appropriate.

The first chapter introduces the concept of whole numbers. Experiences with sets of actual objects form the foundation for understanding whole numbers as the number of objects in a set. Counting forms the foundation for understanding the sequential nature of whole numbers—they occur in a line, they have an order. The number line is introduced as a means to picture the order of the whole numbers. In order to provide a vocabulary for talking about numbers and for writing numbers, base ten numeration is introduced as an application of counting to ten. The chapter concludes with an introduction to measurement, with an emphasis on comparing one quantity to another.

Chapter two introduces the four arithmetic operations on whole numbers. Each operation is given a specific definition and then various interpretations of an operation are presented as theorems. After a thorough discussion of the definition of an operation, and its several manifestations, the essential properties of the operation are described, proved, and applied. These properties form the heart of an understanding of arithmetic and they form the heart of this textbook. These properties can be tedious to learn, but they are indispensible for understanding the methods of computation and the applications. This chapter concludes with an introduction of fractions in the context of dividing objects into equal parts.

Chapter three covers the standard methods for calculating sums, differences, products, and quotients in the base ten numeration system. It begins with a more detailed presentation of our numeration system.

Chapter four contains a brief introduction to the number theory that is needed for a thorough understanding of fractions and their arithmetic. The heart of this chapter is Euclid's algorithm for finding the greatest common factor of two numbers.

Chapter five introduces the rational numbers as an extension of the whole numbers achieved by taking fractions of the unit segment on the number line. In this way the rational numbers become points on the number line and include the points which are the whole numbers. Major attention is paid to understanding how different fractions can represent the same point on the number line. The four arithmetic operations are then extended to the set of rational numbers. Applications of these ideas abound as becomes evident from the number of exercises appearing in this chapter. The chapter concludes with a section on ratio and proportion.

Chapter six describes how the base ten numeration system can be extended to the right of the decimal point so as to include a decimal expansion of many fractions. Most rational numbers have infinite repeating decimals. Arithmetic with infinite decimals requires the idea of rounding off a decimal to some place. Infinite nonrepeating decimals are introduced as a process of making ever more precise the location of some point on the number line. It is shown how an infinite nonrepeating decimal arises in the attempt to find a decimal expansion of the square root of a prime number. The chapter concludes with sections on percentages and exponents.

Chapter seven presents the idea of an integer as a means of expressing a magnitude and direction. After extending the set of whole numbers to include the negative integers, the four arithmetic operations are extended to all of the integers. Although the idea of defining negative integers by extending the number line to the left of zero is simple, and the extension of the arithmetic operations to this larger set is also simple, it remains a challenging task to verify the desired properties of these operations.

Chapter eight presents clock arithmetic as a significant application of school arithmetic. It involves all the concepts of arithmetic, in a setting which requires a thorough rethinking of what things mean. Their study involves serious mathematical thinking. In clock arithmetic, reciprocals of numbers occur without the construction of fractions, and subtraction can always be performed, without the construction of new numbers. Elementary number theory finds interesting applications in determining when numbers have reciprocals or not.

Chapter nine presents an elementary introduction to the public key encryption called RSA encryption, to provide a significant real world application of clock arithmetic.

Neither clock arithmetic nor RSA encryption is part of the K-6 curriculum. Both involve serious, multistep mathematical thought. They also involve large amounts of computation: additions, subtractions, multiplications, divisions with remainder, exponentiation, prime factorization. A large number of middle school children would find clock arithmetic and RSA encryption a lot of fun. Once they grasp the ideas and work a few examples by hand with a calculator, they can write simple routines for carrying out the encryption by computer. A great deal of mathematically significant activity can be based on clock arithmetic. Such activity will never enter the schools if the teachers in these schools have not previously mastered the material themselves. Enrichment courses and gifted student programs should use mathematically significant material such as clock arithmetic and RSA encryption.

My one semester course covers only the first six chapters of the book. It is important to resist the temptation to rush through topics that at first glance seem too easy or too full of tedious detail. Once important details are skipped, students will expect to skip all details and they will have lost their structure of a solid foundation. The integers, clock arithmetic, and RSA encryption, as well as important topics in geometry, probability, and statistics not included in this text, must wait for a second semester course. Ambitious students can study the integers, clock arithmetic, and RSA encryption for extra credit. A recent report of the Conference Board of the Mathematical Sciences, *Mathematical Education of Teachers Project* [**otMS01**], recommends three semester courses in mathematics content for prospective elementary school teachers. I support this recommendation.

This text can also provide prospective and in-service teachers with the first step in building their mathematics background to a level that will enable them to succeed in elementary science courses. Elementary school science programs struggle from many teachers' inadequate background in science. Insufficient math background is, in turn, the reason why many teachers have taken few, if any, science courses. The solid foundation in elementary mathematics provided in this book will prepare teachers for algebra and elementary science courses. A student who has successfully gone through this book will have no trouble going through a book such as *Algebra* by I. M. Gelfand and A. Shen [**GS93**]. Upon the completion of that book, the student would have the mathematical prerequisite for introductory courses in Biology, Chemistry, Geology, and Physics, provided that these courses concentrate on the science to be taught in the schools.

It is my pleasure to acknowledge the influence of several people on this book. The most pervasive influence has come from the students in my Math 266 (formerly Math 366) course over the past five years. Their constant reaction to the way I presented the material resulted in numerous revisions of nearly every paragraph. If this book accomplishes its goal to any degree, it is due to the influence of these students. Their good nature, hard work, and genuine interest in mathematics gave me much-needed incentive to continue writing it. Holly Bernstein provided many valuable suggestions during the semester in which she graded the homework papers for Math 266. It was Roger Howe's 1999 review [**How99**] which introduced me to Liping Ma's book and a subsequent revision in my view on what should be in a mathematics course for elementary education majors. I was also influenced by

Richard Askey's review [**Ask99**] of Ma's book. The same issue of *American Educator* contains an excellent article [**Wu99**] by H. Wu, whose writings, for example [**Wu**, *What is so Difficult about the Preparation of Mathematics Teachers?*], have influenced my thinking on this subject. I have learned much from many conversations on teacher preparation with Tom Parker and Scott Baldridge who are writing the text [**PB00**]. Following their advice, I purchased and read the Singapore mathematics texts for grades 1 through 5 [**Tea99**], whose excellent style of exposition I have tried to emulate in this book. I have benefitted from conversations with Sybilla Beckmann who has recently completed the text [**Bec02**]. She directed me to the web site of Jerry Dancis (jnd@math.umd.edu), which contains valuable information about problem solving in the schools.

For three decades my favorite arithmetic text was Raub's [**Rau77**], first published in 1877 by Porter and Coates, but the Singapore texts show, finally, how it really should be done.

It is a pleasure to thank my wife, Jen, herself an elementary school teacher, and my children, two of whom are elementary school teachers, for their encouragement of this project. Many members of my extended family have entered every aspect of this book.

July 2, 2003, Washington University, St. Louis, MO.

Chapter 1

Counting

1.1. Sets

Set theory and its vocabulary are used in all branches of mathematics. We shall use the language of sets to define whole numbers, the process of counting, and the four operations of arithmetic of whole numbers. This means that the idea of a set is our starting point, something that we already know and understand. We begin with a postulate whose aim is to clarify what we mean by a set.

Postulate 1.1. A *set* consists of designated objects called the *members* or *elements* of the set. The designation must allow us to decide whether or not an object is an element of the set. Two sets being equal means that they consist of exactly the same members.

The elements of a set are designated in a variety of ways. For a set of physical objects at hand, the elements can be exhibited. A set of classroom manipulatives would be an example of this kind. The elements of some sets are designated by a list of words or symbols or pictures, which may either be the elements of the set or name the elements of the set. Eventually we will be concerned with sets of abstractions, such as the set of all whole numbers or the set of all even numbers or the set of all prime numbers.

An example of a set defined by a list is

$$A = \{\square \triangle \bigcirc\}$$

The members of this set are the figures $\square$, $\triangle$, and $\bigcirc$ written on the paper. We use the notation $\square \in A$ to stand for $\square$ *is an element of* A. The notation $\diamond \notin A$ means that $\diamond$ *is not an element of* A.

At an elementary level this set could be designated by the picture

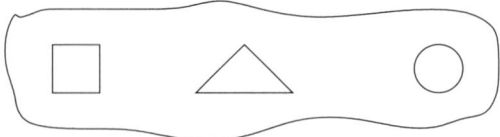

It is important to have something like the loop drawn around the elements of the set to emphasize that the elements of the set are those things contained inside the loop.

Consider the set C consisting of all the coins you have in your right front pants pocket. The elements of this set are physical objects that can be exhibited. If you have no coins in your right front pants pocket (maybe because you don't have a right front pants pocket), then C is the *empty set*, which is the set consisting of no members. It is usually denoted $\emptyset$. In the case when you have no coins in your right front pants pocket, you would say that $C = \emptyset$.

Another example is the set P consisting of all of someone's grandchildren. That description designates the members of P, but we can also list the names of the members of P, as

$$P = \{Nicholas, Joshua, Caitlin, Anthony, Isabella\}.$$

Notice that this list is of the *names* of the members of the set P, not of the members themselves.

Let us agree that the empty set is described as an empty list by $\emptyset = \{\}$.

It is impossible to list all the members of some sets. For example, if $\mathbf{W}$ is the set of all whole numbers, then we cannot list all of its members. When communicating with someone who already knows what the set of all whole numbers is, we could designate it by a list ending with three dots, such as $\mathbf{W} = \{0, 1, 2, \dots\}$, where the three dots indicate continuation of this list *forever* in the manner of the whole numbers as defined below. Another notation we will use is a kind of list, with the designation of what goes into the list coming after a colon. For example, we can write

$$\mathbf{W} = \{n : n \text{ is a whole number}\}$$

which we read as $\mathbf{W}$ is the set of all n such that n is a whole number.

What are the elements of the set $D = \{Leah, Leah, Leah\}$? The definition of D is ambiguous. We could interpret it to mean that D consists of a person named Leah, or of three people each named Leah, or of three copies of the word Leah. This thorny problem of how to distinguish between an object and its name causes constant confusion in discussions of set theory.

Another example of this sort is the set $E = \{1, 1, 2, 2, 2\}$. What would you say are the elements of E?

One way to deal with the naming problem is to attach a secondary name to distinguish different elements with the same name. For example, if the set D above is interpreted to be a set of three people, each named Leah, then we should add their last names to make our designation precise. For example, $D = \{$Leah Jensen, Leah Dutton, Leah Berger$\}$. Following this scheme, if the above defined set E really contains five elements, we could write them as $E = \{1_1, 1_2, 2_1, 2_2, 2_3\}$.

For our purposes, we will try to avoid the problem of confusing the name with the object named by agreeing that any set we define by a list of names or pictures consists of those names and pictures in the list. Under this agreement, the above set $D = \{$Leah, Leah, Leah$\}$ has three members—three copies of the word Leah—and the set $B = \{\square, \square, \square, \square\}$ has four members, the four boxes in this list.

1.1.1. Subsets. All the members of a set can be members of a larger set. Or a given set can be decomposed into smaller sets, such as the set of people in the classroom can be separated into males and females, or into those whose last name initial falls in the range A through J and those whose last name initial falls in the range K through Z. This decomposition of sets forms the foundation for addition and subtraction of whole numbers.

Definition 1.2. A set A is a *subset* of a set B, if every member of A is also a member of B. If A is a subset of B, then write $A \subseteq B$.

If A is a subset of B, but $A \neq B$, then A is a *proper subset* of B, for which we write $A \subset B$.

According to this definition, in order for a set F not to be a subset of a set G, it must have an element which is not an element of G. A subset can be pictured by drawing the elements of the set with a loop around them and then make a dotted loop around the elements of the subset.

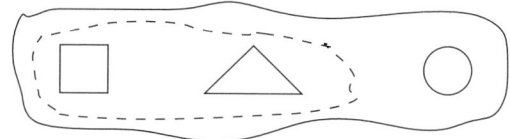

Example 1.3. Any set is a subset of itself, but not a proper subset.

Example 1.4. The set $B = \{2, 3\}$ is a proper subset of $A = \{1, 2, 3\}$, because each element of B is an element of A and $B \neq A$. The set $C = \{1, 2, 5\}$ is not a subset of A, because 5 is an element of C, but not an element of A. Notice that neither A nor B is a subset of C.

Example 1.5. Mathematicians agree that it is convenient to say that the empty set is a subset of any set. This convention is logically consistent with the definition of subset. Namely, suppose A is some given set. By the definition of subset, if $\emptyset$ were not a subset of A, then there must be an element of $\emptyset$ which is not an element of A. But there is no such element because $\emptyset$ has no elements.

Subsets are encountered in the context of classifying objects or people by attributes. Consider the set P of all people in the classroom. The set of people in the classroom whose last name begins with J is a subset of P. What might be the subset of all people in the classroom whose last name begins with X? Other examples would be the subset comprising the left-handed people in the classroom, or the subset comprising the people with red hair in the classroom, or the subset of red-haired, left-handed people in the classroom.

Oral exercise 1.6. Describe some subsets of the set $\{\clubsuit, \spadesuit, \diamondsuit, \heartsuit, a, b, \alpha, \beta\}$.

Example 1.7. List all the subsets of $A = \{1, 2, 3\}$ in a systematic way: all subsets with no elements, all subsets with one element, etc.

No elements	$\emptyset$
One element	$\{1\}, \{2\}, \{3\}$
Two elements	$\{1,2\}, \{1,3\}, \{2,3\}$
Three elements	$\{1,2,3\}$

Counting these subsets, we see that the set A has a total of $1 + 3 + 3 + 1 = 8 = 2 \cdot 2 \cdot 2 = 2^3$ subsets.

Oral exercise 1.8. List all of the subsets of the set $\{\triangle, \square\}$. How many subsets are there?

1.1.2. One-to-one correspondence.

Definition 1.9. A *one-to-one correspondence from* a set A to a set B is an assignment of each element of A to exactly one element of B in such a way that no elements of B are omitted.

A one-to-one correspondence from the set $A = \{\clubsuit, \diamondsuit, \heartsuit, \spadesuit, \square, \triangle\}$ to the set $B = \{\alpha, \beta, \gamma, \delta, \epsilon, \varphi\}$ is

$$\{ \clubsuit \quad \diamondsuit \quad \heartsuit \quad \spadesuit \quad \square \quad \triangle \; \}$$
$$\downarrow \quad \downarrow \quad \downarrow \quad \downarrow \quad \downarrow \quad \downarrow$$
$$\{ \alpha \quad \beta \quad \gamma \quad \delta \quad \epsilon \quad \varphi \; \}$$

A one-to-one correspondence from a set A to a set B can be described as a set of ordered pairs $\{(a,b) : a \in A, b \in B\}$, such that each element of A is in exactly one pair and each element of B is in exactly one pair. Namely, given

1.1. Sets

a one-to-one correspondence from A to B, an element $a \in A$ is assigned to an element $b \in B$, to give the ordered pair (a, b). On the other hand, given the set of ordered pairs, then an element $a \in A$ must be in exactly one such pair (a, b), that tells us to send a to b to make the one-to-one correspondence.

Example 1.10. Any set is in one-to-one correspondence with itself. One such correspondence is obtained by pairing each element with itself.

Example 1.11. If there is a one-to-one correspondence from A to B, then there is a one-to-one correspondence from B to A obtained by reversing the order of each pair. For example, if $A = \{1, 2, 3\}$ and $B = \{\square, \Diamond, \triangle\}$, and if we have the one-to-one correspondence from A to B given by the set of ordered pairs $\{(1, \square), (2, \Diamond), (3, \triangle)\}$, then a one-to-one correspondence from B to A is given by the set of ordered pairs obtained by reversing each pair: $\{(\square, 1), (\Diamond, 2), (\triangle, 3)\}$.

Example 1.12. If there is a one-to-one correspondence from a set A to a set B, and if there is a one-to-one correspondence from the set B to a set C, then there is a one-to-one correspondence from A to C obtained by pairing to an element a of A the element of C which is paired with the element of B which is paired with a. An example will clarify this simple procedure. Suppose A and B are the sets of the preceding example, and suppose that $C = \{\clubsuit, \heartsuit, \spadesuit\}$. Use the one-to-one correspondence from A to B given in the preceding example, and let the one-to-one correspondence from B to C be $\{(\square, \clubsuit), (\Diamond, \heartsuit), (\triangle, \spadesuit)\}$. Then a one-to-one correspondence from A to C is given as follows:

$$
(1.1) \quad
\begin{array}{ccc}
(1, \square) & (\square, \clubsuit) & \rightarrow (1, \clubsuit) \\
(2, \Diamond) & (\Diamond, \heartsuit) & \rightarrow (2, \heartsuit) \\
(3, \triangle) & (\triangle, \spadesuit) & \rightarrow (3, \spadesuit)
\end{array}
$$

If there is a one-to-one correspondence from the set A to the set B, then Example 1.11 shows that there is a one-to-one correspondence from B to A as well.

Definition 1.13. If sets A and B are in one-to-one correspondence, then we say that A and B are *equivalent* sets. We denote equivalent sets by the notation $A \sim B$.

Notice that Example 1.10 shows that any set is equivalent to itself. In particular, if two sets are equal (meaning they have exactly the same members), then they are equivalent. Notice, however, that equivalent sets need not be equal, as $A \sim B$ in Example 1.11, but $A \neq B$. Example 1.12 shows that if $A \sim B$ and if $B \sim C$, then $A \sim C$.

Definition 1.14. If set A is equivalent to set B, then we say that set A has *the same number of elements as* set B. We write $n(A)$ for the set of all sets with the same number of elements as A. We call $n(A)$ the *number of elements of* A.

Strangely enough, the idea of *the same number of elements* does not require the idea of number. In fact, we will use this idea to define whole numbers.

Oral exercise 1.15. Explain the difference between sets being equal and sets being equivalent. Give examples of sets that are equivalent, but not equal.

Definition 1.16. A set is *finite* if its elements can be listed. A set is *infinite* if it is not finite. We agree that the empty set is finite.

By this definition, a set is infinite if any list of its elements is incomplete, that is, it does not contain all of the elements of the set.

This definition is technically unsatisfactory because we have not defined what we mean *to list* the elements of a set. Nevertheless, we have a good enough idea of what it should mean that it seems reasonable to take it as a primitive, undefined concept in our present discussion. We shall give a technically correct definition in the next section in Definition 1.46.

In elementary school mathematics one deals almost exclusively with finite sets, to the extent that it seems pedantic to keep using the adjective finite. But infinite sets do play an important role. We shall see in the next section that the set of all whole numbers, **W**, is infinite. Later we shall study Euclid's argument showing that the set of all prime numbers is infinite.

Definition 1.17. If a finite set A is a proper subset of a set B, or is equivalent to a proper subset of a set B, then we say that the number of elements in A is *less than* the number of elements of B. We also express the same thing by saying that A contains less than B, or that A contains fewer elements than B. Write $n(A) < n(B)$ to denote this (and read it "The number of elements of A is less than the number of elements of B"). We also write $n(B) > n(A)$ to mean that $n(A) < n(B)$. In this case we would also say that B contains more than A.

Example 1.18. To determine whether a room full of children contains the same number of boys as girls, we could pair them off. If each boy is paired with a girl and there are no boys or girls left over, then the set of boys is equivalent to the set of girls and therefore the number of boys (meaning the number of elements in the set of boys) is the same as the number of girls (meaning the number of elements in the set of girls). If there are some boys left over, then the set of girls has been put in one-to-one correspondence with a proper subset of the boys and we conclude that the number of

1.1. Sets

girls is smaller than the number of boys. Observe that we have made the determination without counting.

Oral exercise 1.19. If there are chairs and people in a room, describe how to determine without counting whether there are more, the same number, or fewer chairs than people.

Example 1.20. (1) The set $C = \{\triangle \triangle \triangle\}$ is a proper subset of the set $B = \{\triangle \triangle \triangle \triangle \triangle\}$, so $n(C) < n(B)$.

(2) The empty set is a proper subset of the set C, so $n(\emptyset) < n(C)$.

(3) Which of the following sets contains more?
$$A = \{\square, \square, \square, \square, \square, \square, \square\}$$
$$B = \{\triangle, \triangle, \triangle, \triangle\}$$

Pairing a $\square$ with the $\triangle$ below it, we eventually run out of $\triangle$'s before running out of $\square$'s. From this we conclude that A contains more than B, which we also express as B contains less than A. In symbols, write $n(A) > n(B)$ or $n(B) < n(A)$.

Theorem 1.21. *Suppose A, B, and C are finite sets. If $n(A) < n(B)$ and if $n(B) < n(C)$, then $n(A) < n(C)$.*

Proof. The following example illustrates the proof for the general case. Suppose that

(1.2)
$$A = \{\clubsuit, \diamondsuit, \heartsuit\}$$
$$B = \{a, b, c, d, e\}$$
$$C = \{\spadesuit, \square, \triangle, \bigcirc, \alpha, \beta, \gamma, \delta\}$$

The set of pairs $\{(\clubsuit, a), (\diamondsuit, b), (\heartsuit, c)\}$ defines a one-to-one correspondence from A to a proper subset of B, thus showing that $n(A) < n(B)$.

The pairs $\{(a, \spadesuit), (b, \square), (c, \triangle), (d, \bigcirc), (e, \alpha)\}$ define a one-to-one correspondence from B to a proper subset of C, thus showing that $n(B) < n(C)$.

Combining these pairings as follows

(1.3)
$$\begin{array}{lll} (\clubsuit, a), & (a, \spadesuit) & \to \quad (\clubsuit, \spadesuit) \\ (\diamondsuit, b), & (b, \square) & \to \quad (\diamondsuit, \square) \\ (\heartsuit, c), & (c, \triangle) & \to \quad (\heartsuit, \triangle) \end{array}$$

defines a one-to-one correspondence from A to a proper subset of C, thus showing that $n(A) < n(C)$. $\square$

1.1.3. Set Operations.

Definition 1.22. The *union* of the set A and the set B is the set $A \cup B$ consisting of all the elements of A and all of the elements of B. In symbols, write

$$(1.4) \qquad A \cup B = \{x : x \in A \text{ or } x \in B\}$$

In mathematics the use of the word *or* is always inclusive. An element is in A or in B means it is either in A or in B or in both A and B.

The set $A \cup B$ is the set obtained by putting the elements of B together with the elements of A. This can be illustrated by a picture of a set of children playing and another set of children joining them, or by a picture of a basket of balls (the set A) and another basket of balls (the set B) being dumped together into another basket.

Example 1.23. If A is the set of coins in my right hand and B is the set of coins in my left hand, then $A \cup B$ is the set of all of the coins in my hands.

Example 1.24. The union of a set A and one of its subsets B is again the set A, because the elements of B are already elements of A. Taking all of the elements of A and of B together, gives us exactly the elements of A.

Example 1.25. Set unions come from considering the set of objects satisfying one attribute or another (the connecting word *or* used in the inclusive sense noted above). If A is the set of people in the classroom whose first name begins with J, and if B is the set of people in the classroom whose last name begins with J, then $A \cup B$ is the set of people in the classroom whose first name begins with J or whose last name begins with J. Notice that if Jen Jensen is a student in the classroom, then she is a member of A and a member of B, and she is a member of $A \cup B$.

Definition 1.26. The *intersection* of the set A and the set B is the set of all elements which are members both of A and of B. In symbols, write

$$(1.5) \qquad A \cap B = \{x : x \in A \text{ and } x \in B\}$$

Example 1.27. In Example 1.25, Jen Jensen is a member of $A \cap B$, which is the set of all of the people in the classroom whose first and last names begin with J.

The confusion between names and things named rears its ugly head in the present context. If $A = \{\clubsuit, \square, \bigcirc\}$ and $B = \{\clubsuit, \square, \triangle\}$, then is the $\clubsuit$ or the $\square$ in A the same element, or not, as the $\clubsuit$ or the $\square$ in B? The sets $A \cup B$ and $A \cap B$ depend on the answer to this question.

A common way to resolve this problem is to use what are called universal sets from which we take the elements of the sets we are using. Subsets can

1.1. Sets

be defined by drawing a circle around the desired objects, or by coloring them, within the universal set. In the present case, suppose that $U = \{\clubsuit, \square, \bigcirc, \triangle, \heartsuit\}$ is the universal set and suppose that the above sets A and B are subsets of U. Now we know that $A \cup B = \{\clubsuit, \square, \bigcirc, \triangle\}$ and that $A \cap B = \{\clubsuit, \square\}$.

Set intersections arise naturally in the context of classifying objects according to multiple attributes. A set U could consist of large and small balls, some of which are red, some blue, and some yellow. Set A could be the set of large balls in U and set B could be the set of yellow balls in U, then $A \cap B$ would be the set of large, yellow balls in U.

Oral exercise 1.28. If $B \subseteq A$, then $A \cap B = B$. Explain.

Oral exercise 1.29. If A is the set of women in the classroom and B is the set of left-handed people in the classroom, then describe $A \cap B$ and $A \cup B$.

Definition 1.30. Sets A and B are *disjoint* if they have no element in common.

If A and B are disjoint, then $A \cap B = \emptyset$.

Example 1.31. (1) If $A = \{\clubsuit, \diamond, \heartsuit, \spadesuit\}$ and $B = \{\square, \triangle\}$, then A and B are disjoint, because there is no element which is a member of both of them. (2) If W is the set of women in the classroom and M is the set of men in the classroom, then W and M are disjoint.

Oral exercise 1.32. In oral exercise 1.29 above, how could it happen that A and B are disjoint?

An elementary application of these set operations is to the decomposition of a set into the union of two disjoint subsets. For example, if $U = \{\clubsuit, \square, \bigcirc, \triangle, \heartsuit\}$, then some decompositions of U into the union of two disjoint subsets are

$$\begin{aligned} U &= \{\clubsuit, \square, \bigcirc\} \cup \{\triangle, \heartsuit\} \\ U &= \{\clubsuit, \heartsuit\} \cup \{\square, \bigcirc, \triangle\} \\ U &= \{\} \cup \{\clubsuit, \square, \bigcirc, \triangle, \heartsuit\} \end{aligned} \tag{1.6}$$

Such decompositions lie at the foundation of addition and subtraction of whole numbers.

Definition 1.33. If A is a subset of a set W, then the *complement* of A in W is the subset of W consisting of all elements of W which are not elements of A.

A set is decomposed into a union of two disjoint subsets by a choice of subset. The complement of this subset is then the other set in the decomposition.

Oral exercise 1.34. For the set P of all the people in the classroom, describe some natural decompositions of P into a union of two disjoint subsets.

Theorem 1.35. *The union and intersection satisfy the following properties. If A, B, and C are sets, then*

(1) Commutative property:

  (a) $A \cup B = B \cup A$

  (b) $A \cap B = B \cap A$

(2) Associative property:

  (a) $A \cup (B \cup C) = (A \cup B) \cup C$

  (b) $A \cap (B \cap C) = (A \cap B) \cap C$

Therefore, there is no ambiguity in omitting the parentheses in these expressions and writing $A \cup B \cup C$ and $A \cap B \cap C$, respectively.

(3) If A is any set, then $A \cup \emptyset = A$ (identity property of $\emptyset$).

(4) Order property for union: if $B \subset C$ and A is any set, then $A \cup B \subset A \cup C$.

Proof. (1a) The set $A \cup B$ consists of all the elements of A together with all the elements of B, and this is certainly the same as the set of all the elements of B together with all the elements of A, which is $B \cup A$. Therefore, $A \cup B = B \cup A$.

(1b) The set $A \cap B$ consists of all the elements which are in A and in B, and this is certainly the same as the set of all the elements which are in B and in A, which is $B \cap A$. Therefore, $A \cap B = B \cap A$.

(2a) The set $A \cup (B \cup C)$ consists of all the elements of A and all the elements of $B \cup C$, and this is all the elements of A and all the elements of B and all the elements of C. But this is also the set $(A \cup B) \cup C$. Therefore, $A \cup (B \cup C) = (A \cup B) \cup C$.

(2b) The set $A \cap (B \cap C)$ consists of all the elements which are in A and in $B \cap C$, and this is all the elements which are in A and are also in B and also in C. This is also the set $(A \cap B) \cap C$. Therefore, $A \cap (B \cap C) = (A \cap B) \cap C$.

(3) The set $A \cup \emptyset$ consists of all the elements of A and all the elements of $\emptyset$, which is just all the elements of A, since $\emptyset$ has no elements.

(4) The order property is true because $A \cup B$ is the set of all the elements of A and all the elements of B, and this is contained in the set of all the elements of A and all the elements of C, since $B \subset C$. Therefore, $A \cup B \subset A \cup C$. □

§1.1 Exercises

(1) Is the set $A = \{a, b, c, d\}$ a subset of the set $B = \{a, 1, b, 2, c, 3, d, 4\}$? Is it a proper subset? Explain why or why not. *Ans:* A is a subset of B because every element of A is an element of B. It is a proper subset because B contains at least one element, for example 1, which is not an element of A.

(2) Is the set $A = \{a, b, c, d\}$ a subset of the set $B = \{a, 1, b, 2, 3, 4\}$? Explain why or why not.

(3) List all the subsets of $A = \{a, b, c\}$. How many are there?

(4) Use a diagram to describe a 1:1 correspondence from $A = \{\triangle \square \bigcirc \dagger\}$ to $B = \{a, b, c, d\}$. Indicate your 1:1 correspondence by a set of ordered pairs. What 1:1 correspondence from B to A is defined by your 1:1 correspondence from A to B? Describe it by a set of ordered pairs.

(5) Give a 1:1 correspondence from the set $A = \{a, b, c\}$ to itself. Give another one which is not exactly the same. How many 1:1 correspondences from A to itself are there? Describe each of them with a set of ordered pairs. *Ans:* 6

(6) Which set has more? Explain your answer with a 1:1 correspondence from one set to a proper subset of the other, not by counting.

$$A = \{\triangle, \triangle, \triangle, \triangle, \triangle\}, \quad B = \{\bigcirc, \bigcirc, \bigcirc, \bigcirc, \bigcirc, \bigcirc, \bigcirc\}$$

(7) Which set has less? Explain your answer with a diagram.

$$\{\clubsuit, \diamondsuit, \heartsuit, \spadesuit\} \quad \{a, b, c, d, e, f\}$$

(8) Consider the sets

$$A = \{\diamondsuit, \heartsuit\}, \quad B = \{a, b, c, d\}, \quad C = \{p, q, r, s, t, u, v\}$$

Give a 1:1 correspondence to show that $n(A) < n(B)$ and a 1:1 correspondence to show that $n(B) < n(C)$. Use these 1:1 correspondences to define a 1:1 correspondence from A to a proper subset of C showing that $n(A) < n(C)$.

(9) Is the set of all letters in the English alphabet finite? Explain your answer.

(10) Pictured below is a set U of balls, some large and some small, each ball being red, blue, or yellow as indicated with the letters R, B, and Y, respectively.

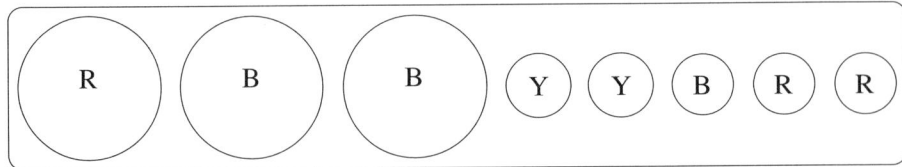

(a) Draw a picture of the set A of small balls in U. Draw a picture of the set B of red balls in U. What is the set of small red balls in U, in terms of set operations on A and B? Draw a picture of this set.

(b) What is the set of balls in U which are small or red, in terms of set operations on A and B? Draw a picture of this set.

(c) If C is the set of blue balls, what is the set $B \cap C$?

(11) Consider again the sets A, B, and C defined in the preceding problem.
   (a) In words, describe the sets $A \cup B$, $B \cup C$, $A \cup (B \cup C)$, and $(A \cup B) \cup C$. Is this consistent with 2(a) of Theorem 1.35?
   (b) In words, describe the sets $A \cap B$, $B \cap C$, $A \cap (B \cap C)$, and $(A \cap B) \cap C$. Is this consistent with 2(b) of Theorem 1.35?

(12) For any set A, what is $A \cup \emptyset$? What is $A \cap \emptyset$?

(13) List the elements of the sets $A = \emptyset \cup \{a\}$, $B = A \cup \{b\}$, $C = B \cup \{c\}$, and $D = C \cup \{d\}$. Explain why $n(\emptyset) < n(A) < n(B) < n(C) < n(D)$ by describing a natural 1:1 correspondence from each onto a proper subset of the next.

(14) For the set $A = \{\triangle \square \bigcirc \dagger\}$, draw a solid loop around a subset B and a dotted loop around a subset C such that $B \cup C = A$ and $B \cap C = \{\square\}$. Can this be done in more than one way? If yes, exhibit another way.

(15) List all possible decompositions of $U = \{\square \triangle \bigcirc\}$ into the union of two disjoint subsets. Remember the cases $\emptyset \cup U$ and $U \cup \emptyset$. How many such decompositions are there? *Ans:* 8.

(16) List all possible subsets of $U = \{\square \triangle \bigcirc\}$. Beside each subset write its complement in U. Explain why the result is a list containing all possible decompositions of U into the union of two disjoint subsets.

(17) List the ways that the set $\{\triangle \square \bigcirc \dagger\}$ can be decomposed into the union of two disjoint subsets. How many ways can this be done? *Ans:* 16.

(18) Let M be the set of all the men in the classroom. Let L be the set of all left-handed people in the classroom. In words describe the sets $F \cup L$, $L \cup F$, $F \cap L$, and $L \cap F$.

(19) Let B be the set of boy children and let C be the set of all children in a certain family. Let A be the set consisting of the parents. In words, describe the sets $A \cup B$ and $A \cup C$ and explain how this description is consistent with (4) of Theorem 1.35.

(20) In flipping a coin, let H denote heads and T denote tails. Flip a penny and a nickel and designate the outcome with an ordered pair of letters, with the first letter denoting the outcome of the penny and the second letter denoting the outcome of the nickel. For example, (T, H) denotes tails on the penny and heads on the nickel.
  (a) List the elements of the set S of all possible outcomes from flipping the pair of coins once. Ans: $S = \{(H,H), (H,T), (T,H), (T,T)\}$.
  (b) If P is the subset of S of all outcomes for which the penny comes up heads, and if N is the set of outcomes for which the nickel comes up heads, what are $P \cap N$ and $P \cup N$? Describe each in words and by listing the elements of each. Ans: $P = \{(H,H), (H,T)\}$, $P \cup N = \{(H,H), (H,T), (T,H)\}$.
  (c) In terms of set operations and set complement applied to S, P, and N, describe the set of all outcomes for which both coins come up tails. Ans: It is the complement of $P \cup N$ in S.

1.2. Whole Numbers

Whole numbers have two essential properties, both of which must be emphasized when the subject is taught. Each whole number represents the number of elements in some set, and the whole numbers are lined up in a specific order. Each whole number has a *successor*, which is the whole number occurring next in line. Zero starts the line, and each whole number other than zero has a *predecessor*, which is the whole number immediately preceding it in line. Here is a definition of the set **W** of whole numbers.

Definition 1.36 (Whole Numbers). The number 0 is defined to be $n(\emptyset)$, the number of elements in the empty set. It is the first whole number, so it has no predecessor. Its successor is defined to be $1 = n(\emptyset \cup \{0\}) = n(\{0\})$, and 0 is the predecessor of 1. Thus, 1 is the number of elements in the set obtained by adding one more element to the preceding set, $\emptyset$. Notice that $0 < 1$, by Definition 1.17, since $\emptyset$ has fewer elements than $\{0\}$. (Not only is there a 1:1 correspondence from $\emptyset$ to a proper subset of $\{0\}$, it **is** a proper subset of $\{0\}$).

The successor of 1 is defined to be $2 = n(\{0\} \cup \{1\}) = n(\{0,1\})$, and 1 is the predecessor of 2. Thus, 2 is the number of elements in the set obtained by adding one more element to the preceding set. Again, Definition 1.17 implies that $1 < 2$.

The successor of 2 is defined to be $3 = n(\{0,1\} \cup \{2\}) = n(\{0,1,2\})$, and 2 is the predecessor of 3 and $2 < 3$. Thus, 3 is the number of elements in the set obtained by adding one more element to the preceding set.

The general step in the definition, after the definition of 0, is that the next whole number is the number of elements in the set of all whole numbers which precede it.

Let **W** denote the set of all *whole* numbers. The set of all whole numbers except 0 is called the set of *natural* numbers and is denoted **N**. Thus, **N** is a proper subset of **W**. The complement of **N** in **W** is the set consisting of the number zero.

By the definition of whole numbers and by Definition 1.17 of inequality, we see that each number is smaller than its successor and larger than its predecessor (except for the case of 0, which has no predecessor). That is,

$$0 < 1 < 2 < 3 < 4 < 5 < 6 < 7 < 8 < 9 < \ldots$$

where the three dots indicate that this goes on for all whole numbers. Combining these inequalities with Theorem 1.21, we conclude that any whole number is smaller than a whole number which occurs later in the course of their definition. For example, $2 < 5$ and $3 < 7$ and so on.

Theorem 1.37. *If m is a natural number, then m is the number of elements in the set of all natural numbers from 1 through m.*

Proof. The proof is illustrated by a specific case, say $m = 5$. In this case the theorem says that $5 = n(\{1, 2, 3, 4, 5\})$. By the Definition 1.36 of whole number, $5 = n(\{0, 1, 2, 3, 4\})$, so the proof of the theorem comes down to showing that there is a one-to-one correspondence from $\{1, 2, 3, 4, 5\}$ to $\{0, 1, 2, 3, 4\}$, which is done by pairing each number of the first set with its predecessor. □

1.2.1. The Number Line. The following process constructs the *number line*, which is a picture of the whole numbers along a line in the order in which they occur. Draw a line, mark a point and label it 0. Mark another point, to the right of 0, and label it 1. Call the line segment from 0 to 1 the *unit line segment*. Go one unit line segment to the right of 1 and mark the end point 2. Continue this indefinitely. For any natural number n, the *number of elements in the set of unit line segments from 0 to n is n.*

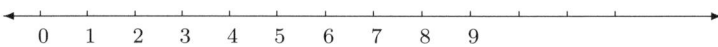

We adopt the philosophy, shared by the Greek mathematicians who wrote Euclid's *Elements*, that every point on this line represents a number. So far we have defined and identified the whole numbers on it.

Theorem 1.38. *(1) If m and n are whole numbers and $m < n$, then m is to the left of n on the number line.*

1.2. Whole Numbers

(2) If m and n are whole numbers and m lies to the left of n on the number line, then $m < n$.

Proof. Each part of this theorem is the converse of the other.

(1) If $m < n$, then the set of unit line segments from 0 to m has fewer elements than the set of unit line segments from 0 to n, so n must be to the right of m.

(2) If m lies to the left of n, then the set of unit line segments from 0 to m must be a proper subset of the set of unit line segments from 0 to n. Thus, $m < n$. □

For future reference, we state an important elementary consequence of this Theorem.

Remark 1.39. If m and n are whole numbers, then exactly one of the following possibilities is true.

(1) $m = n$;

(2) $m < n$;

(3) $m > n$.

In fact, if m is not equal to n, then m and n are two different points on the number line, so one lies to the left of the other one. The number to the left is smaller.

1.2.2. Counting. To count the elements of a set B is to make a one-to-one correspondence from B to the subset of **W** which begins with 1 and contains each number's successor up to some number n. For example, suppose $B = \{\clubsuit, \diamondsuit, \heartsuit, \spadesuit\}$ and suppose that we count the elements of B by pointing to $\clubsuit$ and saying 1, a process we can indicate with the ordered pair $(\clubsuit, 1)$, then pointing to $\diamondsuit$ and saying 2, indicated with the ordered pair $(\diamondsuit, 2)$, then pointing to $\heartsuit$ and saying 3, indicated with the ordered pair $(\heartsuit, 3)$, and finally pointing to $\spadesuit$ and saying 4, indicated with the ordered pair $(\spadesuit, 4)$. This process has created a set of ordered pairs which defines a one-to-one correspondence from B to the set of whole numbers $\{1, 2, 3, 4\}$. By the preceding theorem we know that $4 = n(\{1, 2, 3, 4\})$, and therefore this process of counting shows us that $4 = n(B)$.

Oral exercise 1.40. (1) Count the □'s in each pile.

```
                    □
                  □ □
                □ □ □
              □ □ □ □
            □ □ □ □ □
          □ □ □ □ □ □
        □ □ □ □ □ □ □
```

(2) What comes next? 3, 4, 5,

(3) What comes next: {△, △, △, △, △}, {△, △, △, △}, {△, △, △},

(4) What comes next? 3, 2, 1,

Oral exercise 1.41. The following picture illustrates the arabic numeral and the English word for a number and a set with that many elements. Discuss what can be done with this picture to illustrate the successor and the predecessor of this number.

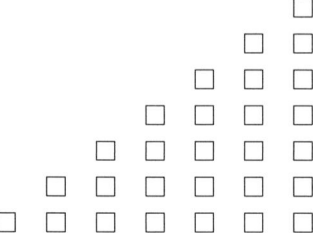

4   four   {♣, ♢, ♡, ♠}

One effective system for counting is to put things into groups of ten. This also provides groundwork for the base ten numeration system.

1.2.3. Numeration. In order to count, we need names, both written and verbal, for the whole numbers. Suppose that we have learned to count the fingers on our hands (thumbs being regarded as fingers in this context). To do this, we must name the numbers 1, 2, 3, 4, 5, 6, 7, 8, 9, and 10, which verbally are one, two, three, four, five, six, seven, eight, nine, and ten. The first whole number is 0, called zero. These numbers, other than 10, are called *digits*, no doubt because of their association with counting our fingers. The symbol 10 already displays a property of our numeration scheme, which we shall now see.

Knowing how to count to ten, we can count larger numbers of objects by putting them into groups of ten, and then counting these groups into groups of ten, and so on. The following examples are visually limited by typesetting requirements. In the classroom these examples should be done with easily manipulated objects, such as soda straws and rubber bands to hold together the groups of ten. Plastic cups can serve as the places for placing the individual straws, the groups of ten, and so on.

Suppose a straw look like this: †. To count the set of straws

† † † † † † † † † † † † †

1.2. Whole Numbers

we count ten of these, put a rubber band around them (denoted by a horizontal line) and move this group of ten to the left, as follows

$$\underline{\dagger\dagger\dagger\dagger\dagger\dagger\dagger\dagger\dagger\dagger} \qquad \dagger\dagger\dagger\dagger$$
$$\text{1 ten} \qquad\qquad 4$$

We have 1 ten (in the ten's cup) and 4 individual straws (in the one's cup). This picture suggests the numeral 14 for this number, where the 1 occupies the ten's place and the 4 occupies the one's place. The English word for this number is fourteen, but a more logical name might be one-ten-four.

If we had begun with a set of ten straws

$$\dagger\,\dagger\,\dagger\,\dagger\,\dagger\,\dagger\,\dagger\,\dagger\,\dagger\,\dagger$$

then after counting a group of ten there would be 1 group of ten and no individual straws remaining, like this

$$\underline{\dagger\dagger\dagger\dagger\dagger\dagger\dagger\dagger\dagger\dagger}$$
$$\text{1 ten} \qquad\qquad 0$$

The set of those remaining is the empty set, in which the number of elements is zero. For this 1 group of ten and 0 we write 10, and this is the numeral for the number ten. A logical name for it could be one-ten-zero. Count the following set of straws

$$\dagger$$

by counting ten of the straws, set this group of ten to the left, and count ten more, and set this group of ten below the first group. There will then be seven objects remaining, as follows.

$$\underline{\dagger\dagger\dagger\dagger\dagger\dagger\dagger\dagger\dagger\dagger}$$
$$\underline{\dagger\dagger\dagger\dagger\dagger\dagger\dagger\dagger\dagger\dagger} \qquad \dagger\dagger\dagger\dagger\dagger\dagger\dagger$$
$$\text{2 tens} \qquad\qquad 7$$

We have 2 tens and 7 remaining, which we write 27 and call twenty-seven, although a more logical name would be two-tens-seven.

For another example, count the set of straws

$$\dagger$$

by counting ten of these, set this group of ten to the left, count ten more and set it below the set of ten already there, and so on until you cannot make any more groups of ten. We then have

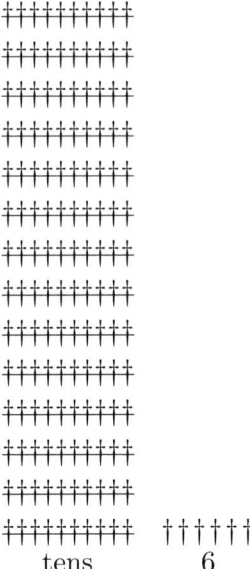

Having more than ten tens, we count the tens into a group of ten, tie it with a rubber band (denoted here with a diagonal line) and move this ten of ten to the left as follows.

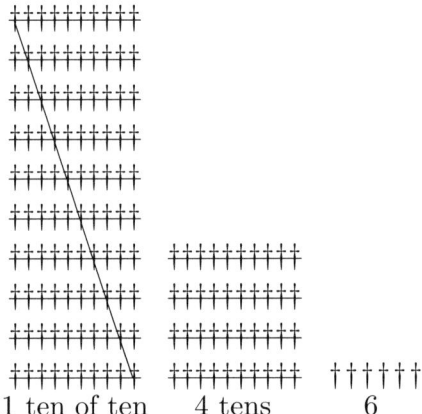

We have 1 ten-of-ten in the ten-of-ten's cup, 4 tens in the ten's cup and 6 straws in the one's cup. The picture suggests the numeral 146 for this number, where the 1 occupies the ten-of-ten's place (hundred's place), the 4 occupies the ten's place and the 6 occupies the one's place. The English word for this number is one-hundred-forty-six, but a more logical name might be one-ten-of-ten-four-tens-six.

1.2. Whole Numbers

Oral exercise 1.42. Use this method of counting into groups of ten to count the number of elements in a set of soda straws, or any other convenient set of objects. Use rubber bands to tie together each group of ten.

1.2.3.1. *Other bases.* Suppose that we count the set of straws

††††††††††††††

by counting them into groups of five, rather than into groups of ten.

$$\begin{array}{cc} \text{\textbardbl\textbardbl\textbardbl} & \\ \text{\textbardbl\textbardbl\textbardbl} & \text{††††} \\ 2 \text{ fives} & 4 \end{array}$$

We have 2 groups of five in the five's cup and 4 straws left in the one's cup. This picture suggests the numeral 24_5, where the subscript 5 indicates that we have now counted into groups of five. Here the 2 occupies the five's place and the 4 occupies the one's place. This is called a *base five* numeral, because it was formed by counting into groups of five. A logical name for it is two-fives-four.

Oral exercise 1.43. Repeat the Oral Exercise 1.42, but now count into groups of five. The resulting numerals are in base five. What would be the logical name for the base five numerals 10_5, 23_5, and 203_5, where the subscript 5 is included to distinguish these numerals from our usual base ten numerals. What is the base five numeral for five?

Oral exercise 1.44. Repeat the same exercise, but now count into groups of two. The resulting numerals are in base two. What should be the logical name for the base two numerals 10_2, 111_2, and 1001_2? What is the base two numeral for two?

The following theorem summarizes this method of naming the numbers. The name of a number is its *numeral*. The digits name the first ten whole numbers: 0, 1, 2, 3, 4, 5, 6, 7, 8, 9, whose English names are zero, one, two, three, four, five, six, seven, eight, and nine, respectively.

Theorem 1.45. *If n is a whole number, then its base ten numeral is obtained by taking a set A of n elements and counting the elements of A into groups of ten. Use a digit to record the number of remaining elements. Write this digit in what is called the one's place.*

Next count the groups of ten into groups of ten. Use a digit to record the number of remaining groups of ten. Write this digit to the left of the first digit, in what is called the ten's place. Continue in this fashion until no more groups of ten can be made.

This method assigns a numeral to each whole number. If two whole numbers are different, then their numerals are different.

The English name for each numeral follows a modification of what we might call the logical English name. These follow the evident pattern of the following table.

numeral	English	logical English
10	ten	one-ten-zero
11	eleven	one-ten-one
12	twelve	one-ten-two
13	thirteen	one-ten-three
15	fifteen	one-ten-five
20	twenty	two-ten-zero
32	thirty-two	three-ten-two
107	one-hundred-seven	one-ten-of-ten-zero-ten-seven

The English name becomes evidently more efficient than the logical English name when we look at an example such as 1718, whose English name is one-thousand-seven-hundred-eighteen and whose logical English name is one-ten-of-ten-of-ten-seven-ten-of-ten-one-ten-eight.

Proof. There are two things to be proved. Does the method assign a numeral to each whole number? Can different whole numbers be assigned the same numeral? The answer to the first question is easily seen to be yes, provided we are convinced that the method reaches a conclusion on any given whole number, especially a very large whole number. Are we sure that we might not end up making groups of ten forever? This point will be addressed in Section 3.1.

As for the second question, if m and n are different whole numbers then their numerals will be different for the following reason. One of these whole numbers must be smaller than the other (see Remark 1.39 above), say $m < n$, and therefore m occurs to the left of n on the number line.

Now m is the number of elements in the set of unit line segments from 0 to m, which is a proper subset of the set of unit line segments from 0 to n. Count the unit line segments from 0 to m into groups of ten, and then continue on to count the unit line segments from 0 to n into groups of ten. In the case pictured, the number of groups of ten is the same for both cases, but the number of unit line segments remaining must be different, because m lies to the left of n. In the case pictured, the number remaining is 3 in the case of m and 8 in the case of n, thus showing that m and n have different numerals.

1.2. Whole Numbers

It can also happen that the number of groups of ten will be greater for the case of n, as pictured below.

After counting into groups of ten, the number of unit line segments remaining might be the same for m and n (as pictured), or it might be different. The essential point now is that we have a different number of groups of ten in the two cases, one such for m and two for n in the picture. Now the process begins anew on the set of groups of ten, and the set for m is a proper subset of that for n. There are two cases as just described. Eventually the number of groups of ten must be the same for both m and n, possibly not until the end when it is zero for both. At that point, the number remaining must be different, as happened in the first case above. □

1.2.3.2. *Finiteness again.* We can now improve Definition 1.16, as promised.

Definition 1.46. A set A is finite if the number of elements of A is some whole number n.

This definition relates to lists as follows. Suppose that A is a finite set whose number of elements is 6, for example. This means that there is a 1:1 correspondence from $\{1, 2, 3, 4, 5, 6\}$ to A, by Theorem 1.37. This correspondence makes a list of the elements of A, for which the first item on the list is the element paired with 1, the second item on the list is the element paired with 2, and so on. We see, then, that *to list* the elements of a set A means to make a 1:1 correspondence from a subset $\{1, 2, 3, \ldots, n\}$ of N, for some natural number n, to A. With this understanding of *to list*, our two definitions of finite are the same.

The next theorem summarizes some basic properties which you long ago internalized, but which we make explicit now in order to be aware of what must be taught about the whole numbers.

Theorem 1.47. *(1) If A is a finite set of whole numbers, then its elements can be arranged in increasing order, from a smallest element to a largest element.*

(2) The set $\mathbf{W}$ of all whole numbers is infinite.

(3) If A is any subset of $\mathbf{W}$ (finite or infinite), then its elements can be put into increasing order from a smallest element. If the subset is infinite, then it will have no largest element.

Proof. (1) The first statement follows from the fact that the whole numbers lie on a line. Namely, suppose that A is a finite set of whole numbers. Then the number of elements in A is some whole number n. Start at 0 on the number line and go to the right until you encounter for the first time an element of A. Continue until you encounter the next element of A. Continue the process n times and you will have encountered all of the elements of A. The first element encountered is the smallest element of A, the second element encountered is the next largest element of A, and so on until the last element encountered is the largest element of A. The order of the encounters orders the elements of A.

(2) The set **W** must be infinite, because if we tried to make a list of the elements of **W**, then we might as well list the elements in increasing order. A finite list must end somewhere, say with the number n, which would be the largest number in the list. But then the successor of n is a whole number not on the list, so our list didn't include all of the elements of **W**. We conclude that no such list can be made, which means that **W** is infinite.

(3) The last statement of the theorem follows by the same arguments. Proceed as in the proof of (1). If A is an infinite set, then the process will never stop, so there will be no largest element in A. □

1.2.4. Summary.

(1) The number of elements in a set relates a whole number to the idea of *how many*.

(2) Counting the elements of a set involves making a 1:1 correspondence from the set to the subset of the natural numbers $\{1, 2, \ldots, n\}$, for some natural number n.

(3) Counting emphasizes the order of the whole numbers. Their order is part of the definition of whole numbers.
    (a) Counting backwards emphasizes each number's predecessor.
    (b) Filling in missing numbers from a list emphasizes the order.

(4) Definition of whole numbers is nearly indistinguishable from counting.

(5) Our numeration system is based on counting into groups of ten. The digits are needed to count to ten and to record how many remain when another group of ten cannot be formed.

§1.2 Exercises

(1) Explain why a one-to-one correspondence from the set $B = \{$Nettie, Robert, Richard, Viola, Ruth, Ralph$\}$ to the set $A = \{1, 2, 3, 4, 5, 6\}$, amounts to counting the elements of B.

§1.2 Exercises

(2) Make up two more exercises like Oral Exercise 1.40 to illustrate the sequential feature of the whole numbers.

(3) Make up another exercise like Oral Exercise 1.41 to illustrate a digit's numeral, English name and a set with that many elements.

(4) Consider the set of straws

(1.7) $$A = \{†\,†\,†\,†\,†\,†\,†\,†\,†\,†\,†\,†\,†\,†\,†\,†\,†\,†\,†\}$$

Give a step by step explanation for how to find the base ten numeral for $n(A)$. What is a logical name for it? What is the base ten numeral for $n(B)$, where $B = A \cup \{†\}$, that is, B is the set obtained from A by putting one more straw into it.

(5) Let A be the set defined in (1.7). Give a step by step explanation for how to find the base five numeral for $n(A)$. What is a logical name for it? *Ans:* three-fives-four. What is the base five numeral for $n(B)$, where $B = A \cup \{†\}$?

(6) Let A be the set defined in (1.7). Give a step by step explanation for how to find the base two numeral for $n(A)$. What is a logical name for it? *Ans:* one-two-of-two-of-two-zero-two-of-two-one-two-one. What is the base two numeral for $n(B)$, where $B = A \cup \{†\}$?

(7) Write in increasing order the base five numerals for the numbers one through the number whose base five numeral is 34_5.

(8) Write in increasing order the base two numerals for the numbers one through the number whose base two numeral is 1011_2.

(9) Fill in the blanks, then explain what ideas are being illustrated by the problem.

$$\{\bigcirc\ \bigcirc\ \bigcirc\} \qquad \{\oplus\ \oplus\ \oplus\ \oplus\}$$

There are ___ plain balls.
There are ___ decorated balls.
There are ___ balls altogether. *Ans:* There are 3 plain balls and 4 decorated balls, both answers obtained by counting the balls in each set. There are 7 balls altogether, determined by counting all the balls in both sets. The problem illustrates counting the number of elements in a set and counting the number of elements in the union of two sets.

(10) Fill in the blanks.

$$\{\bigcirc\ \bigcirc\ \bigcirc\ \bigcirc\} \leftarrow \{\bigcirc\ \bigcirc\ \bigcirc\ \bigcirc\}$$

There are 4 balls in the set.
Add ___ more. So 4 and ___ make ___.
There are _____ balls altogether.

(11) Fill in the blanks.

$$\{\triangle \triangle \triangle\} \quad \{\square \square \square \square\}$$

There are ___ triangles.
There are ___ squares.
There are ___ triangles and squares altogether, so 3 and ___ make ___.

(12) Fill in the blanks.

$$\{\triangle \triangle \triangle \triangle \triangle \triangle\} \quad \{\ \}$$

There are ___ triangles.
There are ___ squares.
There are ___ triangles and squares altogether, so ___ and ___ make ___.

(13) Make up three more exercises, like the preceding three, which show how two numbers make 6.

(14) How many dots are in each triangle?

  (a) How many dots in the next triangle?
  (b) How many dots in the triangle possessing 6 dots on each side?
  (c) How many dots in the triangle possessing n dots on each side, for any $n > 0$?

(15) How many dots in each square?

  (a) How many dots in the next square?
  (b) How many dots in the square possessing 5 dots on each side?
  (c) How many dots in the square possessing n dots on each side, for any $n > 0$?

1.3. Measurement

Any *measurement* is a comparison with a known quantity, called the *unit of the comparison*. This known quantity is arbitrary, but must be known to all those to whom one wants to communicate the comparison. We shall consider here measurements of length, area, volume, weight, capacity (liquids and dry substances such as flour), value (money), angle, and time.

1.3.1. Length.

1.3.1. Length. A *line* extends indefinitely in both directions. It consists of the *points* on it. It is impossible to draw a whole line on a piece of paper. We use arrowheads to indicate the indefinite extension.

A *line segment* is the portion of a line lying between two points on the line, and includes those two points. A line segment is what we actually draw on a piece of paper. To measure the line segment

we compare it to a *reference* line segment such as

It takes 5 of the reference line segments, laid end to end, to cover the line segment being measured.

Referring to the reference line segment as the unit, we say that the longer line segment is 5 units long.

The width of my desk is the width of my two arms outstretched as far as possible. This comparison communicates good information to me, but not to somebody who doesn't know how big I am, or how long my arms are. In order to communcate a measurement, we must compare to a standard unit, such as an inch, foot, yard, meter, or mile. If you are familiar with the length of one foot, then you have a good idea of the width of my desk if I tell you that it is 7 feet wide.

A standard unit is often used because it would be impractical, or impossible, directly to compare the two things under consideration. For example, a big roll of carpet is not taken to your living room to measure out how much is needed to cover the floor. Instead, you would measure the floor with a tape measure, in feet and inches probably, and then go to the carpet store to measure that amount on the roll of carpet.

Sometimes the comparison to an intermediate object is more subtle. Told that Niels is 74 inches tall, you might picture this height in terms of your own height or of someone's height which you know to be close to that. Knowing yourself to be 68 inches tall, you might think Niels is 6 inches taller, or you might compare his height to your brother's, which is about 72 inches, and conclude that Niels is about 2 inches taller than your brother, that is, just a little taller than your brother.

Oral exercise 1.48. Your desk is how many hands wide? (A hand is fingers together, and excluding the thumb).

Once we understand that a measurement is a comparison, then we realize that the unit for the comparison is quite arbitrary, and functions primarily as a tool of communication. If possible, the unit should be on the same scale as the object being measured. You measure your height in inches, not miles, and the distance from St. Louis to Chicago is given in miles, not inches.

Oral exercise 1.49. Measure the width of your desk in inches. Your hand is how many inches wide? Is the last knuckle of your little finger about 1 inch long?

Different units can be compared to each other. In the English system of lengths, the most commonly used units are as follows. An inch is roughly the length of the last knuckle of an adult's little finger.

12 inches	= 1 foot
3 feet	= 1 yard
5280 feet	= 1 mile
1760 yards	= 1 mile

It is common to use two or more of these units at the same time. A length of 50 inches could be described as

$$1 \text{ yard } 14 \text{ inches}$$

which we arrive at by first counting 36 inches along the length, to give 1 yard with 14 inches remaining. On the remaining 14 inches we count 12 inches, which is a foot, and we have 2 inches remaining. Compare this process to that of finding the base ten numeral of a whole number. Then, our length of 50 inches can be expressed as

$$1 \text{ yard } 1 \text{ foot } 2 \text{ inches}$$

Notice the similarity between this expression and a numeral of our numeration system. A major difference is that here we must label each place (yard, foot and inches, respectively), while the place values are understood in our numeration system. Converting from one unit to another involves applications of the operations of arithmetic, which we consider in the next chapter.

Oral exercise 1.50. Express Niels's height in feet and inches.

The metric system of lengths has units which relate to each other by multiples of ten, in the same way as the value of places in our base ten numeration system. The official relation between the English and metric systems is

$$1 \text{ inch } = 2.54 \text{ centimeters (cm)}$$

Here is a table comparing the commonly used units of length in the metric system.

1.3. Measurement

10 millimeters	= 1 centimeter
100 centimeters	= 1 meter
1000 meters	= 1 kilometer

Oral exercise 1.51. How many millimeters in a meter? How many centimeters in a kilometer?

Oral exercise 1.52. Express 123 centimeters in terms of meters and centimeters. Express 1235 millimeters in terms of meters, centimeters, and millimeters.

One should work within the English system or the metric system, and regard conversion between the two as a secondary problem. Children should be taught one system at a time. Conversion from one system to the other is never the main point.

1.3.2. Area. This is a comparison of one region of some surface with another. To begin, we compare the area of a *reference square*, such as

□

with that of regions in the plane which can be built from it, such as these:

The region on the left contains 4 reference squares, so its area is 4 units. The area of the region on the right is 5 units. The area of the region formed by the two regions together is the total number of reference squares, which is 9 units.

Oral exercise 1.53. What is the area of each of the following regions? Which region has the greater area?

If the sides of the unit reference square are each 1 length unit long, then its area is called 1 square length unit. For example, if the sides of the unit squares are 1 inch long, then the areas of the preceding two figures are 11 square inches (written 11 in^2) and 6 in^2, respectively.

Oral exercise 1.54. If the side length of the unit square is 1 cm, what are the areas of the preceding two figures?

Definition 1.55. The *perimeter* of a region is the distance around its boundary.

Usually the perimeter is measured in terms of the side length of the unit square.

Oral exercise 1.56. Suppose that the unit square has side length of 1 ft and that the following regions are made up of unit squares.

(1) What is the perimeter of each region?
(2) Are the areas the same?
(3) Are the perimeters the same?
(4) Can two regions have the same area, but different perimeters?
(5) Can two regions have the same perimeter, but different areas?

1.3.3. Volume. The volume of a solid object is the amount of space it occupies. We measure volume by comparing it to a reference quantity, which we now take to be a *unit cube*, like this:

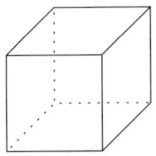

We begin by measuring the volume of solid objects which can be built from unit cubes. These can be drawn on a lattice of dots as follows.

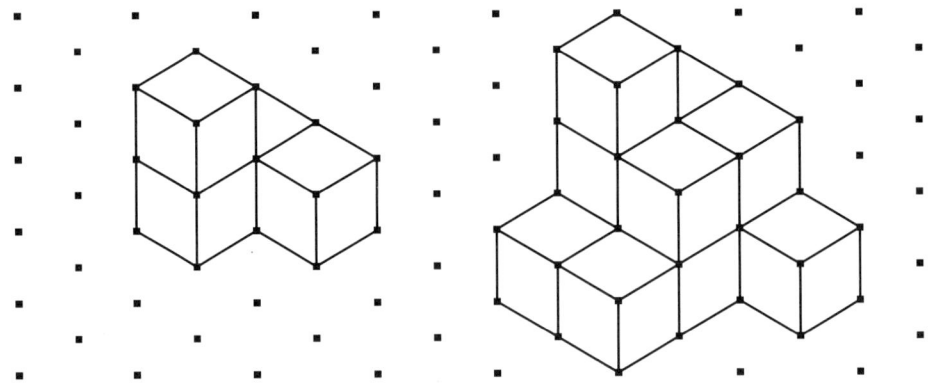

The solid on the left contains 4 unit cubes, so its volume is 4 units. If the unit cube has edge length of 1 ft, then its volume is 1 cubic foot (written 1 ft^3) and this solid has volume of 4 ft^3 because it contains exactly 4 unit cubes.

Oral exercise 1.57. How many unit cubes in the solid on the right? If the length of each edge of the unit cube is 1 cm, what is the volume of this solid in cubic centimeters? The volume of this figure is ambiguous because some of the unit cubes building it are not visible from this view. What are some other possible volumes for this figure?

1.3.4. Weight. The weight of a solid object is the amount of force applied to it by gravity. Important ideas from physics are needed to measure this. To begin with, to measure weight is to compare the weight of one object to that of another. From the physics of levers (think about teeter-totters), we can make a balance scale to compare the weights of two objects. If the arms of the scale are the same length then the weights at the end of each arm are the same if the arms are exactly horizontal.

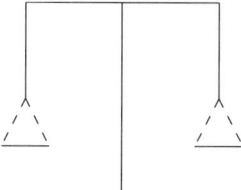

A spring scale works on the principle that the distance that a spring stretches is *proportional* to the force applied to it. This means that if a given force stretches the spring 1 inch, then twice that amount of force will stretch it 2 inches, three times that amount of force will stretch it 3 inches, and so on. Instruction on how to use a scale is the way to introduce the idea of weight.

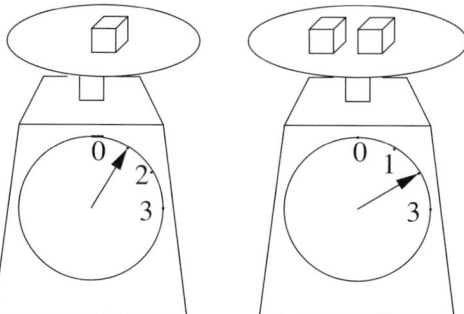

Over the centuries there have been many standard reference weights. In English units, an ounce is a reference weight which is very nearly the weight of ten U.S. pennies. Other units of weight in this system are pounds and tons, which are related as shown in the following table (where the standard abbreviations are also shown). A *stone* is an old-fashioned unit of body weight still used in parts of Great Britain.

16 ounces (oz)	= 1 pound (lb)
14 lbs	= 1 stone
2000 lbs	= 1 ton (T)

Isaac Newton's second law of motion says that the force of gravity on an object is the *mass* of the object times the acceleration of the object when it is dropped (near the surface of the earth and assuming no air resistance). Galileo observed that this acceleration due to gravity is constant. So the weight is this constant acceleration times the mass. The metric system measures the mass, even though it is commonly referred to as weight.

A *gram* is the mass of the water (at 4° C) needed to fill a cube whose edge length is 1 centimeter. Thus, a gram is the mass of a cubic centimeter (cc) of water. A *kilogram* is 1000 grams. It is the mass of the water needed to fill a cube whose edge length is 10 cm.

Oral exercise 1.58. How many kilograms of water are needed to fill a cube whose edge length is 1 meter?

1.3.5. Capacity. We compare liquids, such as water, gasoline, and oil and liquid-like substances such as flour, sugar, and spices by the number of times the substance fills up a specified container. Sometimes we compare their weights; we buy flour and sugar by the pound. Sometimes we compare their volumes, which means we think of their containers as being rectangular solids, or even cubes. For example, water is often sold by the cubic foot.

In the English system, 1 pint of water is the capacity of 1 pound of water. There are 16 ounces of water in 1 pint. Various units are related as follows.

3 teaspoons	= 1 tablespoon
2 tablespoons	= 1 oz of water
8 oz of water	= 1 cup
2 cups	= 1 pint
2 pints	= 1 quart
4 quarts	= 1 gallon
31.5 gallons	= 1 barrel (oil)

Oral exercise 1.59. How many pounds does: one pint of water weigh? One quart of water weigh? One gallon of water weigh?

In the metric system, the capacity measure of 1 gram of water (at $4°C$) is 1 milliliter. Thus, 1 kilogram of water is 1 liter of water, which has the volume of a cube of edge length equal to 10 cm.

1.3. Measurement

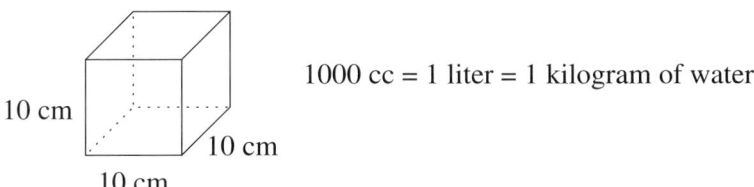

1000 cc = 1 liter = 1 kilogram of water

These relationships among the metric units of volume, weight, and capacity outshine those among the English units. For example, a cubic foot of water weighs approximately 62.5 pounds. Since a gallon of water weighs 8 pounds, there are approximately 7.8 gallons of water in a cubic foot of water.

Oral exercise 1.60. Suppose that we have seven cups of flour sitting before us in individual cups. Count these cups into groups of 2 and pour each of these pairs into a pint jar. How many pint jars have we filled and how many cups remain? Count these pint jars of flour into groups of 2 and pour each pair into a quart jar. How many quart jars are filled and how many pint jars of flour remain? What is the base two numeral for the number of cups of flour that we began with?

1.3.6. Volume by Displacement. If a solid object such as a rock or a brick is submerged in water, it will displace a quantity of water equal to its volume. This fact can be used as a method for measuring volumes. Suppose that a rock is put into an aquarium which is in the shape of a rectangular solid 1 foot wide, 3 feet long, and 2 feet high. Before the rock is put into it, the water is 1 foot high. After the rock is completely submerged, the water is 1 foot 3 inches high. The displaced water is the amount in a rectangular solid 1 foot wide, 3 feet long and 3 inches high.

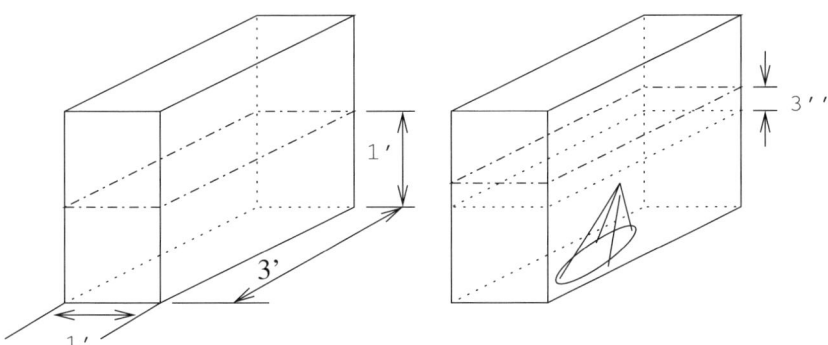

With the tools of arithmetic we will be able to calculate the volume of this displaced water, and this will be the volume of the rock.

1.3.7. Value. In United States currency, 100 cents make one dollar. A penny is worth 1 cent, a nickel is 5 cents, a dime is 10 cents, a quarter is 25 cents, and a half-dollar is worth 50 cents.

Oral exercise 1.61. Suppose we start with a pile of pennies. Count them into piles of 5 and exchange each such pile for 1 nickel. Next count the nickels into piles of 5 and exchange each pile with 1 quarter. At the end of this process you have 3 quarters, 4 nickels and 1 penny. What is the base five numeral for the number of pennies that we started with?

1.3.8. Angles. A *ray* is the half of a line issuing from a point. The point is called the *base* of the ray.

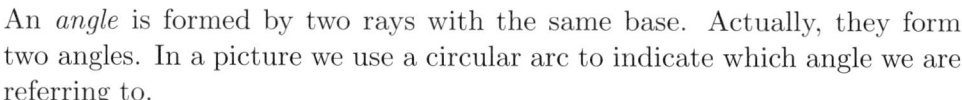

An *angle* is formed by two rays with the same base. Actually, they form two angles. In a picture we use a circular arc to indicate which angle we are referring to.

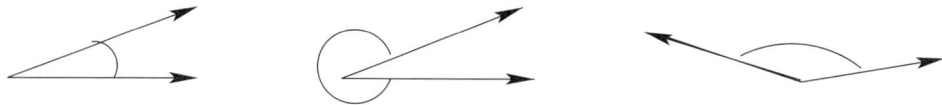

The two rays are called the *edges* and their common base is called the *vertex* of the angle.

If the two edges of an angle go in opposite directions on the same line, then the angle is called a *straight* angle.

Two angles are called *supplements* of each other if they share a common side and the other two sides form a straight line, like this, where angles a and b are supplements of each other.

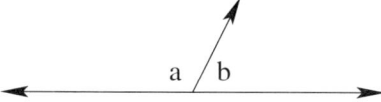

Two angles form a *vertical pair* if their sides combine to form a pair of intersecting lines, like this, where angles a and b form a vertical pair, and also angles c and d form a vertical pair:

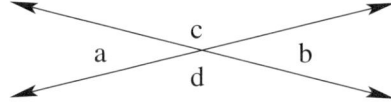

Theorem 1.62. *If angles a and b form a vertical pair, then they are equal to each other.*

Proof. In the above figure, angles a and b are both supplements of angle c, so they must be equal to each other. □

1.3. Measurement

Definition 1.63. A *right* angle is an angle with the property that if its edges are both extended past the vertex, then all four of the angles formed are the same size.

Thus, two right angles make a straight angle and four right angles make the *whole* angle, which is the angle for which the two edges coincide, but we take the long way round from one to the other. We denote a right angle by putting a little square in place of the usual circular arc. Here is a right angle with both sides extended to show the four right angles.

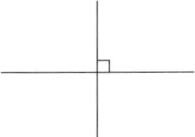

Angles can be compared according to how many of one kind can fit within another. A standard reference angle is one for which exactly 90 of them make a right angle. The reference angle is called 1 degree, written 1°, and thus, the measure of a right angle is 90 degrees, written 90°. We use a protractor to measure angles. Angle measurement should be introduced with instruction on how to use a protractor.

Definition 1.64. A *right triangle* is a triangle for which one of its angles is a right angle. The side opposite the right angle is called the hypotenuse.

A triangle is *isosceles* if two of its sides are equal. In figures, the equal sides are designated by a similar dash through each, as in the following figure.

A triangle is called *equilateral* if all three sides are equal.

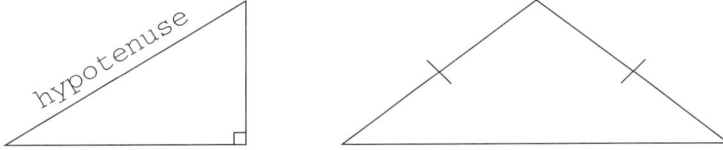

Theorem 1.65. *If a triangle is isosceles, then the angles opposite the equal sides are equal.*

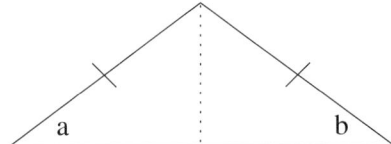

Proof. Shown in the figure is a dotted line which bisects the angle formed by the two equal sides. It is a *line of symmetry* of the triangle, which means that if the triangle is flipped over with this line as the axis of rotation, then the resulting figure will coincide with the original one. Such a flip will interchange the two angles opposite the equal sides. Therefore, they must coincide after the flip, so they are equal to each other. □

Corollary 1.66. *If a triangle is equilateral, then all three angles are equal.*

Oral exercise 1.67. Explain how to prove this Corollary.

1.3.9. Circle. A *circle* is the locus of all points whose distance is the same from the point called its *center*. The distance of each point from the center is called the *radius* of the circle. A *diameter* of the circle is a line segment from a point on the circle through the center to the *opposite* point on the circle. The length of the diameter is 2 radii.

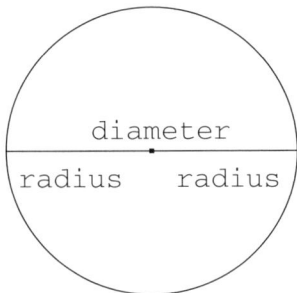

The perimeter of a circle is called its *circumference*. Here is a way to measure the circumference of a circle in terms of its diameter. Find an object which has a good, true circular shape, such as a coffee mug or a piece of heavy (not flexible) plastic pipe. Cut a strip of paper, about a half inch wide and whose length is at least four diameters of the circle. From one end of this paper strip carefully mark four segments each of which is one diameter long. Carefully wrap the paper strip around the mug and mark the circumference on the paper strip. The mark should lie between 3 and 4 diameters, which means that the circumference is between 3 and 4 diameters long.

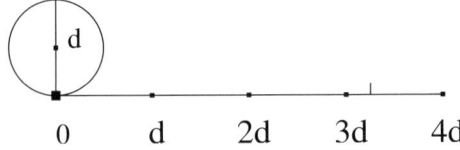

This relationship between the circumference and the diameter is the same for all circles. To see this, imagine a circle drawn on a sheet of rubber. Stretch the rubber to increase the diameter of the circle by a certain amount, say to double it. The stretching will double distances in all directions on the sheet of rubber. In particular, the circumference will also be doubled. Thus, the new circumference measured in terms of the new diameter will be the same as the original circumference measured in terms of the original diameter.

1.3. Measurement

Definition 1.68. The number π is defined to be the number of diameters in the circumference of a circle.

The measurement described above showed us that π is not a whole number. It is between 3 and 4. As we develop our concepts of number we will keep returning to the nature of π.

Oral exercise 1.69. Carry out the above measurement for some circular objects brought to class.

1.3.10. Time. Time is a measure of duration. We measure time by comparing the duration of an event with the duration of some reference event. A unit for measuring time is the *day*, which is the length of time from noon (at a given point on earth) until the next noon (at the same point). Noon at a point is when the sun passes directly over the meridian of that point (think of the sun passing through the plane determined by the meridian). This period of time actually varies a bit, so we use the average of these days over the period of a year, and call this the *Mean Solar Day*, which is the official day. At least, this was the official definition of the time period *one day* until quite recently. Now scientists define our unit of time in terms of the radiation emitted by the element cesium under specified conditions, a very precise standard to use, but one having no easy reference to daily life. For everyday use, the old fashioned definition of day serves our purposes well, and will be the definition used in this text.

The *hour* is then defined by saying that 24 hours make one day. Since there are 24 groups of 15 in the whole angle of 360°, it follows that an hour is the length of time it takes the sun to pass from one meridian to the line of longitude 15° to the west.

Example 1.70. Longitude 90° west (of the meridian through Greenwich, England) passes just to the east of St. Louis and the longitude 15° to the west of that is longitude 105°, which passes through Denver. Thus, it takes the sun exactly one hour to go from noon at St. Louis to noon at Denver.

Define 60 *minutes* to make one hour and 60 seconds to make 1 minute. Commonly used units of time are seconds, minutes and hours. A length of time of 2 hours 14 minutes 23 seconds is written

(1.8) $$2 : 14 :: 23$$

which is very much like a base sixty numeral, because each group of 60 seconds is a minute, indicated as one group of 60 seconds to the left of the :: sign. Each group of 60 minutes is an hour, indicated as one group of 60 minutes to the left of the : sign. But then each group of 24 hours is a day and thus the next place would not follow the base sixty pattern.

A year is the length of time it takes the earth to make 1 complete revolution around the sun. A year is approximately

365 days 5 hours 48 minutes 49 seconds

which, for practical purposes, is approximated by 365 days, with the extra time of 5 : 48 :: 49 temporarily ignored. After four years this ignored time has become just slightly less than one day, which is then added to the year and such years are called *leap* years. A leap year has 366 days and occurs every four years, on the years which are divisible by 4, such as 2004, 2008, 2012 and so on by fours. This adjustment is a bit too much (because not quite a whole day has accumulated in four years), so to adjust for that, every 100 years the leap year is skipped. But that adjusts too far, so every 400 years the leap year is not skipped. The rule is that the years divisible by 100 are not leap years unless they are divisible by 400. Thus, 2000 was a leap year, but 1900, 1800 and 1700 were not. After about 4000 years some further adjustment will be needed as the error will have become nearly a whole day.

A year is divided into twelve months on our Gregorian calendar. The number of days in each month varies, according to the following nursery rhyme found in [**Bak73**]:

> Thirty days hath September,
> April, June and November;
> February has twenty-eight alone;
> All the rest have thirty-one,
> Excepting leap-year, that's the time
> When February's days are twenty-nine.

A simpler version of this rhyme is the following, found in [**Dav48**]:

> Thirty days hath September,
> April, June and November;
> All the rest have thirty-one,
> Excepting February, twenty-eight alone.

This version is easier to remember, but it doesn't remind us of where to put the extra day in a leap year. In any case, the first two lines are the important ones to remember.

There are 7 days in a week and approximately 52 weeks in a year, as can be determined by counting the 365 days of a year into groups of 7 days.

Oral exercise 1.71. How many days more than 52 weeks is a nonleap year?

§1.3 Exercises

(1) Measure in inches the distance you go in two normal paces. What is this distance in feet and inches?

(2) Measure in inches the height of your front door. What is it in feet and inches? In yards, feet and inches?

(3) Use the line segment ____ as the unit to measure the line segment _____.
Hint: mark the length of the unit segment on the edge of a piece of paper and use that to mark units on the longer segment. *Ans:* 7 units.

(4) In terms of the unit line segment in the preceding problem, how much longer than the line segment _____ is the line segment _____? Describe what you measure and how you do it.

(5) (a) Measure _____ in centimeters.
(b) Measure _____ in inches.
(c) Measure the line segment in (a) in inches and in (b) in centimeters.

(6) The following regions are made of 1 inch squares.

(a) Find their areas. *Ans:* 7 in^2 and 7 in^2.
(b) Find their perimeters. *Ans:* 14 in and 16 in.

(7) The following figures are made of 1 ft squares.

(a) Find their areas.
(b) Find their perimeters.

(8) If the unit square has side length of 1 foot, what is its perimeter in feet?

(9) The following solids are made up of 1 inch cubes. Find the volume of each solid. *Ans:* 4 in^3 and 6 in^3.

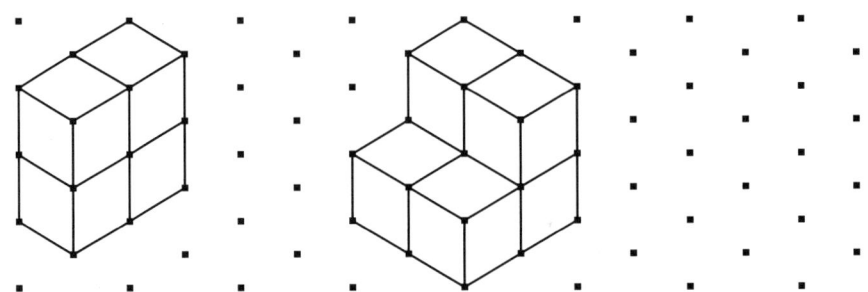

(10) The following solids are made up of 1 foot cubes. Find the volume of each solid.

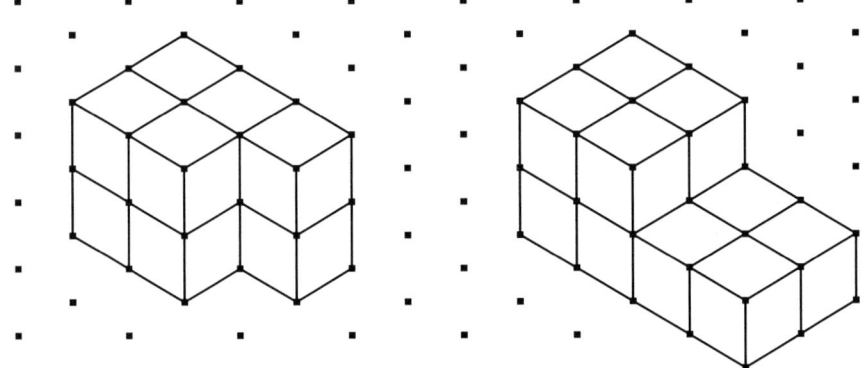

(11) How many 1 ft cubes fit into a 1 yard cube? Draw a picture of it. What is the volume in cubic feet of the cubic yard? *Ans:* 27 cubes. The volume is 27 ft^3.

(12) Imagine that we have some measuring cups, some pint jars and some quart jars. Suppose that we have a bucket of water which exactly fills 9 measuring cups.
  (a) What is the maximum number of pint jars that can be filled? How many full cups will remain?
  (b) What is the maximum number of quart jars that can be filled? How many full pints will remain?
  (c) Express the quantity of water in terms of quarts, pints, and cups.
  (d) What is the base two numeral for the number of cups of water originally in the bucket?

Chapter 2

Whole Number Arithmetic

The four arithmetic operations on whole numbers are given precise definitions. The essential properties of each operation are stated and proved. In the last section the idea of partitive division is extended to the idea of dividing any object into a given number of equal parts.

2.1. Addition

We define addition in terms of the union of disjoint sets.

Definition 2.1 (Addition). If a and b are whole numbers, and if A and B are disjoint sets such that $n(A) = a$ and $n(B) = b$, then the sum $a + b$ is defined to be $a + b = n(A \cup B)$.

Read $a + b$ as "a plus b" or as "b added to a". The numbers a and b in $a + b$ are called the *addends* and the number $a + b$ is called the *sum*.

Example 2.2. To define $3 + 4$, take a set such as $A = \{\dagger\ \dagger\ \dagger\}$, so that $n(A) = 3$ and a set such as $B = \{\ddagger\ \ddagger\ \ddagger\ \ddagger\}$, so that $n(B) = 4$. Then A and B are disjoint sets and $3 + 4 = n(A \cup B) = n(\{\dagger\ \dagger\ \dagger\ \ddagger\ \ddagger\ \ddagger\ \ddagger\}) = 7$.

Example 2.3. Four children are playing trench. Two more children join the game. How many children altogether are then playing trench? Picture the 4 children playing, say with a circle drawn around them to indicate a set of children, and picture the two children arriving to join the game. These two sets of children are disjoint and their union is the set of children now playing trench. Therefore, there are $4 + 2$ children now playing trench.

If $A \sim \mathcal{A}$ and $B \sim \mathcal{B}$, and if $A \cap B = \emptyset$ and $\mathcal{A} \cap \mathcal{B} = \emptyset$, then combine a 1:1 correspondence from A to $\mathcal{A}$ with a 1:1 correspondence from B to $\mathcal{B}$ to produce a 1:1 correspondence from $A \cup B$ to $\mathcal{A} \cup \mathcal{B}$. Therefore, $\mathcal{A} \cup \mathcal{B} \sim A \cup B$ and thus $n(\mathcal{A} \cup \mathcal{B}) = n(A \cup B)$. This shows that in the definition of $a + b$ it does not matter which sets A and B you use as long as each has the correct number of elements. It is this simple fact that lies behind the wide application of addition.

Theorem 2.4. *Addition has the following properties. If $a, b,$ and c are whole numbers, then*

(1) The commutative property says that b added to a is the same number as a added to b:
$$a + b = b + a$$

(2) The associative property says that in adding more than once, the additions can be grouped in pairs in any possible way:
$$a + (b + c) = (a + b) + c$$

(3) The identity property of 0 says
$$a + 0 = 0 + a = a$$

(4) The order property of addition says that a given amount added to unequal numbers leaves the inequalities in the same order: If $b < c$, then for any whole number a,
$$b + a < c + a$$

Proof. In this proof, as for most proofs in this book, we shall present a specific case in such a way as to illustrate the principles of the proof of the general case.

(1) For the commutative property, let us prove that $3 + 5 = 5 + 3$. Take any set of 3 elements, say $A = \{a, b, c\}$ and any set of 5 elements disjoint from A, say $B = \{p, q, r, s, t\}$. Then $A \cup B = B \cup A$ by the commutative property of union, and therefore
$$3 + 5 = n(A \cup B) = n(B \cup A) = 5 + 3$$
Both sums equal the number of elements in the set $\{a, b, c, p, q, r, s, t\}$. This argument illustrates the general case, in contrast to an argument like the following: since $3 + 5 = 8$ and $5 + 3 = 8$, it follows that $3 + 5 = 5 + 3$.

(2) Next we explain why $3 + (5 + 2) = (3 + 5) + 2$. Continue to use the sets A and B and take any set of 2 elements, disjoint from A and B, say $C = \{h, j\}$. Then
$$3 + (5 + 2) = 3 + n(B \cup C) = n(A \cup (B \cup C))$$
$$= n((A \cup B) \cup C) = n(A \cup B) + 2 = (3 + 5) + 2$$

The middle equality is true by the associative property of set union. Both sums equal the number of elements in the set

$$A \cup (B \cup C) = \{a, b, c, p, q, r, s, t, h, j\} = (A \cup B) \cup C$$

Since these two sets are the same, they are certainly equivalent, and that proves the associative property of addition.

(3) Let us prove the identity property for the case $3+0$. Using the same set A above, we have

$$3 + 0 = n(A \cup \emptyset) = n(A) = 3$$

because $A \cup \emptyset = A$. By the commutative property we know that $0+3 = 3+0$, so we conclude also that $0 + 3 = 3$.

(4) Let us prove the order property for the case of adding 2 to both sides of the inequality $3 < 5$. Now $3 < 5$ because $P = \{1, 2, 3\} \subset Q = \{1, 2, 3, 4, 5\}$ and $3 = n(P)$, $5 = n(Q)$. The set $C = \{h, j\}$ above has 2 elements and is disjoint from Q. By the order property for union,

$$P \cup C \subset Q \cup C$$

By definition of addition,

$$3 + 2 = n(P \cup C) \quad \text{and} \quad 5 + 2 = n(Q \cup C)$$

Therefore, $3 + 2 < 5 + 2$ by Definition 1.17. □

Because of the associative property, we may omit the parentheses and write $a+b+c$ to mean either $a + (b + c)$ or $(a + b) + c$. The associative and commutative properties combined imply that the numbers in this sum can be rearranged in any order without changing the sum. Thus

$$a + b + c = a + c + b = b + a + c = b + c + a = c + a + b = c + b + a$$

Example 2.5. Joshua has 3 marbles and his brother Nicholas has 5 marbles. If they put their marbles together in a jar, there will be $3+5$ marbles in the jar, regardless of who puts their marbles into the jar first.

Pairing each of his marbles off with one of Nicholas's, Joshua sees that he has fewer marbles than Nicholas. Anthony gives each of them 4 new marbles. Now Joshua has $3 + 4$ marbles, which is still less than Nicholas's $5 + 4$ marbles, because the new marbles can be paired off and the extra marbles Nicholas had remained unpaired.

This can be illustrated with sets as follows. Before Anthony's gift

```
Joshua   {  O  O  O        }
            ↕  ↕  ↕
Nicholas {  O  O  O  O  O  }
```

After Anthony's gift:

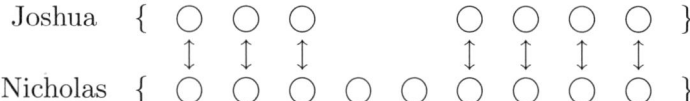

This can be illustrated with line segment diagrams showing the paired parts as follows. Before Anthony's gift:

Joshua ⌊―――3―――⌋
Nicholas ⌊―――――3―――――⌋
         ⌊―――5―――⌋

After Anthony gives them each 4 marbles:

Joshua ⌊―――3―――⌋―――4―――⌋
Nicholas ⌊―――――3―――――⌋―――4―――⌋
         ⌊―――5―――⌋

Example 2.6. Addison has 5 eggs in his basket when he goes to the chicken house to gather eggs. He finds no eggs to add to his basket. He finishes with altogether $5 + 0 = 5$ eggs in his basket.

2.1.1. Addition on the Number Line. To calculate $3 + 5$, start at 3 on the number line and count 5 numbers consecutively to the right. One ends at the number $3 + 5$ because there are 3 unit intervals from 0 to 3 and we go 5 more unit intervals to the right of 3 so the total number of unit intervals from 0 to the end point is the number in the union of these two sets of intervals.

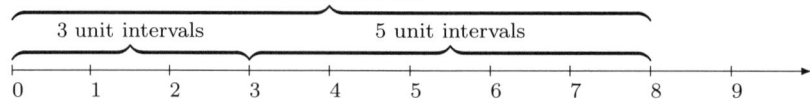

Theorem 2.7 (Addition on the Number Line). *If m and n are whole numbers, then $m + n$ is the point on the number line arrived at by starting at m and going n unit intervals to the right from m.*

Proof. There are m unit intervals from 0 to m on the number line, and there are n different unit intervals going that many to the right of m. By definition of addition, there are $m + n$ unit intervals in the union of these two sets, and therefore $m + n$ is the number represented by the end point of going n units to the right of m. □

The idea of the number line provides a powerful tool for picturing the information contained in a problem, and for explaining what arithmetic operations are needed in order to solve the problem. Following the practice of the Singapore Texts [**Tea99**], we shall develop this technique of line segment diagrams throughout the text.

Example 2.8. Sums arise in comparisons. Rita has a coffee that costs $3 and a sandwich which costs $4 more than the coffee. How much is her bill? If we picture this information as follows

```
    coffee    |―――――|
              $3
    sandwich  |―――――――――――――|  ] ?
                      $4
```

then addition on the number line informs us that the sandwich cost $3 + 4$ dollars and that Rita's whole bill is $3 + (3 + 4)$ dollars.

Oral exercise 2.9. Niels's class has 23 pupils, Leah's has 31, and Rochelle's has 14.

(1) If Niels brings his pupils into Leah's classroom, express as a sum the number of pupils there will be in that classroom. If Leah brings her pupils into Niels's classroom, express as a sum the number of pupils there will be in that classroom. What property of addition is demonstrated by this example?

(2) Niels's class is on the playground for recess, where they are joined by Leah's class and shortly thereafter by Rochelle's class. Express the total number of children on the playground (from these three classes) as a sum. Now suppose that Rochelle had brought her pupils to Leah's room and then the two classes joined Niels's class on the playground. Express the total number of children on the playground as a sum. What property of addition is demonstrated by this example?

2.1.2. Numeration and sums. In our numeration system, after counting a set of objects into groups of ten repeatedly, we end up with the original set decomposed into the union of the set of ones, the set of tens, the set of tens of tens, and so on, and these are disjoint subsets. Therefore, the number of objects is the sum of the ones, the tens, the tens of tens, and so on. The same comment applies to measurements expressed in more than one unit.

Example 2.10. A set of 27 apples can be counted into two groups of ten with 7 apples remaining. The original set of apples is decomposed into the subset of 2 groups of ten and the subset of 7. Therefore,
$$27 = 20 + 7$$
because 2 groups of ten is 20 according to our numeration system. By the same reasoning,
$$483 = 400 + 80 + 3$$
because 4 groups of ten-of-ten is 400 and 8 groups of ten is 80, according to our numeration system.

Example 2.11. A height of 6 feet 2 inches is the union of 6 feet with 2 inches. This becomes a sum if we convert it all to inches. Then 6 feet is 72 inches and this height is $72 + 2 = 74$ inches.

If Morris drives for 1 hour and 10 minutes, stops for 30 minutes to unload his truck, and then drives another 3 hours and 20 minutes, the total time he drove is the sum of 1 hour 10 minutes and 3 hours 20 minutes. Viewed as the union of sets of hours and minutes, this is

$$1 + 3 \text{ hours } 10 + 20 \text{ minutes} = 4 \text{ hours } 30 \text{ minutes}$$

Oral exercise 2.12. (1) Explain why $709 = 700 + 9$.

(2) Explain why $423_5 = 400_5 + 20_5 + 3$.

(3) Explain why $1101_2 = 1000_2 + 100_2 + 1$.

(4) In what sense is 1 hour 10 minutes 3 seconds a sum?

2.1.3. Decomposing Sets. Decomposing a set into the union of a subset with the complement of the subset leads to the idea of decomposing a whole number into the sum of two numbers. For example, if

$$A = \{\dagger \, \dagger \, \ddagger \, \ddagger \, \ddagger\}$$

and $B = \{\dagger \, \dagger\} \subset A$, then the complement of B in A is $B' = \{\ddagger \, \ddagger \, \ddagger\}$ and

$$A = B \cup B'$$

is a union of disjoint sets showing that $5 = 2 + 3$, since A has 5 elements, B has 2 elements, and B' has 3 elements.

Making such decompositions can simplify mental calculations of sums.

Example 2.13. Knowing the decompositions $10 = 6 + 4$ and $5 = 4 + 1$, I can do the following addition by decomposing and making 10:

$$6 + 5 = 6 + 4 + 1 = 10 + 1 = 11$$

Oral exercise 2.14. Decompose and make 10 to find: (a) $5 + 9$, (b) $8 + 6$, (c) $7 + 6$, (d) $9 + 8$.

2.1.4. Angles of a Triangle. Cut out a triangle from a sheet of paper. Tear off two of the vertices and place the pieces beside the remaining vertex to show that the sum of the angles of a triangle is a straight angle. In degree measure of angles, this means that the sum of the angles of a triangle is $180°$.

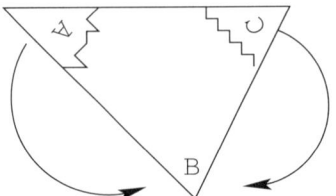 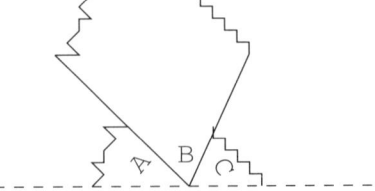

§2.1 Exercises

The object of every exercise is to explain how it is done. For the word problems, use line segment diagrams to illustrate your solution.

(1) Fill in the blanks, then explain what ideas are being illustrated.

$$\{\bigcirc\ \bigcirc\ \bigcirc\}\quad\{\oplus\ \oplus\ \oplus\ \oplus\}$$

There are ___ plain balls.
There are ___ decorated balls.
There are ___ balls altogether. *Ans:* There are 3 plain balls and 4 decorated balls, both answers obtained by counting the balls in each set. There are 7 balls altogether, determined by counting all the balls in both sets. The problem illustrates counting the number of elements in a set and counting the number of elements in the union of two disjoint sets.

(2) Fill in the blanks.

$$\{\bigcirc\ \bigcirc\ \bigcirc\ \bigcirc\}\leftarrow\{\bigcirc\ \bigcirc\ \bigcirc\ \bigcirc\ \bigcirc\}$$

There are 4 balls in the set.
Add ___ more. So 4 and ___ make ___.
There are ___ balls altogether.

(3) Make up a story problem, with picture diagrams, illustrating the order property of addition.

(4) Make up a story problem that illustrates the commutative property of addition.

(5) Illustrate a set as a union of a subset with its complement to demonstrate the decomposition $8 = 5 + 3$.

(6) Use the number line to illustrate $3 + 1$, $3 + 2$, $3 + 3$, and $3 + 0$.

(7) Find all whole numbers a and b for which $a + b = 9$. *Ans:* $0+9$, $1+8$, $2+7$, $3+6$, $4+5$, $5+4$, $6+3$, $7+2$, $8+1$, $9+0$.

(8) Find all whole numbers a and b for which $a + b = 10$.

(9) Decompose and make 10 to find $8 + 7$. Make a picture diagram to illustrate this. *Ans:* $8+7 = 8+(2+5) = (8+2)+5 = 10+5 = 15$, the last step being numeration.

(10) Decompose and make 10 to find $6 + 9$. Cite the role of numeration.

(11) Make a picture of 7 objects and use it to illustrate what number goes in the □ in the equation $5 + □ = 7$. *Ans:* $\{\dagger\ \dagger\ \dagger\ \dagger\ \dagger\ \ \dagger\ \dagger\}$.

(12) Use a picture of a set with 6 elements to illustrate what number goes in the □ in □ + 4 = 6.

(13) Sigrid, Sue, and Jen enter into partnership: Sigrid puts in 6420 dollars; Sue 5130 dollars; and Jen as much as Sigrid and Sue together: how much did Jen put in and how much did all three put in together? *Ans:* Use the line segment diagram

```
        Sigrid        Sue
      |―――――――|    |―――――――|
        $6420         $5130
      |―――――――――――――――――――|
                 Jen
```

(14) Morris bought a house, a farm, a store, and a mill. The house cost 3500 dollars; the farm, 7800 dollars; the store, as much as the house and the farm; and the mill, as much as the house and the store: what was the cost of the store, the cost of the mill, and the cost of everything together? Use a line segment diagram to illustrate your solution.

(15) Tony bought a house and a cafe. The house cost $35000. The cafe cost $15000 more than the house. What was the total cost of the house and the cafe? *Soln:* First find the cost of the cafe.

```
                   35000
  house  |―――――――――――――――――――|
  cafe   |―――――――――――――――――――――――|
                                15000
```

(16) Diane worked for 3 hours and 20 minutes on her homework. After a break of 1 hour and 10 minutes, she worked another 2 hours and 30 minutes on her homework. How much time did she spend actually working on her homework?

(17) John's front yard has an area of 124 ft^2 and his back yard has an area of 235 ft^2. If his yard comprises the front yard and the back yard, what is the area of his yard? *Ans:* Front yard is a set of 124 ft^2 and the back yard is a set of 235 ft^2. The two yards together would be the union of these two sets of square feet, which will have 124 + 235 ft^2.

(18) (Triangle Inequality) Explain why the sum of any two sides of a triangle is greater than the third side. In your argument, you may assume known that the straight line is the shortest path joining two points.

(19) *Theorem*: If a, b, and c are numbers for which the sum of any two of them is greater than the third, then there is a triangle whose sides are of length a, b, and c. Explain the following proof: Assume that $a \geq b$ and $a \geq c$. Draw a line segment of length a. At one end draw a circle of radius b. At the other end draw a circle of radius c. Join a point of intersection of these two circles with the end points of the line segment, as shown:

2.2. Subtraction

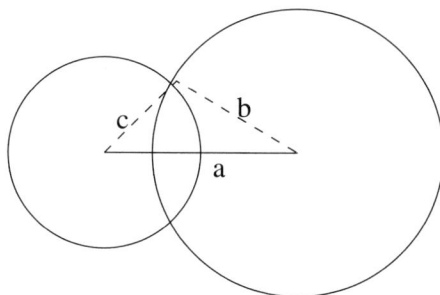

What needs explanation is why the two circles must intersect. Why are the following two situations impossible?

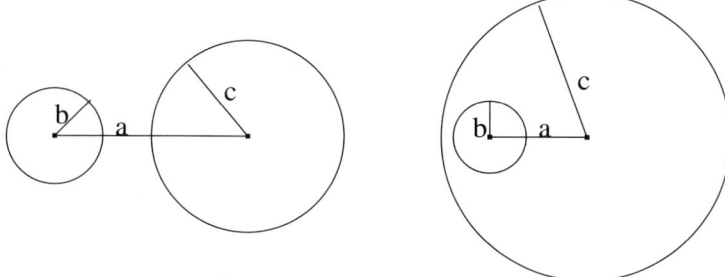

(20) Follow the proof given in the preceding problem to construct with ruler and compass a triangle with sides of length 2 inches, 3 inches and 4 inches. Do these three numbers satisfy the hypothesis of the Theorem in the preceding problem?

2.2. Subtraction

We define subtraction to be the inverse of addition. This can be done in two ways, as *comparative subtraction* or as *take-away subtraction*. Both ways yield the same difference, because addition is commutative.

Definition 2.15 (Comparative subtraction)**.** If a and b are whole numbers and if $b \leq a$, then $a - b$ is that whole number c for which $b + c = a$. In symbols, $a - b = c$ means that $a = b + c$.

In words, $7 - 3$ is what must be added to 3 to get 7. It is the amount 7 is larger than 3. To calculate $7 - 3$, start at 3 and count $\{4, 5, 6, 7\}$ until you reach 7. There are $7 - 3$ numbers in this set.

Definition 2.16 (Take-away subtraction)**.** If a and b are whole numbers and if $b \leq a$, then $a - b$ is that whole number c for which $c + b = a$. In symbols, $a - b = c$ means that $a = c + b$.

In words, $7 - 3$ is the number of elements remaining after 3 elements are taken away from a set of 7 elements. Add back the 3 removed elements and we have again a set of $(7 - 3) + 3$ elements.

The statement $a - b$ is read *a minus b* and the result is called the *difference*. The number a is called the *minuend* and the number b is called the *subtrahend*. In words, the definition of comparative subtraction is

the sum of the subtrahend and difference is the minuend.

The definition of take-away subtraction is

the sum of the difference and subtrahend is the minuend.

Remark 2.17. These definitions say the following. If you state that $275 - 81 = 194$, then you mean that $275 = 81 + 194$ (comparative subtraction) or that $275 = 194 + 81$ (take-away subtraction), and of course both sums are the same because of the commutative property of addition.

Theorem 2.18. *If a and b are whole numbers and $b \leq a$, then $a - b$ in comparative subtraction is equal to $a - b$ in take-away subtraction.*

Proof. If $c = a - b$ for comparative subtraction, then $b + c = a$. But $b + c = c + b$, by the commutative property of addition, and therefore $c + b = a$ and $c = a - b$ for take-away subtraction as well. □

Why is $7 - 3 = 4$? Because $3 + 4 = 7$ (comparative subtraction) and because $4 + 3 = 7$ (take-away subtaction).

Oral exercise 2.19. Why is $9 - 5 = 4$? Why is $30 - 20 = 10$? (Give both the comparative subtraction and take-away subtraction definitions).

Example 2.20 (Comparative subtraction). How much larger than 3 is 5? The question means how much has to be added to 3 to make 5.

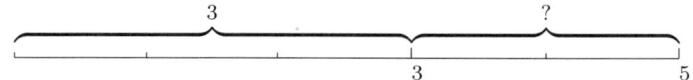

The answer is $5 - 3 = 2$ because $5 = 3 + 2$; that is, $5 - 3 = 2$ must be added to 3 to make 5.

Example 2.21 (Making change). JB pays for a \$2 coffee with a \$5 dollar bill. How much change does he receive? The change is the amount 5 is larger than 2, therefore it is $5 - 2 = 3$, since $2 + 3 = 5$. The old-fashioned way of counting change is based on comparative subtraction. To find the change, start with the value of the purchase and count up to the value paid for it. In this case, start at 2, count the set $\{3, 4, 5\}$ to find that the change is 3 dollars.

2.2. Subtraction

Example 2.22 (Comparative subtraction). If I have 6 pencils and 9 children to whom I want to give a pencil, how many more pencils do I need in order to have 1 pencil for each child?

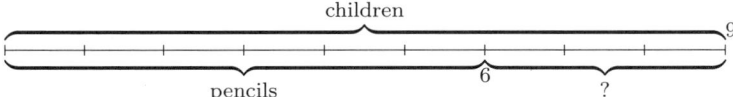

The answer is $9 - 6 = 3$, because $9 - 6$ is how much larger 9 is than 6.

Example 2.23 (Comparative subtraction). It is 7 miles from home to Grandmother's house via Hathaway, which is 3 miles from home. As we drive through Hathaway, how much farther is it to Grandmother's house?

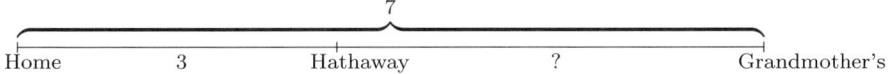

The answer is the number of miles that must be added to 3 to get 7, which is $7 - 3$ because $3 + (7 - 3) = 7$, by definition of comparative subtraction.

Theorem 2.24 (Comparative subtraction on the number line). *If a and b are whole numbers with $a \geq b$, then $a - b$ is the number of unit intervals required to go from b to a on the number line.*

Proof. Since $a \geq b$, we know that a lies to the right of b on the number line. If $c = a - b$, then $a = b + c$ shows that if we start at b and go c units to the right we end at a. □

Here is an illustration of the comparative subtraction $7 - 4 = 3$ on the number line. It shows that $7 - 4$ added to 4 is 7; that is, $4 + (7 - 4) = 7$.

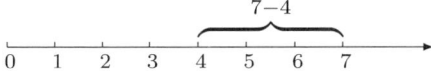

Oral exercise 2.25. Make up a story problem whose solution uses comparative subtraction. Relate its solution to the definition or the number line.

Example 2.26 (Take-away subtraction). If I have 5 marbles and I take away 3 of them, how many are left? If c is the number of marbles left, and if I add to these c the 3 taken away, I'll have the original 5 marbles. That is, $c + 3 = 5$. Therefore, $c = 5 - 3$ is the number of marbles left.

Theorem 2.27 (Take-away subtraction). *If A is a set and if a subset B is removed from A, then $n(A) - n(B)$ elements remain in A. In other words, $n(A) - n(B)$ is the number of elements in the complement of B in A.*

Proof. The idea of the proof can be seen in the proof of the case where $A = \{\square\square\square\square\triangle\triangle\triangle\}$, and B is the subset of all $\square$'s in A. If the elements of B are removed from A, then what remains is the subset $C = \{\triangle\triangle\triangle\}$, which is the complement of B in A. Hence, B and C are disjoint and $A = C \cup B$, which shows that $n(A) = n(C) + n(B)$, which shows that $n(A) - n(B) = n(C)$ in take-away subtraction. $\square$

It is sometimes difficult to see the distinction between comparative subtraction and take-away subtraction. Their representations on the number line, however, are very different. Compare the following theorem to Theorem 2.24.

Theorem 2.28 (Take-away subtraction on the number line). *If a and b are whole numbers with $a \geq b$, then $a - b$ of take-away subtraction is the point on the number line arrived at by starting at a and going b unit intervals to the left.*

Proof. There are a unit intervals from 0 to a. Going b unit intervals to the left from a is to take away b of these unit intervals. The number of unit intervals from 0 to the point reached is the number of unit intervals remaining and is thus $a - b$. $\square$

Here is an illustration of $7 - 4 = 3$ as take-away on the number line. It shows that 4 added back to the $7 - 4$ remaining gives 7, that is $3 + 4 = 7$.

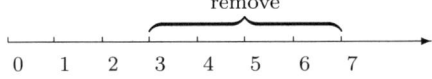

Example 2.29 (Take-away subtraction). If I have a bag of 11 candy bars and I take away 5 of them, there remain $11 - 5 = 6$ candy bars in the bag. If I add the 5 candy bars back to the bag, there will again be $6 + 5 = 11$ candy bars in the bag.

$$\underset{0\ \ 1\ \ 2\ \ 3\ \ 4\ \ 5\ \ 6\ \ 7\ \ 8\ \ 9\ \ 10\ \ 11}{\overset{\text{remove }5}{\overbrace{\qquad\qquad\qquad}}}$$

Example 2.30. Every sum of two numbers is related to two subtraction problems. For example, $8 = 3 + 5$ and therefore $8 - 3 = 5$ and $8 - 5 = 3$. In pictures,

$$\{\square\square\square\square\square\square\square\square\} = \{\square\square\square\} \cup \{\square\square\square\square\square\}$$

Oral exercise 2.31. Make up a story problem whose solution uses take-away subtraction.

2.2. Subtraction

Oral exercise 2.32. Make up a story problem whose solution uses comparative subtraction.

2.2.1. Properties of Subtraction.

Theorem 2.33. *Subtraction has the following properties. In each identity it is assumed that the indicated subtractions are defined.*

(1) The identity property of 0 means that 0 subtracted from any number leaves the number unchanged. If a is any whole number, then

$$a - 0 = a$$

(2) The order property of subtraction asserts that if the same amount is subtracted from both sides of an inequality, the inequality is unchanged:

$$\text{If } a < b, \text{ then } a - c < b - c.$$

(3) The translation property of subtraction asserts that the difference between two whole numbers is unchanged if the same amount is added or is subtracted from each: If a, b, and c are whole numbers, then

$$a - b = (a + c) - (b + c)$$
$$a - b = (a - c) - (b - c)$$

(4) The subtraction associative properties are: If a, b, and c are whole numbers, then

$$(a + b) - c = a + (b - c)$$
$$a - (b - c) = (a - b) + c$$
$$a - (b + c) = (a - b) - c$$
$$a - (b - c) = (a + c) - b$$

Proof. The proof of each statement uses the definition of subtraction, as in Remark 2.17: To check that something is the difference of two numbers, we must show that the sum of that something and the subtrahend equals the minuend.

(1) The identity property of 0, which is $a - 0 = a$ is true because $a = 0 + a$.

We shall use specific numbers to illustrate the idea of the proof of each of the remaining properties.

(2) To prove the order property, consider what happens when we subtract 3 from both sides of the inequality $5 < 9$. On the number line, 5 lies to the left of 9. Think of a rod lying on the number line with its left end at 5 and its right end at 9. Slide the rod 3 units to the left. Then its left end will be at $5 - 3$ and its right end will be at $9 - 3$, so $5 - 3$ lies to the left of $9 - 3$. Hence $5 - 3 < 9 - 3$.

(3) The same kind of argument on the number line proves the translation properties as follows. To prove the first translation property, for the case

$$9 - 5 = (9 + 3) - (5 + 3)$$

think of $9 - 5$ as the number of unit intervals from 5 to 9. This is the number of unit intervals under the brace running from 5 to 9. Slide the brace to the right 3 units, and it runs from $5 + 3$ to $9 + 3$, showing that $(9 + 3) - (5 + 3) = 9 - 5$.

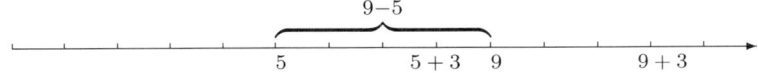

The same argument proves the second translation property, for the case

$$9 - 5 = (9 - 3) - (5 - 3)$$

because now we slide the brace to the left by 3 units instead of to the right.

(4) The subtraction associative properties are proved from the definition of subtraction in the way described in Remark 2.17.

To prove the first subtraction associative property for the case

$$(2 + 5) - 3 = 2 + (5 - 3)$$

the minuend is $(2 + 5)$, the subtrahend is 3, and $2 + (5 - 3)$ is the difference because it plus the subtrahend is

$$2 + (5 - 3) + 3 = 2 + 5$$

is the minuend. Notice that $(5 - 3) + 3 = 5$ by the definition of comparative subtraction.

To prove the second subtraction associative property for the case

$$7 - (5 - 3) = (7 - 5) + 3$$

the minuend is 7, the subtrahend is $(5 - 3)$, and thus, the difference is $(7 - 5) + 3$ because it plus the subtrahend is

$$(7 - 5) + 3 + (5 - 3) = (7 - 5) + 5 = 7$$

is the minuend. Notice that $3 + (5 - 3) = 5$ by definition of comparative subtraction and that $(7 - 5) + 5 = 7$ by definition of take-away subtraction.

To prove the third subtraction associative property for the case

$$7 - (4 + 2) = (7 - 4) - 2$$

the minuend is 7, the subtrahend is $(4+2)$, and thus, the difference is $(7-4) - 2$ because the subtrahend plus this is

$$(4+2) + ((7-4) - 2) = 4 + (2 + ((7-4) - 2)) = 4 + (7-4) = 7$$

which is the minuend.

To prove the fourth subtraction associative property for the case

$$7 - (9 - 4) = (7 + 4) - 9$$

the minuend is 7, the subtrahend is $(9-4)$. The difference is $(7+4) - 9$, because this plus the subtrahend is

$$((7+4) - 9) + (9-4) = (7+4) - (9 - (9-4)), \text{ 2nd sub. assoc. prop.}$$
$$= (7+4) - ((9-9) + 4), \text{ 2nd sub. assoc. prop.}$$
$$= (7+4) - 4 = 7 + (4-4) = 7, \text{ 1st sub. assoc. prop.}$$

which is the minuend. $\square$

Oral exercise 2.34. Explain how the first subtraction associative property, for the case $(2+5) - 3 = 2 + (5-3)$, is illustrated by

$$\{\bigcirc \bigcirc \triangle \triangle \cancel{\triangle}\cancel{\triangle}\cancel{\triangle}\}$$

Oral exercise 2.35. Explain how the second subtraction associative property, for the the case $7 - (5-3) = (7-5) + 3$, is illustrated by

$$\{\triangle \triangle \overbrace{\{\square \square \cancel{\oslash} \cancel{\oslash} \cancel{\oslash}\}}^{\uparrow}\} = \{\triangle \triangle \cancel{\square}\cancel{\square} \bigcirc \bigcirc \bigcirc\}$$
$$= \{\triangle \triangle \cancel{\square}\cancel{\square} \cancel{\oslash} \cancel{\oslash} \cancel{\oslash}\} \cup \{\bigcirc \bigcirc \bigcirc\}$$

Oral exercise 2.36. Explain how the third subtraction associative property, for the case $7 - (4+2) = (7-4) - 2$, is illustrated by

$$\{\triangle \triangle \overbrace{\{\square \square \square \square \bigcirc \bigcirc\}}^{\uparrow}\} = \{\triangle \triangle \overbrace{\{\square \square \square \square\}}^{\uparrow} \cancel{\oslash} \cancel{\oslash}\}$$

Oral exercise 2.37. Explain how the fourth subtraction associative property, for the case of $7 - (9-4) = (7+4) - 9$, is illustrated by

$$\{\triangle \triangle \triangle \square \square \square \square\} - \{\square \square \square \square \square \cancel{\oslash} \cancel{\oslash} \cancel{\oslash} \cancel{\oslash}\} = \{\triangle \triangle\}$$
$$= \{\triangle \triangle \cancel{\square}\cancel{\square}\cancel{\square}\cancel{\square} \cancel{\oslash} \cancel{\oslash} \cancel{\oslash} \cancel{\oslash}\}$$

2.2.2. Mental Math. The translation properties justify mental math calculations of the following type. The stategy is to add or subtract a fixed amount to both minuend and subtrahend in order to simplify the calculation. For example

$$13 - 8 = (13+2) - (8+2) = 15 - 10 = 5$$

where the last step follows from numeration, $15 = 5 + 10$, and the first subtraction associative property

$$15 - 10 = (5 + 10) - 10 = 5 + (10 - 10) = 5$$

Oral exercise 2.38. Explain how to do the following in your head:
(a) $17 - 9$. (b) $13 - 5$.

Oral exercise 2.39. Which subtraction properties are being used to find $15 - 3$ by thinking

$$15 - 3 = (10 + 5) - 3 = 10 + (5 - 3) = 10 + 2 = 12$$

Explain how this is illustrated by

$$\left\{\begin{array}{cccccccccc} \square & \square & \square & \square & \square & \square & \square & \square & \square & \square \\ \square & \square & \not\square & \not\square & \not\square & & & & & \end{array}\right\}$$

Example 2.40. Combining the subtraction associative properties leads to properties like

$$(a + b) - (c + d) = (a - c) + (b - d)$$

as long as $a \geq c$ and $b \geq d$. Here is the proof, with commutative and associative properties of addition used without comment.

$$\begin{aligned}
(a + b) - (c + d) &= ((a + b) - c) - d, \text{ 3rd sub. assoc. prop.} \\
&= ((b + a) - c) - d \\
&= (b + (a - c)) - d, \text{ 1st sub. assoc. prop.} \\
&= ((a - c) + b) - d \\
&= (a - c) + (b - d), \text{ 1st sub. assoc. prop.}
\end{aligned}$$

This new property justifies subtraction place by place, as in

$$37 - 14 = (30 + 7) - (10 + 4) = (30 - 10) + (7 - 4)$$

showing that $37 - 14$ is found by subtracting the one's place $7 - 4$ and subtracting the ten's place $30 - 10$, or rather, $3 - 1$ tens, and adding the results.

2.2.3. Counting Nondisjoint Unions. In the definition of addition, why is such a fuss made about the two sets being disjoint? After all, when teaching addition in the first grade, we would use sets of actual objects such as marbles or blocks in which cases it is difficult to imagine a situation where the two sets being combined are not disjoint. Not long after the introduction of addition and subtraction, you might want to consider the following sort of problem. For the group of people Dona Cragar, Samantha Cragar, Andrea

2.2. Subtraction

Cragar, Nicholas Bartholomew, Joshua Bartholomew, Caitlin Bartholomew, and Anthony Williams, let

$$A = \text{the set of these people whose first name begins with A}$$
$$= \{\text{Andrea Cragar, Anthony Williams}\}$$

and let

$$C = \text{the set of these people whose last name begins with C}$$
$$= \{\text{Dona Cragar, Samantha Cragar, Andrea Cragar}\}$$

How many people in this group have first name beginning with A or last name beginning with C? To answer this question, we first determine that the set of all such people is $A \cup C$. The number of such people is then the number of elements in $A \cup C$, which we write

$$n(A \cup C) = 4$$

This is not the number of elements in A plus the number of elements in C, which is

$$n(A) + n(C) = 2 + 3 = 5$$

because this latter sum counts Andrea Cragar, twice. She is in both sets,

$$A \cap C = \{\text{Andrea Cragar}\}$$

Here is the general result.

Theorem 2.41. *If A and C are sets, then*

(2.1) $$n(A \cup C) = n(A) + n(C) - n(A \cap C)$$

Proof. The following line segment diagram illustrates why the sum $n(A) + n(C)$ counts the elements in the intersection twice, while the number of elements in $A \cup C$ counts them once.

```
A          ⊢─────────┬─── A∩C ───┤
C                    ⊢── A∩C ──┬──────────────┤
A∪C        ⊢─────────┬─── A∩C ──┬──────────────┤
```

□

Example 2.42. An *outcome* of rolling a green die and a red die is the pair of numbers (green first, red second) giving the number of dots on the top face of each die. For how many outcomes will the number up on at least one of the dice be 1 or 2?

Solution: Let G denote the set of all outcomes for which the number on the green die is 1 or 2. Remember that we list the outcome (green, red).

$$G = \{(1,1)\,(1,2)\,(1,3)\,(1,4)\,(1,5)\,(1,6)$$
$$(2,1)\,(2,2)\,(2,3)\,(2,4)\,(2,5)\,(2,6)\}$$

When the green die comes up 1, the red die could be 1, 2, 3, 4, 5 or 6, so there are 6 outcomes for which the green die comes up 1. In the same way, there are 6 outcomes for which the green die comes up 2. Therefore, $n(G) = 6 + 6 = 12$.

Let R denote the set of all outcomes for which the number on the red die is 1 or 2.
$$R = \{(1,1)\,(2,1)\,(3,1)\,(4,1)\,(5,1)\,(6,1)$$
$$(1,2)\,(2,2)\,(3,2)\,(4,2)\,(5,2)\,(6,2)\}$$
By the same counting procedure we conclude that $n(R) = 12$.

G and R are not disjoint. In fact, $G \cap R$ is the set of outcomes for which the number up on both dice is 1 or 2, and we can count to see that there are 4 such outcomes.
$$G \cap R = \{(1,1)\,(1,2)\,(2,1)\,(2,2)\}$$
The problem asks us to find the number of outcomes in G or in R, and this number is $n(G \cup R)$. Theorem 2.41 tells us that
$$n(G \cup R) = n(G) + n(R) - n(G \cap R) = 12 + 12 - 4 = 20$$
is the number of outcomes for which at least one of the dice is a 1 or a 2.

§2.2 Exercises

In problems 3-8, cite the subtraction property used and then give a picture proof of the mathematical statement. Explain how the computation relates to the solution of the story problem.

(1) $4 < 7$ implies that $4 - 2 < 7 - 2$. Caitlin is 7 years old and Anthony is 4. Who is younger now? Who was younger 2 years ago?

Solution: $4 < 7$ means that Anthony is younger now. Two years ago, Anthony was $4 - 2$ years old and Caitlin was $7 - 2$ years old. By the second subtraction translation property, $4 - 2 < 7 - 2$, which means that Anthony was younger than Caitlin two years ago as well.

A picture proof is the following. Each triangle represents 1 year.
Now:
$$\text{Anthony } \{\ \triangle\ \triangle\ \triangle\ \triangle\ \}$$
$$\updownarrow\ \updownarrow\ \updownarrow\ \updownarrow$$
$$\text{Caitlin } \{\ \triangle\ \triangle\ \triangle\ \triangle\ \triangle\ \triangle\ \triangle\ \}$$
Two years ago:
$$\text{Anthony } \{\ \triangle\ \triangle\ \not\triangle\ \not\triangle\ \}$$
$$\updownarrow\ \updownarrow\ \not\updownarrow\ \not\updownarrow$$
$$\text{Caitlin } \{\ \triangle\ \triangle\ \not\triangle\ \not\triangle\ \triangle\ \triangle\ \triangle\ \}$$

§2.2 Exercises

(2) $6-3 = (6+2)-(3+2)$. Caitlin is 6 years old and Anthony is 3. Caitlin is how many years older than Anthony now, and will this difference in age be the same 2 years from now?

(3) $(1+4)-2 = 1+(4-2)$. Joshua sold a baseball card for 1 dollar and he received his allowance of 4 dollars. His mother reminded him that he owed her 2 dollars for the advance she had given him yesterday. Does it make any difference in how much money he ends up with if his mother gives him his full allowance and then he pays her the 2 dollars, or if she deducts the 2 dollars from his allowance and pays him what is left? Explain.

(4) $6-(4-2) = (6-4)+2$. Anthony has 6 dollars and he owes his mother 4 dollars. Today he is to receive his allowance of 2 dollars and he is to pay his debt. His mother offers to let his allowance reduce his debt by 2 dollars and so he must now pay the remainder. Instead, Anthony pays her the 4 dollars he owes her and then asks for his 2 dollar allowance. How much money does Anthony end up with in each case?

(5) $8-(5+1) = (8-5)-1$. Janet has 8 yards of material from which she uses 5 yards to make a dress and 1 yard for a matching sash. How many yards of material are left? Does it make a difference if she sums the amount of material used and subtracts that from the starting amount or if she subtracts each amount used from the amount remaining at each stage?

(6) $5-(6-3) = (5+3)-6$. Anthony has 5 dollars and he owes his mother 6 dollars. Today he receives his allowance of 3 dollars. Describe the two ways he can pay his debt. How much money does he have left?

(7) Roll 2 dice (say, red and green). How many outcomes have a 1 on at least one of the dice? You may list all of the desired outcomes and then count them as a check of your calculation, but then do the calculation as an application of Theorem 2.41. *Ans:* Let A be the set of outcomes with 1 on the red die, let B be the set of outcomes with 1 on green die. Then $A \cap B$ is the set of outcomes with 1 on both dice, and $A \cup B$ is the set of outcomes with 1 on at least one of the dice. Then $n(A \cup B) = n(A) + n(B) - n(A \cap B) = 11$.

(8) Roll 2 dice. How many outcomes have a 5 or 6 on at least one of the dice? You may list all of the desired outcomes and then count them as a check of your calculation, but then do the calculation as an application of Theorem 2.41.

(9) Find $17-4$ by $17-4 = (10+7)-4$ and explain how the subtraction properties justify this. Illustrate with a picture. *Ans:* See Oral exercise 2.39.

(10) Alexi has $43. How much more money does she need to buy a new camera which costs $79?

(11) Nicholas, having 75 cents, earned 27 cents and spent 52 cents: how much had he left? *Solution:*

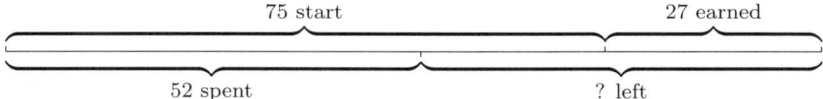

(12) A roll of carpet contained 154 yards; there were sold from it 30 yards, 44 yards, 15 yards, and 18 yards: how much remained?

(13) At an election Robertson received 225,670 votes and Dutton received 212,517 votes. What was Robertson's majority? *Solution:*

Robertson
Dutton

(14) A steam engine model costs $88. It costs $23 more than an airplane model. How much does the airplane model cost?

(15) An *exterior angle* of a triangle is an angle formed by extending outward one side of the triangle. In the figure below, an exterior angle is marked with a question mark. The exterior angle is a supplement of the angle at the same vertex. The other two angles of the triangle are called the *opposite angles* of the given exterior angle.

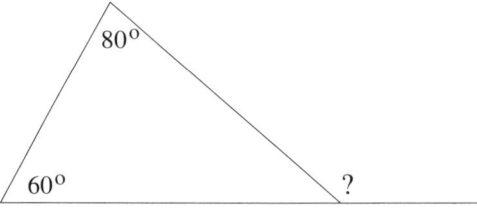

(a) Prove that an exterior angle of a triangle is equal to the sum of its two opposite angles.
(b) How many degrees in the angle questioned? Explain your answer.

(16) Here is a right triangle with two of its exterior angles. Find the number of degrees in the exterior angle X.

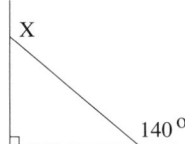

(17) Here is an isosceles triangle. AED and BEC are straight lines. Find the angle α.

2.3. Multiplication

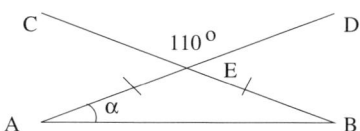

2.3. Multiplication

We define multiplication to be repeated addition.

Definition 2.43 (Multiplication). If a and b are whole numbers, then $a \cdot b$ is the sum $b + \cdots + b$, where a is the number of b's in this sum. We say that this is a *groups* of b. Define $0 \cdot b = 0$, that is, 0 groups of b is 0 by definition.

The statement $a \cdot b$ is read a *times* b or a groups of b, and the resulting number is called the *product*. The number a is called the *multiplier* and the number b is called the *multiplicand*. The multiplier tells how many groups and the multiplicand tells how many in each group.

Definition 2.44. A *multiple* of a number m is any product $m \cdot a = a \cdot m$, where a is any natural number.

In this definition we require a to be nonzero. We do not include 0 among the multiples of any natural number.

Example 2.45. Here are 4 groups of 3 △'s.
$$\{\triangle \triangle \triangle\}, \{\triangle \triangle \triangle\}, \{\triangle \triangle \triangle\}, \{\triangle \triangle \triangle\}$$
In these 4 groups of 3 △'s there are $3 + 3 + 3 + 3 = 4 \cdot 3$ △'s.

Example 2.46. 3 groups of 5 is $3 \cdot 5 = 5 + 5 + 5$.
5 groups of 3 is $5 \cdot 3 = 3 + 3 + 3 + 3 + 3$.

Example 2.47. The multiples of m are the numbers obtained in counting by m. For example, the multiples of 4 are
$$4, 8, 12, 16, 20, 24, 28, 32, 36$$
and so on, which are the multiples $4 \cdot 1, 4 \cdot 2, 4 \cdot 3, 4 \cdot 4, 4 \cdot 5$, and so on.

Example 2.48. $0 \cdot 6 = 0$ by definition. Six groups of 0 is $6 \cdot 0 = 0 + 0 + 0 + 0 + 0 + 0 = 0$. The reason we define $0 \cdot 6 = 0$ is so that $0 \cdot 6 = 6 \cdot 0$.

Example 2.49. $1 \cdot a = a$ for any whole number a, because 1 group of a is a. In the opposite order, $a \cdot 1 = a$, because a groups of 1 is a, by definition of the whole number a. (In that definition we started with the empty set, then added 1 more element a times, at which point a is defined to be the number of elements in this set).

Example 2.50. If Walter has 3 boards, each 2 feet long, then he has 3 groups of 2, which is $3 \cdot 2 = 2 + 2 + 2$ feet of board altogether.

Example 2.51. If Jen plants 5 rows of trees with 3 trees in each row, how many trees does she plant? Each row contains 3 trees and there are 5 rows. Regard each row as a group, then there are 5 groups of 3 trees, which is $3 + 3 + 3 + 3 + 3 = 5 \cdot 3$ trees.

Solve the following Oral exercises in the style of the preceding two Examples. Describe the number of groups and how many in each group and write the answer as a product, but do not compute the product.

Oral exercise 2.52. How many groups of dots are there? How many dots in each group? How many dots are there?

$$\{\bullet \bullet \bullet\}, \{\bullet \bullet \bullet\} \{\bullet \bullet \bullet\}, \{\bullet \bullet \bullet\}$$

Oral exercise 2.53. A school has 13 classrooms, each with 27 students. How many students are in this school?

Oral exercise 2.54. George has 17 haystacks, each containing 23 tons of hay. How many tons of hay has he?

Definition 2.55. A number c is a times as large as b means that $c = a \cdot b$.

Oral exercise 2.56. Joshua has 74 football cards and 5 times as many baseball cards. How many baseball cards has he?

2.3.1. Rectangular Arrays. Consider $4 \cdot 3$, which is the number of elements in 4 groups of 3. Represent each group by a row of 3 dots like this • • •. Make 4 copies of this row of dots, one under the other, to display this 4 groups of 3 dots as a rectangular *array*

$$\begin{matrix} \bullet & \bullet & \bullet \\ \bullet & \bullet & \bullet \\ \bullet & \bullet & \bullet \\ \bullet & \bullet & \bullet \end{matrix}$$

with 4 *rows* (horizontal) and 3 *columns* (vertical). Then $4 \cdot 3$ is the number of elements in this array of 4 groups of 3.

2.3.2. Area and Volume.

Theorem 2.57 (Area of Rectangles). *If a unit of length is chosen and a rectangle has length l units and width w units, where l and w are whole numbers, then the area of the rectangle is $l \cdot w$.*

Proof. The following case illustrates the general proof. Consider a rectangle whose length is 3 units and whose width is 5 units. Then the rectangle is an array of unit squares for which there are 3 rows, each containing 5 unit squares.

2.3. Multiplication

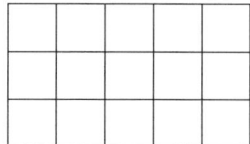

It contains 3·5 unit squares. The area of the rectangle is 3·5 unit squares. □

Theorem 2.58 (Volume of Rectangular Solid). *If a rectangular solid has width w, length l, and height h, all measured in the same length unit, then its volume is $h \cdot (w \cdot l)$ cubic length units. In words, the volume is the height times the area of the base rectangle whose width is w and length is l.*

Proof. The proof of the general case is illustrated by the case where the height $h = 2$, the width $w = 3$, and length $l = 4$. The rectangular solid

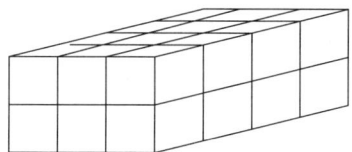

consists of 2 horizontal layers of unit cubes, where each layer contains an array of 3 rows with 4 unit cubes in each row. Each layer contains $3 \cdot 4$ unit cubes. Thinking of each layer as a group, we see that the solid consists of 2 groups of $3 \cdot 4$ unit cubes, which is $2 \cdot (3 \cdot 4)$ unit cubes. The volume of this rectangular solid is $2 \cdot (3 \cdot 4)$ unit cubes. □

2.3.3. Cartesian Product of Sets.
Here is a more formal way to describe rectangular arrays.

Definition 2.59 (Cartesian Product). The Cartesian product of a set A with a set B, written $A \times B$, is the set of all ordered pairs whose first entry is any element of A and whose second entry is any element of B. In symbols,
$$A \times B = \{(a,b) : a \in A, b \in B\}$$

Example 2.60. The elements of $A \times B$ can be displayed as a rectangular array. For example, if $A = \{1,2,3,4\}$ and $B = \{1,2,3\}$, then the elements of $A \times B$ are displayed in the array

(2.2)
$$\begin{array}{ccc} (1,1) & (1,2) & (1,3) \\ (2,1) & (2,2) & (2,3) \\ (3,1) & (3,2) & (3,3) \\ (4,1) & (4,2) & (4,3) \end{array}$$

These pairs show the standard way of addressing the entries of an array. The first number in the pair is the row number and the second number is the column number. For example, entry $(3,2)$ is in row 3 and column 2.

Theorem 2.61. *If A and B are finite sets, then $n(A \times B) = n(A) \cdot n(B)$.*

Proof. For each element a of A there is a row of pairs whose first entry is a and whose second entries run through the entries of B. Therefore, the elements of $A \times B$ can be arranged in an array consisting of $n(A)$ rows each with $n(B)$ elements. This is $n(A)$ groups of $n(B)$. Hence, $n(A \times B) = n(A) \cdot n(B)$. □

Example 2.62. If $A = \{\clubsuit \heartsuit \dagger\}$ and $B = \{\triangle \square\}$, then $A \times B$ consists of the ordered pairs in the array

$$(\clubsuit, \triangle) \quad (\clubsuit, \square)$$
$$(\heartsuit, \triangle) \quad (\heartsuit, \square)$$
$$(\dagger, \triangle) \quad (\dagger, \square)$$

Regarding each row as a group, we see that $A \times B$ consists of $n(A)$ groups of $n(B)$, so that $n(A \times B) = n(A) \cdot n(B)$.

Theorem 2.63 (Multiplication Properties). *If a, b, and c are whole numbers, then:*

(1) $a \cdot b = b \cdot a$. (Commutative property).

(2) $a \cdot (b \cdot c) = (a \cdot b) \cdot c$. (Associative property).

(3) $a \cdot 1 = 1 \cdot a = a$. (Identity property of 1).

(4) $a \cdot (b + c) = a \cdot b + a \cdot c$. (Distributive property over addition).

(5) If $b \geq c$, then $a \cdot (b - c) = a \cdot b - a \cdot c$. (Distributive property over subtraction).

(6) If $b > c$ and if $a \neq 0$, then $a \cdot b > a \cdot c$. (Order property).

Proof. (1) The commutative property is true because $a \cdot b$ is the number of elements in an array comprising a rows with b elements in each row, while $b \cdot a$ is the number of elements in the same array viewed as comprising b columns with a elements in each column. This change of viewpoint can be achieved by rotating the array through $90°$, as here:

1	2	3
4	5	6
7	8	9
10	11	12
13	14	15

1	4	7	10	13
2	5	8	11	14
3	6	9	12	15

A formal explanation of the commutative property can be given in terms of the Cartesian product. If A and B are sets such that $n(A) = a$ and

2.3. Multiplication

$n(B) = b$, then a 1:1 correspondence from $A \times B$ to $B \times A$ is given by

$$(a, b) \leftrightarrow (b, a), \quad \text{for any } a \in A, b \in B$$

(This is a formal description of interchanging rows and columns in an array by rotating it through 90°.) Therefore,

$$\begin{aligned} n(A) \cdot n(B) &= n(A \times B), \quad \text{by Theorem 2.61} \\ &= n(B \times A), \quad \text{by } A \times B \sim B \times A \\ &= n(B) \cdot n(A), \quad \text{by Theorem 2.61} \end{aligned}$$

(2) The associative property is proved by calculating the volume of a rectangular solid in two ways.

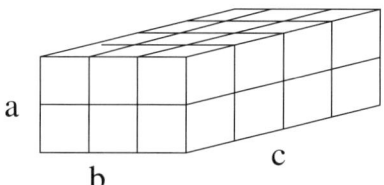

Think of this rectangular solid as comprising a horizontal layers, each layer consisting of an array of b rows with c unit cubes in each row. Then each layer contains $b \cdot c$ unit cubes, and a such layers contain $a \cdot (b \cdot c)$ unit cubes, which is the number of cubes in the rectangular solid.

Now think of the rectangular solid as comprising c vertical layers, each layer consisting of a rows with b unit cubes in each row. Each layer contains $a \cdot b$ unit cubes and so c layers contain $c \cdot (a \cdot b)$ unit cubes, and this equals $(a \cdot b) \cdot c$ by the commutative property. This is the number of cubes in the rectangular solid. We conclude that $a \cdot (b \cdot c) = (a \cdot b) \cdot c$.

A formal explanation of the associative property can be given in terms of Cartesian products. If A, B, and C are sets containing a, b, and c elements, respectively, then a 1:1 correspondence from $A \times (B \times C)$ to $(A \times B) \times C$ is given by

$$(x, (y, z)) \leftrightarrow ((x, y), z), \quad \text{for any } x \in A, y \in B, z \in C.$$

Then

$$\begin{aligned} a \cdot (b \cdot c) &= a \cdot n(B \times C) = n(A \times (B \times C)), \quad \text{by Theorem 2.61} \\ &= n((A \times B) \times C), \quad \text{by } A \times (B \times C) \sim (A \times B) \times C \\ &= n(A \times B) \cdot c = (a \cdot b) \cdot c, \quad \text{by Theorem 2.61} \end{aligned}$$

(3) In Example 2.49 above we have explained the identity property of 1.

(4) To prove $a \cdot (b + c) = a \cdot b + a \cdot c$, consider two arrays, one comprising a rows with b elements in each row, and the other comprising a rows with c elements in each row. Putting these arrays side by side, we obtain an

array comprising a rows with $b+c$ elements in each row. The large array contains $a \cdot (b+c)$ elements, and it is the union of the original two arrays, so it contains $a \cdot b + a \cdot c$ elements.

Here is a picture of $3 \cdot (5+2) = 3 \cdot 5 + 3 \cdot 2$.

$$
\begin{array}{ccccccc}
\bullet & \bullet & \bullet & \bullet & \bullet & \triangle & \triangle \\
\bullet & \bullet & \bullet & \bullet & \bullet & \triangle & \triangle \\
\bullet & \bullet & \bullet & \bullet & \bullet & \triangle & \triangle
\end{array}
$$

(5) To prove the distributive property of multiplication over subtraction, we must prove that
$$a \cdot b = a \cdot c + a \cdot (b-c)$$
because the definition of comparative subtraction says that this is the meaning of $a \cdot b - a \cdot c = a \cdot (b-c)$. The proof is

$$
\begin{aligned}
a \cdot c + a \cdot (b-c) &= a \cdot (c + (b-c)), \quad \text{Dist. Prop.} \\
&= a \cdot b, \quad \text{by definition of comp. sub.}
\end{aligned}
$$

which is what we needed to show. A picture of $3 \cdot (5-2) = 3 \cdot 5 - 3 \cdot 2$ is

$$
\begin{array}{ccccc}
\triangle & \triangle & \triangle & \cancel{\triangle} & \cancel{\triangle} \\
\triangle & \triangle & \triangle & \cancel{\triangle} & \cancel{\triangle} \\
\triangle & \triangle & \triangle & \cancel{\triangle} & \cancel{\triangle}
\end{array}
$$

(6) To prove the order property of multiplication, consider an array comprising a rows each with b elements and another array with a rows each with c elements. If $b > c$, then each row of the first array has more elements than each row of the second array, so the second array can be put into 1:1 correspondence with a proper subset of the first array. Therefore, the first array has more elements, that is, $a \cdot b > a \cdot c$. $\square$

The associative property allows us to omit parentheses when we write a product of more than two numbers. For example, we write $a \cdot b \cdot c$ to mean either $a \cdot (b \cdot c)$ or $(a \cdot b) \cdot c$, which are equal by the associative property. The associative and commutative properties combined allow us to write this product in any order. Thus

$$a \cdot b \cdot c = a \cdot c \cdot b = b \cdot a \cdot c = b \cdot c \cdot a = c \cdot a \cdot b = c \cdot b \cdot a$$

The commutative and distributive properties imply that

(2.3)
$$\begin{aligned} (b+c) \cdot a &= b \cdot a + c \cdot a \\ (b-c) \cdot a &= b \cdot a - c \cdot a \end{aligned}$$

We shall refer to these two identities as distributive properties as well. In this formulation, the size of the group is a throughout. In applications we

2.3. Multiplication

shall frequently illustrate these with line segment diagrams in which the line segment unit represents the group. For example,

$$6 \cdot 3 + 2 \cdot 3 = (6+2) \cdot 3$$

because

$6 \cdot 3 = 6$ groups of 3:

plus

$2 \cdot 3 = 2$ groups of 3:

is

$6 + 2$ groups of 3:

And

$$6 \cdot 3 - 2 \cdot 3 = (6-2) \cdot 3$$

because

$6 \cdot 3 = 6$ groups of 3:

If the solution of a problem involves b groups of a, then $a \cdot b$ does not usually provide an explanation of the solution of the problem. For example, if Dale has 3 boards, each 2 feet long, how many feet of board does he have? To solve this, we regard each board as a group of 2 feet, and so having 3 boards means that he has 3 groups of 2 which is $3 \cdot 2 = 2 + 2 + 2$ feet of board. The answer is also given by $2 \cdot 3 = 3 + 3$, but the problem cannot be explained adequately in terms of 2 groups of 3.

Compare the commutative property of addition to the commutative property of multiplication. We know that if A and B are disjoint sets with $a = n(A)$ and $b = n(B)$, then $a + b = b + a$ because $A \cup B = B \cup A$; these two unions are equal. But $a \cdot b = b \cdot a$ because $A \times B \sim B \times A$, and these two sets are not equal. For this reason, $a \cdot b = b \cdot a$ is *less obvious* than $a + b = b + a$. One should say that $a \cdot b$ is equal to $b \cdot a$, rather than that $a \cdot b$ is the same as $b \cdot a$.

Example 2.64. A carton contains 7 glazed doughnuts and 3 times as many cake doughnuts. How many cake doughnuts does it contain? How many doughnuts does it contain altogether?

Solution: The carton contains $3 \cdot 7$ cake doughtnuts. The following line segment diagram

glazed
cake

shows us that the carton contains 4 units with 7 doughnuts in each unit for a total of $4 \cdot 7$ doughnuts altogether.

We can also think of the carton as containing the union of the set of glazed doughnuts with the set of cake doughnuts, for a total of $7 + 3 \cdot 7$ doughnuts altogether. The two answers are the same by the distributive property of multiplication,

$$7 + 3 \cdot 7 = 1 \cdot 7 + 3 \cdot 7 = (1 + 3) \cdot 7 = 4 \cdot 7$$

since $1 + 3 = 4$.

Oral exercise 2.65. Margaret bought her mother a present which cost \$14. Her sister Martha paid twice as much for the present she bought. How much did Martha spend? How much did Margaret and Martha spend altogether? What line segment diagram would you make to illustrate the solution?

Example 2.66. Cecelia made 27 Danish cookies. She made 4 times as many Scotch cookies as Danish cookies. How many Scotch cookies did she make? How many more Scotch cookies than Danish cookies did she make?

Solution: She made $4 \cdot 27$ Scotch cookies. By comparative subtraction, she made $4 \cdot 27 - 27$ more Scotch cookies than Danish cookies. By the distributive property, this is

$$4 \cdot 27 - 1 \cdot 27 = (4 - 1) \cdot 27 = 3 \cdot 27$$

more Scotch cookies than Danish cookies. The line segment diagram

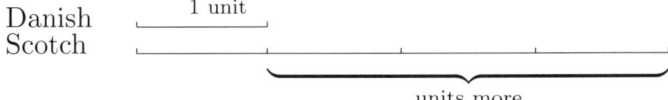

shows us directly that she made 3 units more of Scotch, which is $3 \cdot 27$ more Scotch cookies than Danish cookies. Here 1 unit = 27 cookies.

Oral exercise 2.67. Nels has 7 cows. He has 5 times as many sheep as cows. Give two methods to solve each problem and explain how the distributive property is used to see that both methods give the same answer. Describe what line segment diagram to use. One unit is how many animals?

(1) How many sheep and cows does he have altogether?

(2) How many more sheep than cows does he have?

2.3.4. Numeration and products. After counting a set of objects into groups of ten and putting the remaining objects in the one's place, then counting the groups of ten into groups of ten and putting the remaining tens in the ten's place, and so on, we end up with a decomposition of the original set into the union of disjoint sets: the set of ones in the one's place, the set of tens in the ten's place, the set of tens of tens in the ten of ten's

2.3. Multiplication

place, and so on. From this we have concluded in Section 2.1 that the number of objects in the original set is the sum of the number of ones, the number of tens times 10, the number of ten of tens times $10 \cdot 10$, and so on. For example, a set of 487 objects has

$$(2.4) \qquad 487 = 4 \cdot 10 \cdot 10 + 8 \cdot 10 + 7$$

objects. The numeral for $4 \cdot 10 \cdot 10$ is 400, the numeral for $8 \cdot 10$ is 80, so we can write Equation (2.4) as

$$(2.5) \qquad 487 = 4 \cdot 100 + 8 \cdot 10 + 7$$

Oral exercise 2.68. Write the following numbers as sums in the form of Equations (2.4) and (2.5): (a) 24. (b) 496. (c) 703. (d) 2978. (e) 3004.

2.3.5. Expansions. To *expand out* a product of two sums, or of two differences, or of a sum times a difference, means to apply the distributive property twice as follows:

$$(2.6) \quad \begin{aligned} (a+b) \cdot (c+d) &= (a+b) \cdot c + (a+b) \cdot d, \quad \text{Distributive Prop.} \\ &= (a \cdot c + b \cdot c) + (a \cdot d + b \cdot d), \quad \text{Dist. Prop. twice} \\ &= a \cdot c + b \cdot c + a \cdot d + b \cdot d \end{aligned}$$

which can be pictured as areas, the initial product of the two sums being the area of the rectangle of side lengths $a+b$ and $c+d$, and the four final products being the areas of the four rectangles composing this rectangle.

	c	d
a	a·c	a·d
b	b·c	b·d

An important special case of where we use expansions is when one of the numbers is a multiple of 10, such as in

$$6 \cdot 17 = 6 \cdot (10 + 7) = 6 \cdot 10 + 6 \cdot 7$$

or in

$$\begin{aligned} 14 \cdot 19 &= (10+4) \cdot (10+9) = (10+4) \cdot 10 + (10+4) \cdot 9 \\ &= 10 \cdot 10 + 4 \cdot 10 + 10 \cdot 9 + 4 \cdot 9 \end{aligned}$$

which form the basis of our algorithm for computing products.

Another important expansion is of the square of a sum

$$(2.7) \qquad (a+b)^2 = a^2 + 2 \cdot a \cdot b + b^2$$

Recall that squaring means multiplying the number by itself: $a^2 = a \cdot a$.

Oral exercise 2.69. Cite a justification for each step in the following expansion.

$$(3+5)^2 = (3+5) \cdot (3+5) = (3+5) \cdot 3 + (3+5) \cdot 5$$
$$= 3 \cdot 3 + 5 \cdot 3 + 3 \cdot 5 + 5 \cdot 5 = 3 \cdot 3 + 3 \cdot 5 + 3 \cdot 5 + 5 \cdot 5$$
$$= 3 \cdot 3 + 2 \cdot (3 \cdot 5) + 5 \cdot 5 = 3^2 + 2 \cdot 3 \cdot 5 + 5^2$$

This expansion is illustrated by

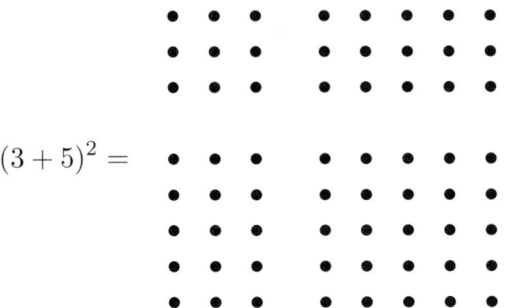

$(3+5)^2 =$

Oral exercise 2.70. Give the justification for each step in the expansion of a sum times a difference, where $a \geq b$.

(2.8)
$$(a+b) \cdot (a-b) = (a+b) \cdot a - (a+b) \cdot b$$
$$= a \cdot a + b \cdot a - (a \cdot b + b \cdot b)$$
$$= a \cdot a + ((b \cdot a - a \cdot b) - b \cdot b)$$
$$= a \cdot a - b \cdot b$$

Example 2.71. An application of formula (2.7) is

$$21^2 = (20+1)^2 = 20 \cdot 20 + 2 \cdot 20 \cdot 1 + 1 \cdot 1 = 400 + 40 + 1 = 441$$

Our observations above about the numeration system allow us to calculate

$$2 \cdot 20 = 2 \cdot 2 \cdot 10 = 4 \cdot 10 = 40$$

and

$$20 \cdot 20 = 2 \cdot 10 \cdot 2 \cdot 10 = 4 \cdot 10 \cdot 10 = 400$$

Since $19 = 20 - 1$ and $21 = 20 + 1$, an application of formula (2.8) is

$$21 \cdot 19 = (20+1) \cdot (20-1) = 20 \cdot 20 - 1 \cdot 1 = 400 - 1 = 399$$

Oral exercise 2.72. Find: (a) 11^2. (b) 30^2. (c) 31^2. (d) $31 \cdot 29$.

2.3.6. Pythagorean Theorem. Recall from Definition 1.64 that the hypotenuse of a right triangle is the side opposite the right angle.

Theorem 2.73 (Pythagorean Theorem). *If a right triangle has sides of length a, b, and c, with c the length of the hypotenuse, then $a^2 + b^2 = c^2$.*

2.3. Multiplication

The theorem says that the area of the square whose side is the hypotenuse is the sum of the areas of the squares whose sides are the other two sides of the triangle.

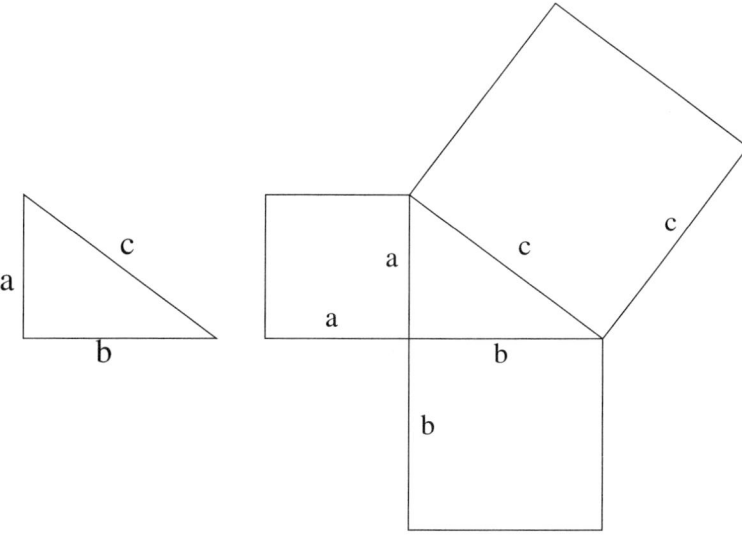

Proof. There are many proofs of this theorem, many of which can be found on the world wide web. Assume that $a \leq b$.

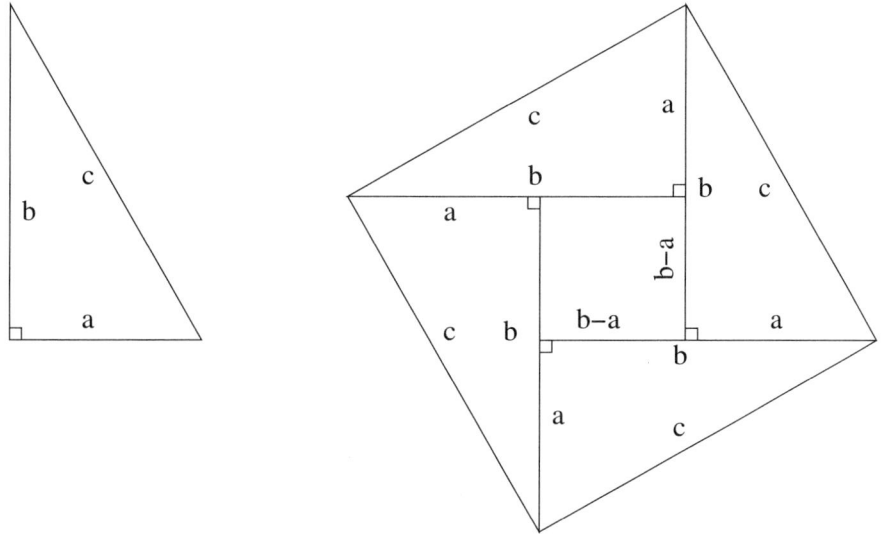

The figure on the right is a square, because the sum of the angles of a triangle is 180°, so the sum of the two non-right angles of a right triangle must be 90°. This square has side length c, so its area is c^2. It is composed of 4 copies of the right triangle and a square of side length $b - a$, so its area is

also given by
$$4 \cdot ((a \cdot b)/2) + (b-a)^2 = 2 \cdot a \cdot b + b^2 - 2 \cdot b \cdot a + a^2 = a^2 + b^2$$
Therefore, the area of this square being the same no matter how we calculate it, we conclude that $c^2 = a^2 + b^2$. □

2.3.7. Rates. A rate tells us how many are in each group. For example, if a car travels at 60 mph, then in 1 hour it will go 60 miles. The hour defines the group and there are 60 miles in each group. If you maintain that speed for 3 hours, then you will go 60 miles in each of 3 hours. The distance you will have travelled in these 3 hours is 3 groups of 60 miles, which is $3 \cdot 60 = 180$ miles.

If it takes me 5 minutes to grade 1 paper, then my time is being consumed at the rate of 5 minutes per paper. At this rate, how long will it take me to grade 13 papers? The answer is 13 groups of 5 minutes, which is $13 \cdot 5 = 65$ minutes.

If a recipe calls for 3 cups of flour per batch, then 4 batches would require 4 groups of 3 cups, which is $4 \cdot 3 = 12$ cups of flour.

Oral exercise 2.74. Diane can walk 3 miles per hour. At this rate, how far will she walk in 1 hour? In 5 hours?

Oral exercise 2.75. If it takes Janet 3 minutes to plant a petunia, how long will it take her, at this rate, to plant 6 petunias?

Example 2.76 (Rate of Separation). Leah and Ragna start at the same point and travel in opposite directions, Leah at the rate of 60 miles per hour, Ragna at the rate of 75 miles per hour. What is their rate of separation? How far apart are they in 3 hours?

Solution:
In 1 hour:

In 1 hour Leah goes 60 miles.

In 1 hour Ragna goes 75 miles in the opposite direction.

In 1 hour they have separated by $60 + 75$ miles.

Their rate of separation is $60 + 75$ mph.

In 3 hours they are $3 \cdot (60 + 75)$ miles apart.

Example 2.77 (Rate of Approach). Ruth's house is 20 miles from Clarence's house. Ruth walks at the rate of 3 mph. Clarence walks at the rate of 5 mph. If they both start walking toward the other's house at the same time, what is their rate of approach? How far apart are they after 2 hours?

2.3. Multiplication

Unitary method:

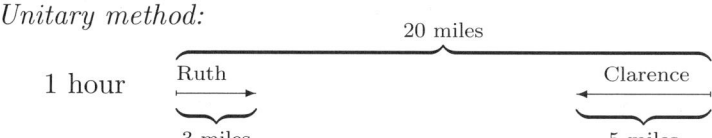

In 1 hour Ruth walks 3 miles.

In 1 hour Clarence walks 5 miles.

In 1 hour they walk $3 + 5$ miles altogether.

Their rate of approach is $3 + 5$ mph.

After 2 hours they have walked $2 \cdot (3 + 5)$ miles altogether.

After 2 hours they are $20 - (2 \cdot (3 + 5))$ miles apart.

2.3.8. Counting. Many counting problems can be carried out by counting the number of groups and then the number in each group, so that the total is the product.

An *outcome* of rolling two dice (a green and a red one) is the number up on each die. How many outcomes are there? If we list these two numbers in order, say with the number on the green die first and the number on the red die second, then an outcome is an ordered pair of numbers. For example, if we roll the dice and 2 is up on the green die and 5 is up on the red die, then this outcome is indicated by the ordered pair $(2, 5)$. The following array lists all possible outcomes.

$$
\begin{array}{cccccc}
(1,1) & (1,2) & (1,3) & (1,4) & (1,5) & (1,6) \\
(2,1) & (2,2) & (2,3) & (2,4) & (2,5) & (2,6) \\
(3,1) & (3,2) & (3,3) & (3,4) & (3,5) & (3,6) \\
(4,1) & (4,2) & (4,3) & (4,4) & (4,5) & (4,6) \\
(5,1) & (5,2) & (5,3) & (5,4) & (5,5) & (5,6) \\
(6,1) & (6,2) & (6,3) & (6,4) & (6,5) & (6,6)
\end{array}
$$

The first row of this array comprises all the outcomes for which the green die is 1. There are 6 outcomes in this row, because when the green die is 1, there are 6 possible values for the red die. Each row, then, contains the six possible values for the red die for a given value on the green die. The array shows that there are 6 groups of 6 outcomes, for a total of $6 \cdot 6 = 36$ outcomes. Each group is an outcome of the green die, and the group contains all possible outcomes of the red die.

For another example, suppose a penny is flipped and a die is rolled. An outcome is an ordered pair whose first entry is heads H or tails T on the penny and whose second entry is 1, 2, 3, 4, 5, or 6 on the die. The total set of outcomes can be organized into 2 groups of 6 for a total of $2 \cdot 6 = 12$ outcomes.

Oral exercise 2.78. Make an array of the possible outcomes from flipping a penny and rolling a die.

2.3.9. Permutations. A *permutation* of a set is an ordering of the elements of the set. Many important applications require a count of the number of permutations of a set. We can solve this problem by studying a few cases. A set $A = \{a\}$ of just one element has only 1 permutation. A set $B = \{a, b\}$ of two elements has the two permutations a, b and b, a and you should convince yourself that there are no other permutations of this set. Thus, a set of two elements has 2 permutations, which we have determined by listing, or enumerating, the permutations and then counting them.

Consider next the permutations of a set $C = \{a, b, c\}$ of three elements. We can enumerate these in a systematic way as follows.

$$\begin{array}{ll} a, b, c & a, c, b \\ b, a, c & b, c, a \\ c, a, b & c, b, a \end{array}$$

The scheme is to choose one of the elements of C as the first element in the permutation and then list all permutations beginning with that element. The result is an array in which the number of rows is the number of elements in the set and the number of columns is the number of permutations of the remaining elements, which in this case is the number of permutations of a set of two elements. We conclude then that the number of permutations of a set of three elements is $3 \cdot 2$, which from our scheme of enumeration is the number of elements in C times the number of permutations of a set with one fewer element.

Definition 2.79. For a natural number n, define n factorial, written $n!$, to be the product of all the natural numbers from 1 up through n. We agree to define $0! = 1$.

For example,

(2.9)
$$\begin{aligned} 2! &= 1 \cdot 2 = 2 \\ 3! &= 1 \cdot 2 \cdot 3 = 2! \cdot 3 = 6 \\ 4! &= 1 \cdot 2 \cdot 3 \cdot 4 = 3! \cdot 4 = 6 \cdot 4 = 24 \\ 5! &= 1 \cdot 2 \cdot 3 \cdot 4 \cdot 5 = 4! \cdot 5 = 24 \cdot 5 = 120 \end{aligned}$$

Oral exercise 2.80. Find: (a) $6!$. (b) $7!$. (c) $8!$. (d) What is the rule for expressing $(n+1)!$ in terms of $n!$?

Theorem 2.81. *If a set has n elements, then it has $n!$ permutations.*

Proof. At the beginning of §2.3.9 we showed that a set with 1 element has only $1 = 1!$ permutation, a set with 2 elements has $2 = 2!$ permutations and a set with 3 elements has $3 \cdot 2! = 3!$ permutations. We show now that a set of 4 elements has 4 times as many permutations as a set of 3 elements.

There are 4 elements in the set, each of which can be the first element in a permutation. Label the rows of an array by these elements. In each row list the permutations for which that element is first. The number of such permutations is the number of permutations of the remaining 3 elements, and there are 3! of these. The entries of this array consist of all the permutations of the set of 4 elements. Our array has 4 rows and 3! columns for a total of $4 \cdot 3! = 4!$ entries, and these are all the permutations of the set of 4 elements.

That argument can be repeated to show that a set of 5 elements contains $5 \cdot 4! = 5!$ permutations, and then the same argument shows that a set of 6 elements contains $6 \cdot 5!$ permutations, and so on. By our definition of the whole numbers, every natural number will be arrived at eventually by this line of reasoning. This completes the proof of the theorem. □

§2.3 Exercises

In these problems, indicate the arithmetic operations to be done, but do not carry out any computations. For example, if the answer to a problem is $7 \cdot (8 - 3) + 6$, then leave it in this form.

(1) Illustrate $3 \cdot 5$ as an array for which each row is a group. Use the array to explain why $5 \cdot 3 = 3 \cdot 5$. Illustrate $3 \cdot 5$ with a line segment diagram in which each segment represents a group. Explain how this latter illustration is related to illustrating $5 + 5 + 5$ on the number line.

(2) There are 6 girls in Rochell's second grade class. There are 3 times as many boys as girls in her class. Use the line segment diagram to illustrate your solution to each of the following problems.

girls $\overbrace{\qquad}^{6}$
boys $\underbrace{\vdash\!\!-\!\!-\!\!-\!\!\dashv}_{1\ \text{unit}}\!\!\vdash\!\!-\!\!-\!\!-\!\!\dashv\!\vdash\!\!-\!\!-\!\!-\!\!\dashv$

(a) One unit equals how many children?
(b) How many units of boys are in her class? How many boys?
(c) How many more units of boys than girls in her classroom? How many more boys than girls?
(d) How many units of students altogether in her classroom? How many students altogether?

(3) Ashley received 5 marbles as a present. The next day she played marbles at school and doubled the number of marbles she had. The day after she played again and tripled the number of marbles she started the day with. At home that second day she said, "It's too bad I didn't triple the number of marbles on the first day, and then double on the second day, because then I would have had more marbles to double on the second day and I would have ended up with more marbles". Use the associative and commutative properties of multiplication to explain why Ashley's reasoning is incorrect.

(4) George measured a rectangular red rug and found it to be 7 feet wide and 12 feet long. He measured a rectangular blue rug and found it to be twice as wide and twice as long as the red rug. The area of the blue rug is how many times greater than the area of the red rug? Illustrate your answer.

(5) In preparation for a 3-month stay abroad, Ralph calculates how much money to put in his account to cover the automatic debits for his monthly payments of: mortgage of 400 dollars, electricity of 100 dollars, gas of 40 dollars, and sewer of 10 dollars. Explain why he finds the correct amount either by adding together the four monthly payments and multiplying this sum by 3 (that is, $3 \cdot (400 + 100 + 40 + 10)$) or by multiplying each monthly payment by 3 and adding these products (that is, $3 \cdot 400 + 3 \cdot 100 + 3 \cdot 40 + 3 \cdot 10$).

(6) Leah can peel potatoes at the rate of 30 per hour. Ragna can peel potatoes at the rate of 40 per hour. How many potatoes can they peel together in 3 hours? Illustrate your solution with a line segment diagram.

(7) Ruth's house is 20 miles from Clarence's house. Ruth walks at the rate of 3 mph. Clarence walks twice as fast as Ruth. If they both start walking towards the other's house at the same time, how far apart will they be in two hours? Use a line segment diagram to illustrate your solution by the unitary method.

(8) On Friday Rita washed 7 dozen eggs. On Saturday she washed 15 eggs. How many eggs did she wash altogether?

(9) Newell bought 7 bulls. He bought 10 times more heifers than bulls. How many heifers did he buy? How many bulls and heifers did he buy? How many more heifers than bulls did he buy? Use line segment diagrams to illustrate your solutions.

(10) A carton of ice cream costs $3. Joe buys 5 cartons of ice cream. He gives the clerk a $20 bill. How much is the change?

§2.3 Exercises   75

(11) If a nickel and a dime are flipped, how many possible outcomes are there? Make an array of the outcomes.

(12) A penny, a nickel, and a dime are flipped. List the outcomes in a systematic way that shows your method for counting them. What is the total number of possible outcomes? *Ans:* $2 \cdot 2 \cdot 2$.

(13) A red die, a white die, and a blue die are rolled. What is the total number of possible outcomes? *Ans:* $6 \cdot 6 \cdot 6$.

(14) Use the ideas in the computation

$$2 \cdot (1+2+3) = (1+2+3+1+2+3) = (1+3)+(2+2)+(3+1) = 3 \cdot 4$$

to verify that $2 \cdot (1+2+3+4) = 4 \cdot 5$. What is the correct formula for $2 \cdot (1+2+3+4+5+6+7+8+9+10)$? Prove it.

(15) Use numeration facts and multiplication properties to explain

$$4 \cdot 27 = 4 \cdot (20+7) = 4 \cdot (2 \cdot 10 + 7) = (4 \cdot 2) \cdot 10 + 4 \cdot 7$$

(16) How many seconds in a minute? How many seconds in an hour? How many seconds in a day?

(17) How many 1 cm cubes fit into a 1 meter cube? What is the volume in cubic centimeters of a 1 meter cube?

(18) A bakery sold 8 blueberry bagels. It sold 5 times as many plain bagels as blueberry bagels. How many plain bagels did it sell? How many bagels did it sell altogether? How many more plain bagels than blueberry bagels did it sell? Use a line segment diagram to illustrate your solutions.

(19) Prove that the set $D = \{a, b, c, d\}$ has 4! permutations by enumerating the permutations in an array with 4 rows, each row consisting of all the permutations of a set of 3 elements. Explain why the number of elements in each row is 3!.

(20) In §1.3.9 we determined that the circumference of a circle is between 3 diameters and 4 diameters. If the radius of the circle is 5 inches, explain why the circumference must be less than 40 inches. It must be more than how many inches?

(21) Explain the following proof of the Pythagorean Theorem. The square on the right is formed from 4 copies of the right triangle. Explain why the four-sided figure inside has right angles at each of its vertices. Calculate the area of the square on the right in two ways to prove the theorem.

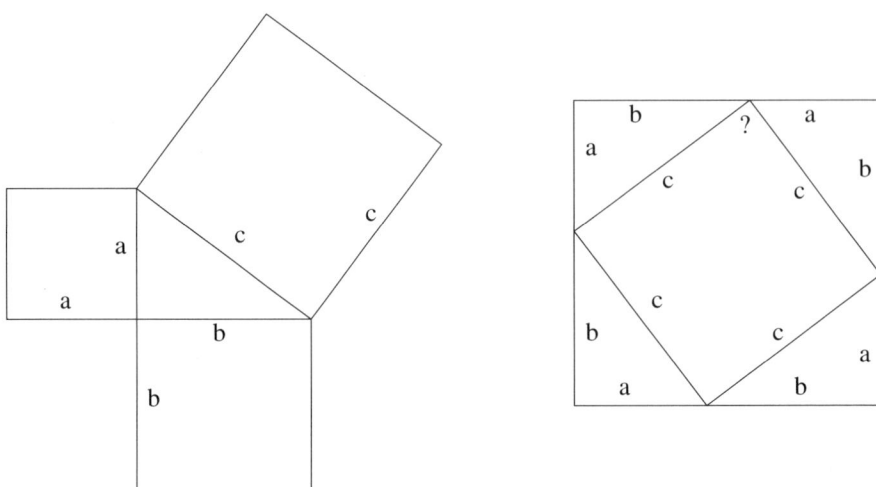

2.4. Division

Division of whole numbers is defined as the inverse operation to multiplication. Division is to multiplication as subtraction is to addition. As in the case of subtraction, there are two kinds of division. If the divisor is the number of groups, it is partitive division. If the divisor is the number in each group, it is measurement division. The two kinds of division produce the same quotient because multiplication is commutative.

Remember that in our definition of multiplication, $a \cdot b$ means a groups of b. The multiplier a is the number of groups and the multiplicand b is the number in each group.

Definition 2.82 (Partitive division)**.** If a and b are whole numbers, then $a \div b$ is the whole number q such that $b \cdot q = a$.

Thus, b groups of q makes a. The partitive division question is: b groups of what makes a? For example, divide 12 children into 4 equal groups. How many in each group? There are 3 children in each group, because 4 groups of 3 is $4 \cdot 3 = 12$.

Definition 2.83 (Measurement division)**.** If a and b are whole numbers, then $a \div b$ is the whole number q such that $q \cdot b = a$.

Thus, q groups of b makes a. The measurement division question is: How many b's in a? For example, divide 12 children into equal groups of 4. How many groups are there? There are 3 groups, because 3 groups of 4 is $3 \cdot 4 = 12$.

In the statement $a \div b = q$, a is called the *dividend*, b is called the *divisor*, and q is called the *quotient*. Such a whole number q might not exist in some cases. That is, for whole numbers a and b, there might not be any whole number q such that $a = b \cdot q$. In such a case, $a \div b$ is not defined (as a whole number, but see §2.4.3 on Long Division below). If the quotient exists (as a whole number), we say that a is *divisible* by b and we say that b is a *factor* of a.

We use the same notation for both kinds of division, because both operations yield the same quotient.

Theorem 2.84. *If a and b are whole numbers, then $a \div b$ in partitive division is equal to $a \div b$ in measurement division.*

Proof. A whole number $q = a \div b$ in partitive division means that $a = b \cdot q$. The commutative property of multiplication says that $b \cdot q = q \cdot b$, which means that $q = a \div b$ in measurement division. □

"So what's the point of two definitions of division?" students often ask. The point is that the two different kinds of division answer two different kinds of questions. Partitive division specifies the number of groups, b, and asks how many are in each equal group. Measurement division specifies the number in each equal group, b, and asks how many groups.

The following exercise demonstrates the difference between a measurement and a partitive problem. Suppose that we have a pile of 12 soda straws.

Measurement Problem: How many children can each receive 4 straws? To solve this problem, give 4 straws to Nicholas, then 4 to Joshua, then 4 to Caitlin, at which point the straws are gone. We conclude that 3 children can each receive 4 straws.

Partitive Problem: Divide the straws equally among 4 children. How many straws will each child get? To solve this problem, first give each child 1 straw, then give each child a second straw, then give each child a third straw, at which point the straws are gone. We conclude that each child will receive 3 straws.

Each multiplication gives rise to a measurement and a partitive division problem. Multiplication and division should be studied together. For example, $7 \cdot 9 = 63$ gives rise to the measurement problem "How many 9's in 63?" and to the partitive problem "7 of what is 63?".

Division by 0 is not defined. The reason for this is that if we try to define $5 \div 0$ to be some whole number q, then this would mean that $0 \cdot q = 5$, which is impossible since $0 \cdot q = 0$ for any whole number q. Should we define $0 \div 0 = 0$, since $0 \cdot 0 = 0$? But $0 \cdot q = 0$ and $q \cdot 0 = 0$, for any whole number q, means that any whole number q would satisfy the definition of partitive or measurement division for $0 \div 0$. If everything is the answer, then the question is pointless. We agree that $0 \div 0$ is undefined.

Example 2.85 (Partitive division problems). These are problems which can be reduced to the question: "b of what is a?". The answer is the number of elements in each of b equal groups.

(1). Ella paid $12 for 4 pounds of butter. What was the cost of 1 pound of butter?

1 unit = cost of 1 pound of butter. 4 units = \$12.

4 of what is 12? $12 \div 4 = 3$, because 4 groups of 3 is $4 \cdot 3 = 12$.

1 pound of butter costs \$3.

(2). An orchard is to have 42 trees divided equally among 6 rows. How many trees in each row?

1 unit = number of trees in each row. 6 units = 42 trees.

6 of what is 42? $42 \div 6 = 7$, because 6 groups of 7 is $6 \cdot 7 = 42$.

There are 7 trees in each row.

(3). Leah ran 20 miles, which was 5 times farther than Ragna ran. How far did Ragna run?

1 unit = number of miles Ragna ran. 5 units = 20 miles.

5 of what is 20? $20 \div 5 = 4$, because 5 groups of 4 is $5 \cdot 4 = 20$.

Ragna ran 4 miles.

(4). If 9 batches of a recipe require 72 cups of raisins, how many cups of raisins are needed for a single batch?

1 unit = number of cups in 1 batch. 9 units = 72 cups.

9 of what is 72? $72 \div 9 = 8$, because 9 groups of 8 is $9 \cdot 8 = 72$.

There are 8 cups of raisins in a single batch.

Example 2.86 (Measurement division problems). These are problems which can be reduced to the question: "How many b's in a?". The answer is the number of groups of b's in a.

2.4. Division

(1). How many pieces 2 feet long can Morty cut from a board which is 12 feet long?

$\underbrace{\qquad}_{2 \text{ ft}}\underline{\qquad\qquad\qquad\qquad\qquad}_{12 \text{ ft}}$

How many 2's in 12? $12 \div 2 = 6$, because 6 groups of 2 is $6 \cdot 2 = 12$.

(2). At 4 dollars per ball, how many baseballs can JB buy for $28?

$\underbrace{\qquad}_{\$4}\underline{\qquad\qquad\qquad\qquad\qquad}_{\$28}$

How many 4's in 28? $28 \div 4 = 7$, because 7 groups of 4 is $7 \cdot 4 = 28$.

(3). An agent can process 9 claims in 1 hour. How many agents (working at the same rate) are needed to process 54 claims in 1 hour?

$\underbrace{\qquad}_{9 \text{ claims}}\underline{\qquad\qquad\qquad\qquad\qquad}_{54 \text{ claims}}$

How many 9's in 54? $54 \div 9 = 6$, because 6 groups of 9 is $6 \cdot 9 = 54$.

(4). If a recipe requires 7 cups of flour, how many batches can be made from 56 cups of flour?

$\underbrace{\qquad}_{7 \text{ cups}}\underline{\qquad\qquad\qquad\qquad\qquad}_{56 \text{ cups}}$

How many 7's in 56? $56 \div 7 = 8$, because 8 groups of 7 is $8 \cdot 7 = 56$.

Oral exercise 2.87. Reduce each problem to the basic partitive or measurement division question. State the answer in the style of the preceding examples.

(1). Divide 70 children into 7 equal teams. How many children are there on each team?

$\underbrace{\qquad}_{? \text{ kids}}\underline{\qquad\qquad\qquad\qquad\qquad}_{70 \text{ kids}}$

(2). There are 24 flowers. To put 4 flowers in each vase, how many vases are needed?

$\underbrace{\qquad}_{4 \text{ flowers}}\underline{\qquad\qquad\qquad\qquad\qquad}_{24 \text{ flowers}}$

(3). Divide 18 petitfours equally among 6 plates. How many petitfours go on each plate?

$\underbrace{\qquad}_{? \text{ petitfours}}\underline{\qquad\qquad\qquad\qquad\qquad}_{18 \text{ petitfours}}$

(4). At 8 dollars each, how many hats can you buy for 32 dollars?

$\underbrace{\qquad}_{\$8}\underline{\qquad\qquad\qquad}_{\$32}$

(5). If Cecilia can buy 6 equally priced hats for $66, how much does each hat cost?

(6). How many weeks in 21 days?

Oral exercise 2.88. These problems require more than one step. Reduce any division step to the basic partitive or measurement question.

(1) If 5 dolls cost $45, how much will 7 dolls cost?

      ? $ $45

1 unit = 1 doll. 1 doll costs ___?

7 dolls cost ___?

(2). Janet bought 18 pencils. She bought 3 times as many pencils as pens. If the pencils and pens each cost $8, how much did she pay altogether?

    pencils   18

    pens

1 unit = number of pens. 1 unit costs ___?

4 units cost ___?

Remember that our goal is to explain how to do problems, not just solve the problems. A line segment diagram illustrates the difference between a partitive and a measurement division problem. The partitive division problem requires finding the value of 1 segment, which I call the unit. This is the *unitary method*.

2.4.1. Properties of Division. Notation frequently used for a divided by b is:

$$a \div b = a/b = \frac{a}{b}$$

Theorem 2.89 (Division Properties). *If a, b, and c are whole numbers, then*

(1) $(a \cdot b)/c = a \cdot (b/c)$, (associative property).

(2) $a/(b \cdot c) = (a/b)/c$, (associative property).

(3) If $a \neq 0$ and $b < c$, then $b/a < c/a$, (order property).

(4) If $0 < b < c$, then $a/b > a/c$, (order property).

(5) $(a + b)/c = a/c + b/c$, (distributive property over addition).

(6) If $a > b$, then $(a - b)/c = a/c - b/c$, (distributive property over subtraction).

It is assumed that any division indicated is defined.

2.4.1.1. *Interpretations and examples of the properties.* An example of the associative property (1) is

$$(3 \cdot 20)/5 = 3 \cdot (20/5)$$

2.4. Division

In terms of partitive division, it says that to divide $3 \cdot 20$ into 5 equal parts results in the same size parts as when we divide 20 into 5 equal parts and take 3 of these, as pictured here:

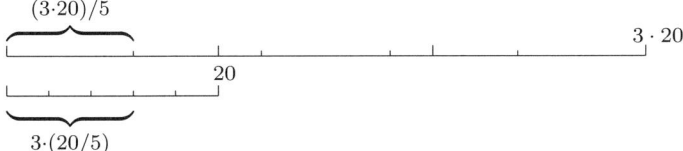

In terms of measurement division, $(3 \cdot 20)/5 = 3 \cdot (20/5)$ says that the number of 5's in $3 \cdot 20$ is equal to 3 times the number of 5's in 20, as pictured here:

Associative property (1) can simplify the calculation in something like
$$(87 \cdot 121)/11 = 87 \cdot (121/11) = 87 \cdot 11$$

Oral exercise 2.90. An example of associative property (2) is
$$72/(4 \cdot 3) = (72/4)/3$$

Explain how the line segment diagram below illustrates that in terms of partitive division; this property says that dividing 72 into $4 \cdot 3$ equal parts results in the same size parts as dividing 72 into 4 equal parts and then dividing each of these parts into 3 equal parts.

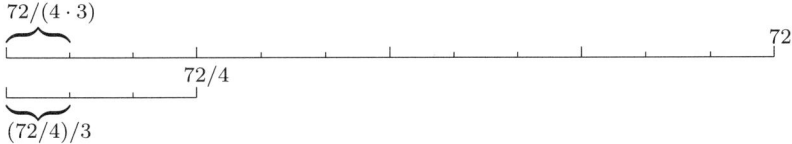

Oral exercise 2.91. Explain how the following line segment diagram illustrates that in terms of measurement division, $72/(4 \cdot 3) = (72/4)/3$ says that the number of $4 \cdot 3$'s in 72 is equal to the number of 3's in $72/4$.

Associative property (2) can simplify the calculation in something like
$$180/(4 \cdot 9) = (180/4)/9 = 45/9 = 5$$

Order property (3) says that *for the same divisor, the smaller dividend yields the smaller quotient.* An example is
$$10/5 < 15/5, \text{ since } 10 < 15$$

It says that if we divide 10 equally among 5, then each will get less than if we divide 15 equally among 5. It also says that there are fewer 5's in 10 than there are in 15.

Oral exercise 2.92. Order property (4) says that *for the same dividend, the smaller divisor yields the larger quotient.* An example is

$$30/3 > 30/5, \text{ since } 3 < 5$$

Give the partitive and measurement interpretations of this inequality.

(5) An example of the distributive property over addition is

$$(15 + 25)/5 = 15/5 + 25/5$$

Partitive: $15 + $25 divided equally among 5 people gives each the same as dividing $15 equally among these 5 and then dividing $25 equally among them:

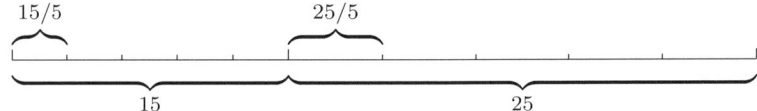

Measurement: The number of 5's in $15 + 25$ is the number of 5's in 15 plus the number of 5's in 25:

(6) An example of the distributive property over subtraction is

$$(28 - 16)/4 = 28/4 - 16/4$$

Partitive: Mom divides $28 equally among her 4 children, so each receives $28/4$ dollars. They divide a $16 snack bill equally among them, so each pays $16/4$ dollars and each has $28/4 - 16/4$ dollars left. Or, Mom pays the $16 snack bill from the $28 and divides the $28 - 16$ dollars left equally among the 4 children, giving each child $(28 - 16)/4$ dollars. Either way, the kids come out the same:

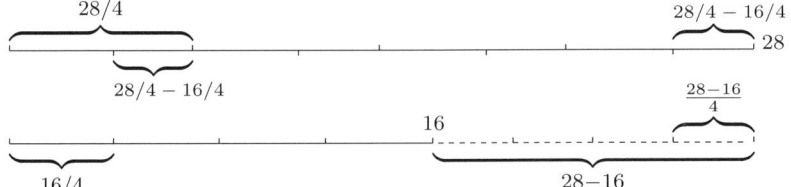

Measurement: The number of 4's in $28 - 16$ is equal to the number of 4's in 28 minus the number of 4's in 16.

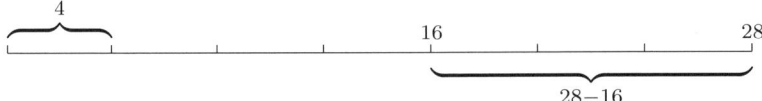

Formal proofs of the division properties. We check a division problem by multiplying the quotient by the divisor and seeing whether that product is the dividend. Such a check is an application of the definition of

2.4. Division

partitive division. In symbols, to check that $a \div b = q$, one verifies that $b \cdot q = a$.

(1) To check this associative property, multiply the expression on the right side of the equation times c to see if you get $a \cdot b$. We do this using properties of multiplication as follows.

$$(a \cdot (b/c)) \cdot c = a \cdot ((b/c) \cdot c), \text{ associative property of mult.}$$
$$= a \cdot b, \text{ definition of } b/c$$

(2) To check this associative property, multiply the expression on the right side of the equation by $b \cdot c$ to see if you get a.

$$(b \cdot c) \cdot ((a/b)/c) = b \cdot (c \cdot ((a/b)/c)), \text{ assoc. prop. of mult.}$$
$$= b \cdot (a/b), \text{ definition of } (a/b)/c$$
$$= a, \text{ definition of } a/b$$

(3) This order property is proved using the order property for multiplication. Multiplying b/a and c/a by a will preserve their order. Thus, their order is the same as $a \cdot (b/a) = b$ and $a \cdot (c/a) = c$, which is $b < c$. Therefore, $(b/a) < (c/a)$.

(4) This order property is proved using the order property for multiplication: since $b < c$, then $a \cdot b < a \cdot c$. We also use the simple facts that $b/b = 1$ and $c/c = 1$. Now

$$a/c = (a \cdot 1)/c = (a \cdot (b/b))/c$$
$$= ((a \cdot b)/b)/c \text{ first assoc. prop.}$$
$$= (a \cdot b)/(b \cdot c) \text{ second assoc. prop.}$$
$$< (a \cdot c)/(b \cdot c) \text{ first order property}$$
$$= (a \cdot c)/(c \cdot b) = ((a \cdot c)/c)/b \text{ second assoc. prop.}$$
$$= (a \cdot (c/c))/b \text{ first assoc. prop.}$$
$$= (a \cdot 1)/b = a/b$$

from which we conclude that $a/c < a/b$.

(5) To prove the distributive property over addition, we show that it satisfies the definition of measurement division by using the distributive property of multiplication over addition to calculate

$$(a/c + b/c) \cdot c = (a/c) \cdot c + (b/c) \cdot c = a + b$$

which checks that $a/c + b/c = (a+b)/c$.

(6) The proof of the distributive property over subtraction goes in the same way using the distributive property of multiplication over subtraction.

$$(a/c - b/c) \cdot c = (a/c) \cdot c - (b/c) \cdot c = a - b$$

which checks that $a/c - b/c = (a-b)/c$. □

2.4.2. Order of operations. In a line of additions, subtractions, multiplications, and divisions, the standard *order of operation* is left to right, multiplication and division followed by addition and subtraction. Operations enclosed in parentheses are done first. For example,

$$20/4/5 = (20/4)/5 = 1$$
$$20/4 \cdot 5 = (20/4) \cdot 5 = 25$$
$$20/4 + 5 = 5 + 5 = 10$$
$$20 + 8/4 \cdot 3 = 20 + 2 \cdot 3 = 20 + 6 = 26$$

Oral exercise 2.93. Use Theorem 2.89 to restate each of the following in a form which simplifies the calculation.

(1) $(7 \cdot 15)/3$

(2) $(18 \cdot 43)/6$

(3) $45/(5 \cdot 3)$

(4) $42/7/3$

Oral exercise 2.94. Use the order properties of division to determine and justify the following.

(1) Which is larger, $63/7$ or $77/7$?

(2) Which is larger, $288/9$ or $288/8$?

Example 2.95 (Combining rates). Ruth and Clarence live 16 miles apart. Ruth walks 3 mph and Clarence walks 5 mph. At 2:00 p.m. each leaves home and starts walking towards the other's house. How long until they meet?

Solution: They meet when they have walked a combined distance of 16 miles.

They start at 2:00 pm.

After 1 hour their combined distance is $3 + 5$ miles.

Their combined rate is $3 + 5$ mph.

The number of hours until they meet is the number of $3 + 5$'s in 16.

They meet after $16/(3 + 5)$ hours.

Oral exercise 2.96 (Catch up). Viola starts walking at noon at the speed of 4 mph. Her brother Ralph notices that she has forgotten her math book.

2.4. Division

At 1 o'clock he starts jogging at 6 mph to catch her. How long does it take for him to catch her?

In 1 hour Viola has gone how many miles?

When Ralph starts, how far ahead is Viola?

```
   0                         ?
   |_____|
 Ralph                      Viola
```

After 1 more hour how far has Ralph gone? How far has Viola gone?

Viola
```
         0              4       ?  4+4
         |_____|_____⌢___|
 Ralph                          6
```

After 1 hour the distance between them is $4 - (6 - 4)$ miles. Explain.

In 1 hour Ralph has caught up by $6 - 4$ miles. Explain.

Ralph's rate of catch up is ____ mph.

Ralph catches up with Vioia after ____ hours.

2.4.3. Long Division. In measurement division, $8 \div 2 = 4$ means that $8 = 4 \cdot 2 = 2 + 2 + 2 + 2$. This means that measurement division $8 \div 2 = 4$ is the number of times 2 can be subtracted from 8 until 0 is reached. In detail

$$\begin{aligned}
0 &= 8 - (2 + 2 + 2 + 2) \\
&= 8 - (2 + (2 + 2 + 2)), & \text{associative property of addition} \\
&= (8 - 2) - (2 + 2 + 2), & \text{(6) of Theorem 2.33} \\
&= ((8 - 2) - 2) - (2 + 2), & \text{same two reasons} \\
&= (((8 - 2) - 2) - 2) - 2, & \text{same two reasons}
\end{aligned}$$

The number of times b can be subtracted from a is the quotient of the measurement division $a \div b$. How much can be subtracted from a for a total of b times is the quotient of the partitive division $a \div b$.

Division as repeated subtraction leads us naturally to the idea of division with remainder. For example, $9 \div 2$ is not defined, because there is no whole number whose product with 2 is 9. Thinking of measurement division, however, we can ask how many groups of 2 are in 9. This is equivalent to asking how many times 2 can be subtracted from 9, until we reach 0 or until an amount less than 2 remains. Asking the question in this way leads to a solution. We have

(2.10)
$$\begin{aligned}
9 - 2 &= 7, & 1 \text{ time} \\
(9 - 2) - 2 &= 7 - 2 = 5, & 2 \text{ times} \\
((9 - 2) - 2) - 2 &= 5 - 2 = 3, & 3 \text{ times} \\
(((9 - 2) - 2) - 2) - 2 &= 3 - 2 = 1, & 4 \text{ times}
\end{aligned}$$

Therefore, 2 can be subtracted from 9 a total of 4 times, at which point there is only 1 remaining. This information is summarized by writing $9 = 4 \cdot 2 + 1$. We say that $9 \div 2$ has *quotient* 4 and *remainder* 1.

Theorem 2.97 (Long Division). *If a and b are whole numbers and if $b \neq 0$, then there are unique whole numbers q, for quotient, and r, for remainder, such that*

(2.11) $$a = q \cdot b + r, \text{ where } 0 \leq r < b$$

We say that $a \div b$ has *quotient* q with *remainder* r. The quotient is the number of b's in a, which is the same as the number of times b can be subtracted from a.

If equation (2.11) is written in the form

$$a = b \cdot q + r$$

then the quotient q is the number of elements in each group when a elements are divided equally among b groups. Since $r < b$, there are not enough elements in the remainder to give one to each group.

Proof. The proof is seen by starting with a set of a elements and counting its elements into groups of b. The process will stop when there remain too few elements to make another group of b. The quotient is the number of groups of b and the remainder is the number left over.

We illustrate the proof for the case of $47 \div 8$. Suppose we have 47 soda straws. Count them into a group of 8. There remain $47 - 8 = 39$ straws. Count these into a group of 8. There remain $39 - 8 = 31$ straws. Count these into a group of 8. There remain $31 - 8 = 23$ straws. Count these into a group of 8. There remain $23 - 8 = 15$ straws. Count these into a group of 8. There remain $15 - 8 = 7$ straws. There are now too few straws to make another group of 8. We have made 5 groups of 8, so 5 is the quotient of $47/8$, and there are 7 straws remaining, so 7 is the remainder of $47/8$ and we write $47 = 8 \cdot 5 + 7$. This counting process finally ends, because after each group is made there remain fewer straws than before. This is the same reasoning we used to see that any whole number has a base ten numeral. □

If you plant 8 trees in a row, how many rows can you make with 47 trees? In practical terms, you might take 8 trees and plant a row, then take another 8 trees and plant another row. Continue doing this until there are too few trees for another row. The orchard will look like this, with the

2.4. Division

remainder as a sixth row which contains only 7 trees.

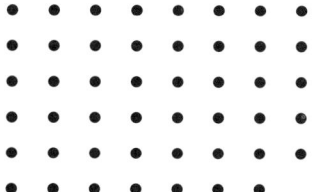

The counting process amounts to repeated subtractions. On the number line, repeated subtractions of 8 from 47 looks like

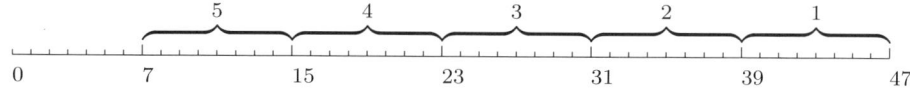

The quotient is the number of braced 8's. The remainder is the amount left.

Oral exercise 2.98. Explain how the quotient of $47 \div 8$ is the number of times 8 can be added to itself before surpassing 47. Illustrate on a number line. Describe the remainder on this picture.

Example 2.99 (Numeration). We carry out a sequence of long divisions by 10 when we find the base ten numeral of a whole number. Counting into groups of ten is the process of measurement division by ten. The remainder is the digit which goes into that place. When dividing by ten, the possible remainders are any whole numbers less than ten. The digits are needed to represent each possible remainder.

Oral exercise 2.100. Explain how finding the base five numeral of seventeen soda straws involves long divisions by five. What are the possible remainders? What digits are needed in base five numeration?

The order properties (3) and (4) of Theorem 2.89 extend, with slight modification, to quotients in long division. These order properties are used to estimate the quotient at each step of the long division algorithm.

Theorem 2.101 (Order properties of quotients). *Suppose a, b, and c are whole numbers and that $a \neq 0$.*

(1). If $b < c$, then the quotient of b/a is less than or equal to the quotient of c/a.

(2). If $0 < b < c$, then the quotient of a/b is greater than or equal to the quotient of a/c.

In (1) it is possible to have equality. For example, the quotient of $90/20$ equals the quotient of $93/20$. It is also possible to have inequality. For example, the quotient of $76/20$ is less than the quotient of $93/20$.

In (2) it is possible to have equality. For example, the quotient of 54/20 is equal to the quotient of 54/25. It is also possible to have inequality. For example, the quotient of 64/20 is greater than the quotient of 64/25.

Proof. (1). The assumption is that b lies to the left of c on the number line. Then the quotient of b/a is the number of a's in b, which is less than or equal to the number of a's in c.

(2). The quotient of a/b is the number of b's in a, which is greater than or equal to the number of larger c's there are in a.

□

Oral exercise 2.102. (1) Which is larger, the quotient of 87/12 or the quotient of 87/10? What is the quotient of 87/10?

(2) Which is larger, the quotient of 114/20 or the quotient of 114/25? What is the quotient of 114/25?

Oral exercise 2.103. Explain how to use a set of 54 soda straws to illustrate that the quotient of 47/5 is less than the quotient of 54/5.

§2.4 Exercises

In each word problem, explain your solution. Use line segment diagrams to illustrate your solutions. If a division $m \div n$ is required, identify the problem as measurement division, by rephrasing it as "How many of n in m?", or as partitive division, by rephrasing it as "n of what is m?".

(1) One bag of sugar costs $3. How many bags of sugar can be bought with $15?

(2) Dona has $28. She has 4 times as much money as Samantha. How much money has Samantha?

(3) Andrea and Leah together have $21. Andrea has twice as much money as Leah has. How much money does Leah have?

(4) Niels and Sophie go to the park. The lake paddle boats can be rented for $4 an hour. If Niels has $12, for how many hours can he rent a paddle boat?

§2.4 Exercises

(5) Patricia, Rosemary, and Peggy go to the park. The lake paddle boats can be rented for $4 an hour. They rent a paddle boat for 6 hours and share the cost equally. How much does each pay?

(6) Anthony put 15 apples equally on 3 plates. How many apples on each plate?

(7) Joshua put 15 apples into bags at 3 apples per bag. How many bags did he use?

(8) Leah divides her class of 27 pupils into 3 equal groups. How many in each group?

(9) Share 12 cherries equally between 2 children. How many cherries does each child get?

(10) Sue has 14 caramels. She gives 2 caramels to each child. How many children receive caramels?

(11) There are 20 coins. Put 5 coins in a set. How many sets are there?

(12) Follow the proof of Theorem 2.89 to prove that $(3 \cdot 20)/5 = 3 \cdot (20/5)$.

(13) Follow the proof of Theorem 2.89 to prove that $36/(4 \cdot 3) = 36/4/3$.

(14) Follow the proof of Theorem 2.89 to prove that $87/8 < 92/8$.

(15) Three identical shirts cost $15. How much do 5 of these shirts cost?

(16) Bethany saves $7 a day in order to buy a shirt which costs $40. How many days before she can buy the shirt?

(17) Follow the proof of Theorem 2.89 to prove that $93/17 > 93/20$ and that $85/22 > 85/25$.

(18) Shopping around, Ragna discovers that Big J is selling a package of 6 socks at the same price that Little J is charging for a package of 8 socks. Which store has the lower price per pair of socks? Cite the division property used.

(19) Use an array to illustrate the quotient and remainder of $12 \div 5$, from the partitive point of view and from the measurement point of view.

(20) Use the number line to illustrate the quotient and remainder of $17/5$, from the point of view of repeated addition and from the point of view of repeated subtraction.

(21) Explain how to use a set of 47 soda straws to illustrate the quotient and remainder of $47/9$.

(22) The product of two numbers is 4312695. One of the numbers is 1205. What is the other? Explain what calculation must be made, but don't make it.

(23) The product of three numbers is 107100. One of the numbers is 42, another 34. What is the third? Explain what calculation must be made, but don't make it.

(24) One side of a rectangle is 5 inches long. The area of the rectangle is 35 in². What is the perimeter of the rectangle?

(25) One side of a rectangle is 3 cm long. The perimeter of the rectangle is 16 cm. What is the area of the rectangle?

(26) Molly sold 200 plain bagels. She sold 4 times as many plain bagels as onion bagels. How many onion bagels did she sell?

(27) Rita sold 30 jars of jelly on Friday. She sold 6 times as many jars of jelly on Friday as she sold on Sunday. How many jars of jelly did she sell on Sunday?

(28) Newell packed 250 ears of corn into bags with 5 ears in each bag.
  (a) How many bags of corn were there?
  (b) He sold all the corn at $2 a bag. How much money did he receive?

(29) Andre bought 70 doughnuts for 5 cents apiece. She packed the doughnuts in cartons with 10 doughnuts in each carton. She then sold all the doughnuts for $1 per carton. What was her profit?

(30) In a calculus class of 27 students there are 3 more men than women. How many women are there? *Hint:* 2 units = ___?

    men
    women

(31) In a box there are red marbles and blue marbles. There are 5 times as many red marbles as blue marbles. There are 20 more red marbles than blue marbles. How many marbles of each color are there?

(32) Ardis walks 3 mph and Dick walks 4 mph. They live 14 miles apart. Each leaves home at noon and begins walking towards the other's house. At what time do they meet? How far from Ardis's house are they when they meet?

(33) Gavin has $20 and Nels has none. They both start jobs at the same time. Gavin makes $5 an hour and Nels makes $7 an hour. After how many hours will they both have the same amount of money?

(34) How many degrees in each angle of an equilateral triangle? Explain your answer.

2.5. Dividing the Whole

Partitive division considers the problem of dividing a set of objects into a given number of equal parts. Whole objects also can be divided into a given

2.5. Dividing the Whole

number of equal parts. We can divide a cake, a trip, a quantity of money, or an amount of time into equal parts. The emphasis is on **equal** parts.

When an object is divided into n equal parts, then each part is called $\frac{1}{n}$ of the object. The name for $\frac{1}{n}$ is the *ordinal* name for the natural number n, except for $n = 2$ when $\frac{1}{2}$ is called a *half*, and for $n = 4$ when $\frac{1}{4}$ is also called a *quarter*. So, $\frac{1}{3}$ is a third, $\frac{1}{5}$ is a fifth, $\frac{1}{8}$ is an eighth, $\frac{1}{15}$ is a fifteenth, etc.

Definition 2.104. The *fraction* $\frac{m}{n}$ of an object is the amount obtained by dividing the object into n equal parts and taking m of these parts.

The number m on the top of $\frac{m}{n}$ is called the *numerator* and the number n on the bottom is called the *denominator*. A fraction $\frac{m}{n}$ is often written m/n to simplify the typography.

If we divide an object into n equal parts and take n of these parts, we obtain the whole object. Thus

$$n \text{ of } \frac{1}{n} \text{ make } 1 \text{ whole.}$$

Divide a circle into 2 equal parts. Each part is a *half* circle, which we write $\frac{1}{2}$ circle. Two half circles make 1 whole circle. Divide a paper circle into 2 equal parts by folding it through the center. The fold line will be a diameter.

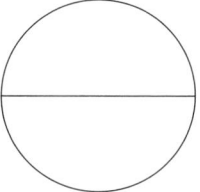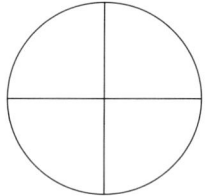

Divide a circle into 4 equal parts. Each part is a *quarter* circle or a *fourth* of a circle, which we write $\frac{1}{4}$ circle. Three of these equal parts is 3 fourths, also called 3 quarters, and written $\frac{3}{4}$ of a circle. Four quarter circles make 1 whole circle. Divide a paper circle into four equal parts by folding it through the center, then fold the half circle again, through the center, perpendicular to the first fold.

A rectangular sheet of paper can be divided into 2 equal parts by folding it in the following ways.

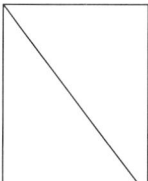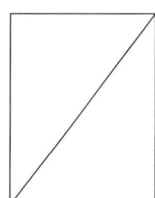

In the first picture, bring the top edge over to coincide with the bottom edge. The resulting crease, which is called a horizontal *line of symmetry*, divides the rectangle into two equal parts. In the second picture, bring the left edge over to coincide with the right edge, thus making the vertical crease, which is a vertical *line of symmetry*. The last two pictures are based on the fact that a diagonal of a rectangle divides it into two equal parts. This can be verified by cutting along the diagonal and superimposing one part over the other after rotating one part through 180°. A diagonal is a line of symmetry only when the rectangle is a square.

Both diagonals together divide a rectangle into 4 equal parts. Each part is called a *quarter* of the rectangle. It is also called a *fourth* of the rectangle. Fold a rectangular piece of paper along a vertical line of symmetry to divide it into two equal parts, then fold it again along the horizontal line of symmetry. This will divide the rectangle into 4 equal parts.

The two diagonals with the vertical and horizontal lines of symmetry altogether divide the rectangle into 8 equal parts. Each part is an $\frac{1}{8}$ of the rectangle.

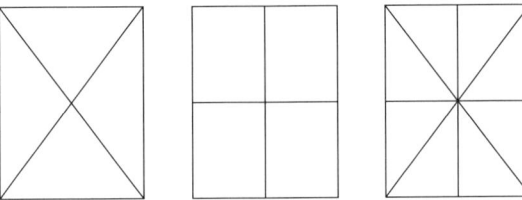

Oral exercise 2.105. In the following picture, $\frac{2}{3}$ of the circle is shaded. This means that how many out of the how many equal parts are shaded?

What fraction of the rectangle is shaded? This means that how many out of the how many equal parts are shaded?

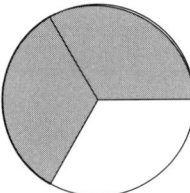 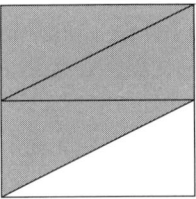

In taking a fraction of an object, it must be divided into **equal** parts.

Oral exercise 2.106. For the following figures, answer the question and explain your answer.

2.5. Dividing the Whole

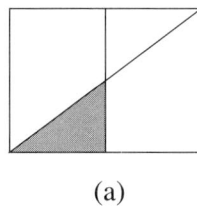
(a)

(b)

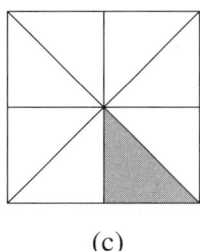
(c)

(1) Is the shaded region of Figure (a) a $\frac{1}{4}$ of the figure?

(2) Is the shaded region of Figure (b) a $\frac{1}{3}$ of the figure?

(3) Is the shaded region of Figure (c) an $\frac{1}{8}$ of the figure?

Example 2.107. A fraction of a fraction of a whole can be done in either order with the same result. If we take $\frac{2}{3}$ of $\frac{4}{5}$ of an object (a rectangle in the illustration), first divide it into 5 equal parts (by vertical strips) and take 4 of these (shown with stripes)

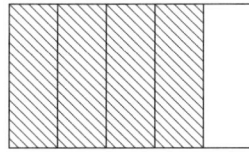

and then divide the whole rectangle into 3 equal parts (by horizontal strips) and take 3 of these (shown with opposite stripes) in such a way that the first striped part has been divided into 3 equal parts, as follows:

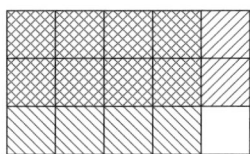

The cross-hatched part is $\frac{2}{3}$ of $\frac{4}{5}$ of the whole rectangle. The two divisions have divided the whole into 15 equal parts, and the cross-hatched part comprises 8 of these. Thus, $\frac{2}{3}$ of $\frac{4}{5}$ of the whole is $\frac{8}{15}$ of the whole. It is evident from the picture that $\frac{4}{5}$ of $\frac{2}{3}$ of the whole is the same amount.

Oral exercise 2.108. Consider the following figures.

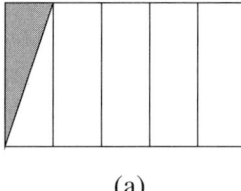
(a)

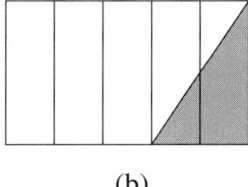
(b)

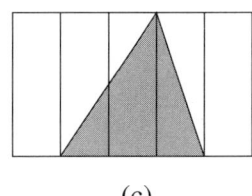
(c)

(1) Explain why the shaded region of Figure (a) is $\frac{1}{10}$ of the figure.

(2) Explain why the shaded region of Figure (b) is $\frac{1}{5}$ of the figure.

(3) Explain why the shaded region of Figure (c) is $\frac{3}{10}$ of the figure.

Two of the shaded regions in Figure (a) make one of the small rectangles, each of which is $\frac{1}{5}$ of the whole rectangle. It takes ten of the shaded regions to make the whole rectangle, so the shaded region is $\frac{1}{10}$ of the whole rectangle. We see then that $\frac{2}{10}$ of the whole rectangle is the same amount as $\frac{1}{5}$ of the whole rectangle. Two different fractions of a whole can produce the same amount.

2.5.1. Area of a Rectangle. In §1.3.2 we defined the area of a rectangle as the number of unit squares needed to build it. We can extend this definition to include half squares.

A square is a special case of a rectangle, so the ways described above to divide a rectangle into 2 equal parts apply to squares. Each part is called a *half square*, written $\frac{1}{2}$ square. A half square formed by a diagonal is easy to see when building figures. It is useful to use them with whole squares to measure areas. Here are some examples drawn on graph paper to make it easy to count the squares and half squares.

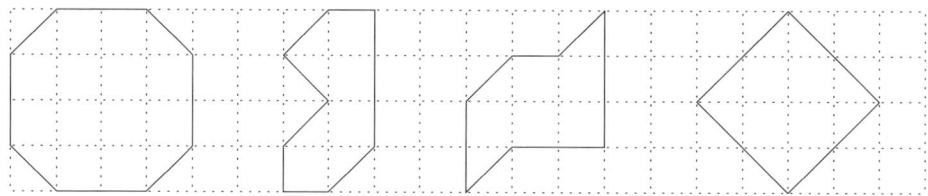

Oral exercise 2.109. How many unit squares in each of these regions? Remember that 2 half squares make 1 unit square.

2.5.2. Area of a Triangle. The following ideas can be observed with triangles cut from paper. Start with a triangle and orient it on the paper so that one of its sides is at the "bottom". Call this side the *base* of the triangle. The choice is arbitrary as to which side you call the base.

Definition 2.110. The rectangle *associated* to a triangle with a chosen base is the rectangle one side of which is the triangle's base and whose opposite side is contained in a line passing through the triangle's opposite vertex. The height of this rectangle is called the *height* of the triangle with reference to its chosen base.

Here are two examples of triangles and their associated rectangles. The base chosen for each triangle is the side at the bottom.

2.5. Dividing the Whole

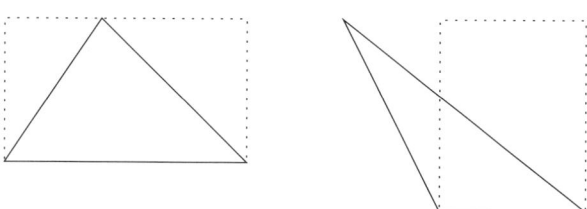

By cutting the triangle and rearranging the parts, we can verify that a triangle has half the area of its associated rectangle. Here are some examples. In each case the base is the side at the bottom.

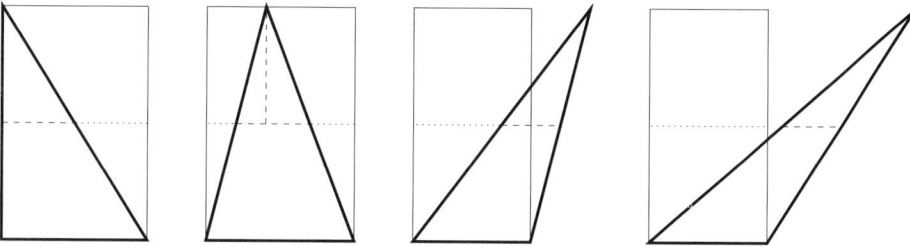

The first cut is through the horizontal line half-way up the associated rectangle. The next move is to rotate the top piece one way or the other to fill in an empty portion of the bottom half of the associated rectangle. The next cut, if needed, will be along a vertical edge of the associated rectangle.

Here is how they look after the triangle is cut and rearranged.

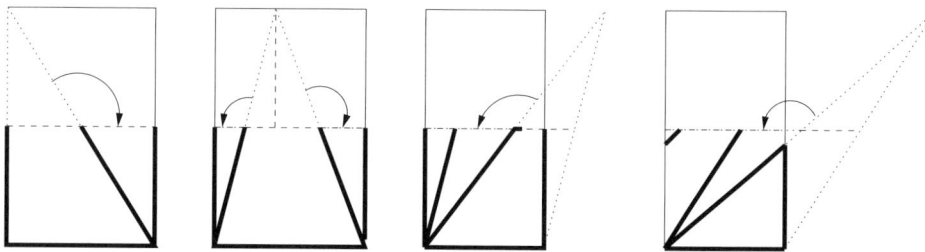

Oral exercise 2.111. Describe how to cut and rearrange each triangle above so that it fills in the bottom half of its associated rectangle.

Oral exercise 2.112. On a sheet of graph paper use a straight edge to draw a triangle. Choose a base for the triangle and draw the associated rectangle. Use scissors to cut the triangle so that the pieces can be rearranged to fill half the associated rectangle.

Theorem 2.113 (Area of a Triangle). *The area of a triangle is half the area of an associated rectangle. Therefore, the area of a triangle is half the product of a base times the height.*

Proof. The above examples illustrate how the proof should go. From the cases shown, you should be able to describe the exact procedure for the cutting and rearranging. With more knowledge of the properties of vertical angles and congruent triangles, we could explain why these procedures will always work. As an introduction to the subject, cutting and rearranging triangles gives adequate insight into the proof. □

Oral exercise 2.114. What is the area of the following shaded triangle? What side do you regard as the base? Describe its associated rectangle for this base.

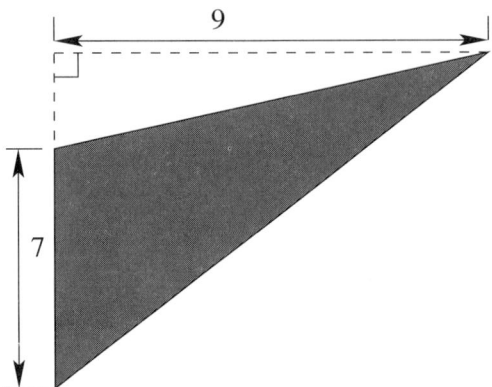

2.5.3. Dividing a Line Segment. Consider the problem of dividing a line segment into n equal parts, for any natural number n. The following method of doing this is based on Euclid's ruler and compass construction of the division of a line segment into n equal subintervals. We do this by laying the line segment on a grid of $n + 1$ equally spaced parallel lines. For example, to divide the line segment

into five equal parts, set it on six equally spaced parallel lines in such a way that one end lies on the bottom line and the other end lies on the fifth line above that, like this.

The points where the parallel lines intersect the line segment will divide it into 5 equal subsegments:

Each equal subsegment is a fifth of the given line segment. Three of these equal subsegments would be 3 fifths, which we write as $\frac{3}{5}$ of the given line segment. Five of these fifths make 1 whole line segment.

2.5. Dividing the Whole

Oral exercise 2.115. Repeat the measurement of the circumference of a circle in terms of its diameter, as described in §1.3.9. On the strip of paper on which you have marked four diameters, use the parallel lines above, or some other you have available, to divide the last diameter segment into 7 equal parts. As a result of your measurement, is the circumference more or less than 3 and $\frac{1}{7}$ diameters?

Archimedes of Syracuse, who lived from 287 - 212 BC, proved that the circumference of a circle is greater than 3 and $\frac{10}{71}$ diameters and less than 3 and $\frac{1}{7}$ diameters. In terms of the number π of Definition 1.68, Archimedes proved that

(2.12) $$3\frac{10}{71} < \pi < 3\frac{1}{7}$$

In subsequent chapters we shall look further into the nature of π. Section 6.5 gives an indication of how Archimedes obtained his estimates. An excellent article on π is available on the web at

`http://turnbull.mcs.st-and.ac.uk/history/HistTopics`
                        `/Pi_through_the_ages.html`

This article is kept in the *MacTutor History of Mathematics Archive*, which is on the Turnbull web server operated by the School of Mathematics and Statistics of the University of St. Andrews Scotland. Turnbull's url is

`http://turnbull.mcs.st-and.ac.uk/history/`

2.5.4. Line Segment Diagrams. A line segment divided into equal parts can illustrate the information of a problem and show us how to solve it.

Example 2.116. Nettie returns from the market with a bag of apples, oranges, and plums. Half the fruit in the bag are apples. There are 3 times as many apples as oranges. There are 3 fewer oranges than plums. How many of each fruit are in the bag?

Solution: Draw a line segment to represent all the fruit in the bag. The apples are then represented by half of this line segment. Divide each half into 3 equal parts, then 1 of these parts represents the oranges. The rest of the line segment must represent the plums. Now the line has been divided into 6 equal parts, and 1 of these parts must represent 3 pieces of fruit, because 1 of these parts represents the number fewer oranges than plums that there are.

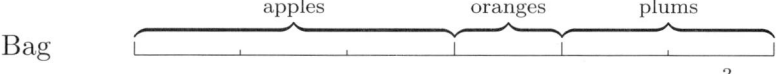

§2.5 Exercises

Use line segment diagrams to explain your solutions.

(1) In a classroom of 21 children, $\frac{1}{7}$ of them are left-handed: how many are left-handed?

(2) In a classroom of 21 children, 4 of them have red hair: what fraction of the children have red hair?

(3) Find the area of the shaded triangle. Trace this triangle and on it draw the triangle's associated rectangle whose dimensions you know.

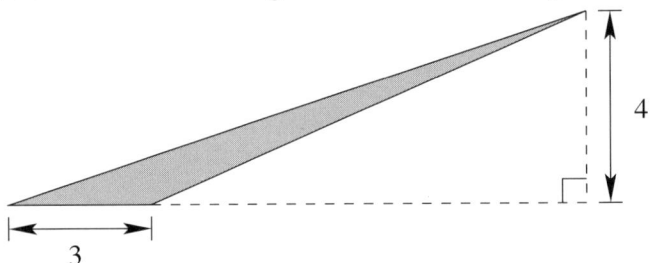

(4) Find the area of the shaded triangle. If the side length of each unit square is 1 foot, what is the area in square feet?

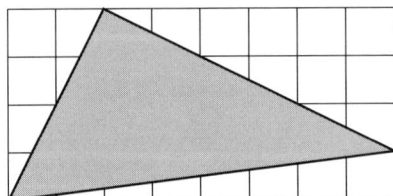

(5) A box contains red marbles, white marbles, and blue marbles. Four-sevenths of the marbles in the box are red. There are twice as many red marbles as white marbles. There are 5 more white marbles than blue marbles. Find the number of marbles of each color in the box. *Hint:* Complete the following line segment diagram. How many segments represent red marbles? How many segments represent white marbles? How many segments represent blue marbles? One subsegment represents how many marbles?

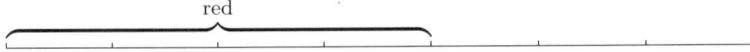

(6) Three-fifths of the people in a bus are children. The rest are adults. There are twice as many children as female adults. There are 10 more female adults than male adults. Find the number of children, female adults, and male adults on the bus.

§2.5 Exercises 99

(7) At a summer camp there are children from Montana and from Missouri. $\frac{3}{5}$ of the children are from Montana. $\frac{2}{3}$ of the Montana children are boys. $\frac{3}{4}$ of the Missouri children are girls. There are 10 more Montana girls than Missouri boys. How many Missouri girls are in the camp?

(8) Each grid square is a unit square. Find the area of each figure.

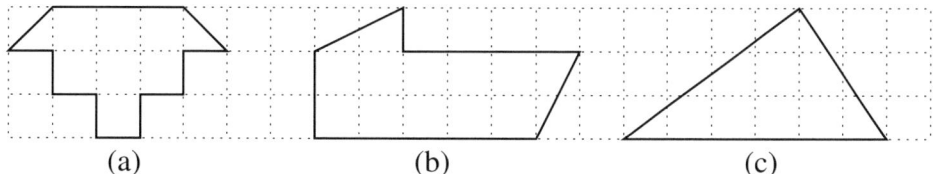

(a) (b) (c)

(9) Inscribe a regular hexagon in a circle, as shown. Prove that the hexagon comprises 6 equilateral triangles. *Hint:* Why is the angle at the center $\frac{1}{6}$ of 360°? Why is the triangle isosceles? Why are the angles at the circle equal? What must be the sum of these two angles? Why are all of the angles equal?

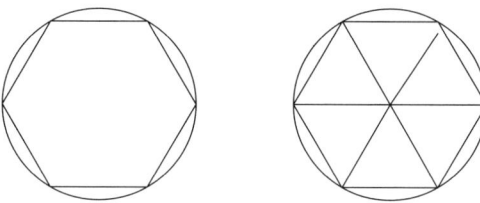

Use this fact to prove that the circumference of the circle is greater than 3 diameters.

Chapter 3

Whole Number Computation

Methods of computation are based on an ingenious numeration system called the Hindu-Arabic system to base ten. The computation algorithms of arithmetic combine the properties of this numeration system with the properties of the arithmetic operations. The algorithms cannot be understood without an understanding of both sets of properties. As we shall see, the algorithms do not depend on the choice of base. Trying the algorithms in a base other than ten serves to test one's understanding of the algorithms.

Grade school children must learn how to compute sums, differences, products, and quotients. Their efforts will be aided if they are taught how the methods work. Teachers must know how to compute and why the methods work, and they must also know how to explain the procedures and how to explain why they work.

Each algorithm uses the basic facts of arithmetic: addition and multiplication tables for the digits. These facts **must be memorized**. Without a firm command of the basic facts, a student finds the algorithms too cumbersome to be of use. Such a student will lose interest in learning the algorithms and will consequently lose the insight study of the algorithms provides into the properties of the arithmetic operations. This lack of insight will adversely affect the student's ability to apply arithmetic to solve problems.

3.1. Numeration

The base ten numeral of a number is built by repeated applications of long division. As we saw in Section 1.2.3, the long division is carried out by counting into groups of ten.

Example 3.1. The base ten numeral for the number thirteen is 13, which means 1 element in the ten's place and 3 elements in the one's place. The element in the ten's place is 1 group of ten elements of the one's place. If A is a set with thirteen elements, we begin by regarding them as thirteen elements in the one's place. Composing these elements into groups of ten, the division algorithm applied to $13 \div 10$ tells us that the result is 1 group of ten and 3 elements remaining. The 1 group of ten then becomes an element of the ten's place. The numeral is completed now because each place has fewer than ten elements. The picture is

$$\begin{array}{cc} \text{ten's} & \text{one's} \\ & \bullet \\ & \bullet \\ \bullet & \bullet \end{array}$$

where the elements of the ten's place are groups of ten elements of the one's place. In terms of the arithmetic of the division algorithm, we have

$$\text{thirteen} = 13 = 1 \cdot 10 + 3$$

Definition 3.2 (Base ten numeration). The Hindu-Arabic system to base ten uses the symbols $0, 1, 2, 3, 4, 5, 6, 7, 8, 9$ for the whole numbers less than ten. These symbols are called *digits*. They represent the number of elements in the place they occupy. New places are created from right to left. The first place is called the one's place. The elements of any place to the left of the one's place are groups of ten elements of the adjacent place to its right. If there are more than 9 elements in a place, then they are composed into groups of ten which become elements in the adjacent place to the left. In base ten, the numeral for ten is 10. From right to left the places are named one's place, ten's place, hundred's place, thousand's place, ten-thousand's place, and so on. The pattern in the names is that of powers of ten. Hundred is $100 = 10 \cdot 10$. Thousand is $1{,}000 = 10 \cdot 10 \cdot 10$. Ten-thousand is $10{,}000 = 10 \cdot 10 \cdot 10 \cdot 10$, and so on.

The basic features of this numeration system are its ten digits and the fact that the elements of a place to the left of the one's place are groups of ten of the elements of the adjacent place to its right. A numeral is read from left to right by saying the digit in a place followed by the place's name, with the modification at the ten's place dictated by the English language. For example, 237 is 2 hundred thirty-seven, and 4,735 is 4 thousand 7 hundred

3.1. Numeration

thirty five. If the English language were to follow the numeration more literally, these numerals would be called 2 hundred 3 ten 7 and 4 thousand 7 hundred 3 ten 5, respectively. Notice that the base ten numeral of ten is 10. The digits are needed to name the remainders when dividing by ten.

Oral exercise 3.3. (1) How many thousands, hundreds, tens, and ones in 6? in 30? in 875? in 7904?

(2) How many ten-thousands, thousands, hundreds, tens, and ones in in 60? in 600? in 6000? in 300? in 8750? in 79040?

(3) Explain the distinctions being made here in the meaning of the words number, digit, and numeral. In the present context it is important to distinguish between the thing (the number) and its name (the numeral). After mastering the numeration system, one may drop these distinctions.

Example 3.4. To find the base ten numeral of the number twenty-four, consider a set with twenty-four elements. Initially, these elements are the elements of the one's place. Since there are more than 9 elements, compose them into groups of ten. By the division algorithm applied to twenty-four divided by ten, we have twenty-four $= 2 \cdot 10 + 4$, which means that there are 2 groups of ten with 4 elements remaining. The 2 groups of ten become 2 elements in the ten's place and the 4 elements remaining stay in the one's place. All places now have fewer than ten elements, so the final numeral is 24, which can be pictured as

```
       ten's   one's
                •
                •
         •      •
         •      •
```

In terms of the arithmetic of the division algorithm we have

$$\text{twenty-four} = 24 = 2 \cdot 10 + 4$$

Each of these examples can be illustrated with physical objects such as soda straws, coins, or one of the commercial products made for this purpose. For example, in the preceding example, start with twenty-four soda straws. Compose them into groups of ten and bind each group with a rubber band, so that these bound bundles of ten become the elements of the ten's place. The result will be two groups of ten and 4 individual straws remaining.

From soda straws one can move on to coins, which add a relevant step of abstraction to the process. For the preceding example, start with twenty-four pennies and compose these into groups of ten. Exchange each group of ten for a dime, so that the dimes become the elements of the ten's place. The result for the preceding example is 2 dimes and 4 pennies.

Example 3.5. Find the base ten numeral of one-hundred-twenty-seven. Since one-hundred-twenty-seven $= 12 \cdot 10 + 7$, a set of one-hundred-twenty-seven elements can be composed into twelve groups of ten with 7 remaining. The twelve groups of ten are put into the ten's place and are now twelve elements in the ten's place. At this stage the picture would be

```
ten's   one's
  •
  •
  •
  •
  •     •
  •     •
  •     •
  •     •
  • •   •
  • •   •
```

Each element in the ten's place is a group of ten elements of the one's place. There are more than 9 elements in the ten's place, so we continue.

From twelve divided by ten we conclude that twelve $= 1 \cdot 10 + 2$, which means that the twelve elements in the ten's place can be composed into 1 group of ten with 2 remaining. Put the 1 group of ten into the hundred's place, and leave the 2 remaining in the ten's place. In summary, the original set has been composed into 1 element in the hundred's place, 2 elements in the ten's place and 7 elements in the one's place. The resulting numeral is 127, which can be pictured as

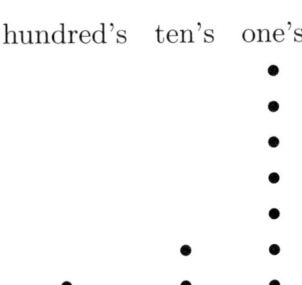

Each element in the hundred's place is a group of ten elements of the ten's place and each element of the ten's place is a group of ten elements of the one's place. As a consequence, each element of the hundred's place is ten groups of ten, which is a hundred, of the elements of the one's place.

3.1. Numeration

Combining the two applications of the division algorithm and using the distributive property, we can express this construction of the numeral as

$$127 = 12 \cdot 10 + 7 = (1 \cdot 10 + 2) \cdot 10 + 7 = 1 \cdot 10 \cdot 10 + 2 \cdot 10 + 7$$

Oral exercise 3.6. Describe how to use soda straws to demonstrate the preceding example. Describe how to use pennies, dimes, and dollars to demonstrate the same thing.

3.1.1. Other bases. Any number greater than one can be used as the base in the Hindu-Arabic numeration system. All computers internally use the base two numeration system.

Definition 3.7 (Base b numeration). Suppose b is a whole number larger than 1. The Hindu-Arabic system to base b uses symbols $0, 1, 2, 3, \ldots$ for the whole numbers less than b. These symbols are called digits. They represent the number of elements in the place they occupy. New places are created from right to left. The first place is called the one's place. The elements of any place to the left of the one's place are groups of b elements of the adjacent place to its right. If there are b or more elements in a place, then they are composed into groups of b which become elements in the adjacent place to the left. From right to left the places are named one's place, b's place, $b \cdot b$'s place, $b \cdot b \cdot b$'s place, $b \cdot b \cdot b \cdot b$'s place, and so on.

The basic features of this numeration system are its b digits and the fact that the elements of a place to the left of the one's place are groups of b elements of the adjacent place to its right. A numeral is read from left to right by saying the digit in a place followed by the place's name. Notice that the base b numeral of b is 10, which is called b. The digits are needed to name the remainders when dividing by b.

If b is less than ten, then the usual digits less than b can be used as the base b digits. If b is greater than ten, then the usual digits can be used together with newly invented digits for the numbers between 9 and b. In this book we shall restrict our examples to the bases two, five, and ten.

Example 3.8. Find the base five numeral of seventy-three. For base five we use the digits $0, 1, 2, 3, 4$. Remember that the base five numeral for five is 10_5, where the subscript 5 is used to remind us that this is a base five numeral. Consider a set with seventy-three elements. Count these elements into groups of five, to obtain fourteen such groups with 3 remaining, as confirmed by the long division

$$\text{seventy-three} = \text{fourteen} \cdot 10_5 + 3$$

The 3 for the remaining elements is put in the one's place and the fourteen groups of five become elements of the five's place. The present situation is pictured as follows.

five's   one's

• •
• • •
• • • •
• • • •
• • • •

Count the fourteen groups of five into groups of five, to obtain two groups of five of five and 4 groups of five remaining, as confirmed by the long division

$$\text{fourteen} = 2 \cdot 10_5 + 4$$

The 4 for the remaining elements is put in the five's place and the two groups of five of five are placed in the next place to the left, which is the $10_5 \cdot 10_5 =$ twenty-five's place. All places now have fewer than five elements, so the base five numeral of seventy-three is 243_5, which is pictured as

twenty-five's five's one's

and read 2 twenty-fives 4 fives 3. Combining the two long divisions with the distributive property of multiplication, we have

$$\text{seventy-three} = 243_5 = (\text{fourteen} \cdot 10_5) + 3$$
$$= ((2 \cdot 10_5 + 4) \cdot 10_5) + 3 = 2 \cdot 10_5 \cdot 10_5 + 4 \cdot 10_5 + 3$$

Oral exercise 3.9. (1) Describe how to use soda straws to demonstrate the preceding example.

(2) Describe how to use pennies, nickels, and quarters to demonstrate the same example.

(3) Why should we not use the symbol 5 for the number five when we are using the base five numeration system?

Example 3.10. Find the base two numeral of six. For base two we use the digits $0, 1$. The base two numeral for two is 10_2, where the subscript 2 is used to remind us that this is a base two numeral. Consider a set with six elements. Initially, these elements are the elements of the one's place. Compose them into groups of two. By long division applied to six divided by two, we can make three groups of two with none remaining. These three groups of two become three elements in the two's place. The present

3.1. Numeration

situation is pictured as

$$\begin{array}{cc} \text{two's} & \text{one's} \\ \bullet & \\ \bullet & \\ \bullet & \end{array}$$

Next, we compose the three elements in the two's place into groups of two. By long division, three $= 1 \cdot 10_2 + 1$, we obtain 1 group of two and 1 remaining. The group of two becomes an element of the $10_2 \cdot 10_2 =$ four's place. Because every place has fewer than two elements, we have completed the construction and the base two numeral of six is 110_2. This numeral is read 1 four 1 two 0. It is pictured as

$$\begin{array}{ccc} \text{four's} & \text{two's} & \text{one's} \\ \bullet & \bullet & \end{array}$$

and written

$$\text{six} = 110_2 = (\text{three} \cdot 10_2) + 0 = ((1 \cdot 10_2 + 1) \cdot 10_2) = 1 \cdot 10_2 \cdot 10_2 + 1 \cdot 10_2 + 0$$

Oral exercise 3.11. Describe how to use soda straws to demonstrate the preceding example.

Example 3.12. Here is the process we have been using, put into schematic form, for finding the base two numeral of thirteen. For conciseness of notation we have used base ten numerals to write the intermediate steps.

$13 \div 2 = 6$ with remainder 1, which goes in the 1's place.

Then $6 \div 2 = 3$ with remainder 0, which goes in the 2's place.

Then $3 \div 2 = 1$ with remainder 1, which goes in the $2 \cdot 2$'s place.

Then $1 \div 2 = 0$ with remainder 1, which goes in the $2 \cdot 2 \cdot 2$'s place.

The process stops here because the last quotient was 0.

The resulting base two numeral of thirteen is

$$1101_2 = 1 \cdot (2 \cdot 2 \cdot 2) + 1 \cdot (2 \cdot 2) + 0 \cdot 2 + 1$$

Example 3.13. The preceding example used a scheme which works to determine the numeral in any base for any number. Here is how it works to find the base five numeral for one-hundred-forty-three.

$$\begin{array}{rcl} 143 \div 5 & = & 28, \text{ with remainder } 3 = \text{ digit in 1's place} \\ 28 \div 5 & = & 5, \text{ with remainder } 3 = \text{ digit in the 5's place} \\ 5 \div 5 & = & 1, \text{ with remainder } 0 = \text{ digit in the } 5 \cdot 5\text{'s place} \\ 1 \div 5 & = & 0, \text{ with remainder } 1 = \text{ digit in the } 5 \cdot 5 \cdot 5\text{'s place} \end{array}$$

and the process ends when the quotient becomes 0. Therefore, the base five numeral for one-hundred-forty-three is 1033_5, that is,

$$1033_5 = 1 \cdot (5 \cdot 5 \cdot 5) + 0 \cdot (5 \cdot 5) + 3 \cdot 5 + 3$$

The above scheme must end because at each stage the quotient is smaller than the previous quotient, so that eventually the quotient becomes zero, at which point the process is complete.

Oral exercise 3.14. (1) How many thirty-twos, sixteens, eights, fours, twos, and ones in the base 2 numeral 110001_2? in 100101_2? in 100001_2? in 101010_2?

(2) How many six-hundred-twenty-fives, hundred-twenty-fives, twenty-fives, fives, and ones in the base 5 numeral 34231_5? in 24011_5? in 42103_5?

(3) What is the logical way to say the numbers 1010_2 and 2032_5?

(4) How many digits are needed for base twelve? What would you suggest be used as the digits for ten and eleven? What is the base twelve numeral for twelve?

3.1.2. Order. The definition of whole number implies that they occur in an order which is illustrated by the number line. From this definition we easily verify that the order of the digits is

$$0 < 1 < 2 < 3 < 4 < 5 < 6 < 7 < 8 < 9$$

The next number is 10, from which we conclude that

$$\text{any digit} < 10$$

Then, by the order property of addition,

$$10 + \text{any digit} < 10 + 10 = 20, \quad 20 + \text{any digit} < 20 + 10 = 30$$

and so on up to 100. It follows that any two-digit number is less than 100. Then $100 < 100 + 100 = 200$, and $6 < 8$ implies that $600 = 6 \cdot 100 < 8 \cdot 100 = 800$, and so on. Following this line of argument, we see that $527 < 819$ because

$$527 = 500 + 27 < 500 + 100 = 600 < 800 < 800 + 19 = 819$$

where we use the order property of addition and of multiplication.

Oral exercise 3.15. In answering the following questions, explain how you reached your answer and how you know that your answer is correct.

(1) Which is larger, 8 or 14? *Ans:* $8 < 14$ because $8 < 10$ and $10 < 10 + 4 = 14$.

(2) Which is larger, 23 or 18?

(3) Which is larger, 359 or 571?

(4) Which is larger, 7538 or 7419?

Theorem 3.16 (Order of Whole Numbers). *If a and b are unequal whole numbers, then $a < b$ precisely when their base ten numerals are different and*

3.1. Numeration

in the first place from the left where they differ, the digit of a in that place is less than the digit of b in that place.

In applying this theorem to two numerals of different lengths, places to the left of the first digit (on the left) are to be regarded as containing the digit 0. For example, according to the theorem, $859 < 2301$ because the first place from the left at which they differ is the thousand's place in which 859 has a 0 and 2301 has a 2 and $0 < 2$.

Proof. The essence of the proof is illustrated by the following example. The numbers 56,287 and 56,341 differ at the third place from the left. To determine the order of these two numbers, we use the meaning of base ten numeration together with the order properties of addition and multiplication.
$$56287 = 50000 + 6000 + 200 + 87 < 50000 + 6000 + 200 + 100$$
$$= 50000 + 6000 + 300 < 50000 + 6000 + 300 + 41 = 56341$$
The essential point of this proof is that
$$287 = 200 + 87 < 200 + 100 = 300 < 300 + 41 = 341$$
The hundred's place is the first place where the two numerals differ (reading from the left). Then 287 must be smaller than 341 because it is smaller than 300, which is no larger than any three digit numeral beginning with a 3 in the hundred's place. The values of the digits in the places to the right of the hundred's place are not relevant for determining the order of these two numbers. $\square$

Oral exercise 3.17. Which is larger: 297,411 or 29,768? 3324_5 or 3241_5? 1011101_2 or 111111_2?

The following theorem, which is nearly the same as Theorem 1.45, summarizes the essential facts about the Hindu-Arabic numeration system. It is stated for base ten, but it is true, as stated, for any base $b > 1$.

Theorem 3.18 (Numeration). *(1) If a whole number is given, then it has a unique numeral in the base ten numeration system.*

(2) If a numeral in the base ten numeration system is given, then it represents a whole number.

Proof. (1) Any given whole number has a base ten numeral determined by repeated applications of long division by ten. The process reaches a conclusion because after each application of long division by ten the quotient becomes smaller. Consequently the quotient must become 0 after a number of steps which can be no more than the size of the number we started with. The number of times long division by ten must be applied is the number of digits that will be in the numeral, since the remainder at each step is a digit in the numeral. When we say that the long divisions by ten come to a conclusion, we are saying that any number has only a finite number of digits in its Hindu-Arabic numeral.

The uniqueness follows from Theorem 3.16, which implies that two different numerals always represent two different numbers in the order described by that theorem. In particular, then, one number cannot have two different numerals (in a single base).

(2) A base ten numeral represents the number of elements in the set determined as follows. The one's place of the numeral indicates the number of individual elements, the ten's place the number of groups of ten of these elements, the hundred's place the number of groups of ten of ten of these elements, and so on. The number of elements in this set is a whole number. It is the whole number represented by the given numeral. □

The number of digits in a base b numeral is the smallest number of times b can be multiplied times itself to yield a product greater than the number. This can be seen by following the process of applying long division. For example, to determine the number of digits in the base two numeral of forty-three, we compute $2 \cdot 2 = 4$, $2 \cdot 2 \cdot 2 = 8$, $2 \cdot 2 \cdot 2 \cdot 2 = 16$, $2 \cdot 2 \cdot 2 \cdot 2 \cdot 2 = 32$, and one more product by 2 will be 64, which is greater than forty-three. Therefore, the base two numeral of forty-three has six digits. Compare this to the process of finding this numeral:

$$43/2 = 21, \text{ r} = 1 \text{ is digit in the one's place}$$
$$21/2 = 10, \text{ r} = 1 \text{ is next digit to the left}$$
$$10/2 = 5, \text{ r} = 0 \text{ is next digit to the left}$$
$$5/2 = 2, \text{ r} = 1 \text{ is next digit to the left}$$
$$2/2 = 1, \text{ r} = 0 \text{ is next digit to the left}$$
$$1/2 = 0, \text{ r} = 1 \text{ is next digit to the left}$$

Therefore, forty-three= 101011_2, which has six digits, the remainders from the six divisions by two.

The product of a number b multiplied by itself n times is called b to the power n, which, for those who already know exponents, is written b^n. The above rule can be restated as: the number of digits in the base b numeral of a is the smallest power of b which is larger than a. In the preceding proof, when we say that the long divisions by b reach the point where the quotient is 0 and thus terminates, we are saying that no matter what number a is given to us, some power of the base b will be greater than a.

§3.1 Exercises

(1) Find the base ten numeral for 1010_2 and for 232_5.

(2) Find the base ten numeral for 110110111_2 and for 403321_5.

(3) Find the base 2 numeral for seventy-five. Give a systematic explanation of the process.

(4) Find the base 5 numeral for seven-hundred-thirty-seven. Give a systematic explanation of the process.

(5) (a) Write the base ten numeral for the following sums and products:
   (i) $7000 + 500 + 80 + 6$
   (ii) $6 \cdot 1000 + 7 \cdot 100 + 3 \cdot 10 + 5$
   (iii) $5 \cdot 10 \cdot 10 \cdot 10 \cdot 10 + 4 \cdot 10 \cdot 10 + 8 \cdot 10$
   (b) Express 27043 as a sum of digit multiples of powers of ten.

(6) Give the proof of Theorem 3.16 for the numbers 3587 and 3488.

(7) Give the proof of Theorem 3.16 for the numbers 1034_5 and 1342_5.

(8) Give the proof of Theorem 3.16 for the numbers 100101_2 and 101001_2.

(9) How many digits are in the base 2 numeral of eighty-five? Determine the answer by finding the smallest power of 2 which is greater than eighty-five, then verify your answer by finding this numeral.

(10) If $1 < b$, then $b = b \cdot 1 < b \cdot b$, by the order property of multiplication. Continue this argument to prove that the powers of b increase with the powers: $b < b \cdot b < b \cdot b \cdot b < \ldots$ Go up to the fourth power of b.

3.2. Computing Sums

The first step in computing sums is to learn the *basic addition facts*, which are sums $a + b$, where a and b are digits in the base ten numeration system. Emphasize that $a + b = b + a$ so as to reduce the number of addition facts to be memorized.

Each addition fact is connected to two subtraction facts. If $a + b = c$, then $c - a = b$ and $c - b = a$. All of these can be learned together. For example,

$$3 + 5 = 8, \quad 8 - 3 = 5, \quad 8 - 5 = 3$$

The sum of some digits is again a digit. For example, $2+3 = 5$ and $3+6 = 9$. The sum of other digits is no longer a single digit. For example, $3 + 7 = 10$ and $3 + 9 = 12$. Sums of digits (and their corresponding differences) must be memorized. Making combinations of ten can help us to derive them and

help us recall those we momentarily forget. For example,
$$5 + 8 = 3 + 2 + 8 = 3 + 10 = 13 \text{ or } 9 + 4 = 9 + 1 + 3 = 10 + 3 = 13$$
Making combinations of ten should be part of this activity.

Oral exercise 3.19. (1) What combinations of two digits sum to 10? What are the corresponding differences?

(2) Use combinations making 10 to find the sums $8 + 7$, $5 + 6$, $5 + 7$, and $9 + 9$.

(3) What combinations of two digits sum to 5?

(4) Use combinations making 5 to find the sums $2 + 4$, $3 + 3$, $4 + 3$, and $4 + 4$.

(5) What combinations of two digits sum to 2? What are the addition facts in base two?

Now consider sums of digit multiples of 10. For example, $20+50$ means 2 tens plus 5 tens, which is $2 + 5 = 7$ tens, namely 70. Therefore, $20 + 50 = 70$ is obtained by adding the digits in the ten's place. Another example is $30 + 60$, which we regard as 3 tens plus 6 tens, which is $3 + 6 = 9$ tens, which is 90. As in the first example, $30 + 60 = 90$ is obtained by adding the digits in the ten's place. This rule for adding multiples of 10 is justified by the distributive property.
$$20 + 50 = 2 \cdot 10 + 5 \cdot 10 = (2 + 5) \cdot 10 = 7 \cdot 10 = 70$$
$$30 + 60 = 3 \cdot 10 + 6 \cdot 10 = (3 + 6) \cdot 10 = 9 \cdot 10 = 90$$

Next consider the sum $20 + 7$. This is 27 by the very meaning of the base ten numeration system. This is the case for the sum of any digit multiple of 10 with a digit. Another such example is $30 + 5 = 35$.

We are now ready for sums such as $13 + 4 = 17$. We can write this out in what we shall call a *formal explanation* as

$$\begin{aligned} 13 + 4 &= 10 + 3 + 4, \text{ by the meaning of 13} \\ &= 10 + 7, \text{ by the addition table} \\ &= 17, \text{ by the numeration system} \end{aligned}$$

Here, and in the following examples, we shall use the associative and commutative properties of addition and multiplication without comment. It is not much harder to do sums such as

$$\begin{aligned} 25 + 64 &= 20 + 5 + 60 + 4, \text{ meaning of numeration} \\ &= 20 + 60 + 5 + 4, \\ &= 80 + 9, \text{ addition facts} \\ &= 89, \text{ meaning of numeration} \end{aligned}$$

Looking back at this formal explanation, we see that we found the sum $25 + 64$ by adding the digits in the ones place $5 + 4 = 9$, and adding the

3.2. Computing Sums

digits in the tens place $2+6=8$ and then summing the result. Writing the computation in columns, we can keep track of the places, and add the digits in each place.

(3.1)
$$\begin{array}{r} 25 \\ +64 \\ \hline 89 \end{array}$$

When the numbers are aligned in columns, the sum is found by adding the digits in each column.

3.2.1. Carrying. Another step is required if the sum in some place is 10 or more, as in the next example.

(3.2)
$$\begin{aligned} 17+28 &= 10+7+20+8, \text{ numeration} \\ &= 10+20+7+8, \text{ assoc., comm.} \\ &= 30+15, \text{ addition facts} \\ &= 30+10+5, \text{ compose group of 10} \\ &= 40+5=45, \text{ addition facts and numeration} \end{aligned}$$

This example shows that when you add the ten's place first, then you end up having to add the ten's column again after composing groups of ten in the one's place. In order to avoid adding a given place more than once, we work from right to left. In the present example, add the digits in the one's place, obtaining $7+8=15$. Compose units of ten in the result, $15=1\cdot 10+5$, and add the 1 group of ten to the ten's place, leaving the remaining 5 elements in the one's place. Next add the digits in the ten's place, including this new element carried over from the one's place, obtaining $1+1+2=4$, which is smaller than 10, so we stop. In columns, with *margin notes*, this goes as follows. Add digits in the one's place:

$$\begin{array}{r} 17 \\ +28 \\ \hline 15 \quad 7+8=15 \text{ in the one's place} \end{array}$$

In the one's place, $15 = 1\cdot 10+5$, so there is 1 group of ten, to be carried to the ten's place (denoted with a small 1 above the ten's place) and 5 remains in the one's place. Then add the digits in the ten's place:

$$\begin{array}{r} \overset{1}{1}7 \\ +28 \\ \hline 45 \quad 1+1+2=4 \text{ in the ten's place} \end{array}$$

which agrees with the formal explanation in equations (3.2).

Definition 3.20 (Carrying). If there are a elements in a given place of a base ten numeral, then *composing groups of ten* of these means to divide a by ten to find the quotient, which is the number of groups of ten. This quotient is then *carried* to the next place to the left. The remainder is the number of elements remaining in the given place.

Theorem 3.21 (Addition algorithm). *If the sum is to be found of a finite set of numbers written in base ten numeration, then: Add the digits in the one's place. Compose groups of ten in this sum. Leave the remainder in the one's place and carry the number of such groups of ten to the ten's place. Sum the digits in the ten's place, compose groups of ten in this sum, carry the number of groups of ten to the next place to the left, and leave the remainder in the ten's place. Repeat the process on the next place to the left. Continue through all places which are occupied by a nonzero digit.*

Proof. The proof comes from the formal explanation, which uses the commutative and associative properties of addition and multiplication and the distributive property together with the base ten numeration system. We illustrate it here with the example of $48 + 75$.

$$\begin{aligned} 48 + 75 &= 4 \cdot 10 + 8 + 7 \cdot 10 + 5, \text{ numeration} \\ &= 4 \cdot 10 + 7 \cdot 10 + 8 + 5, \text{ assoc. and comm. properties} \\ &= 4 \cdot 10 + 7 \cdot 10 + 13, \text{ sum digits in one's place} \\ &= 4 \cdot 10 + 7 \cdot 10 + 1 \cdot 10 + 3, \text{ compose groups of ten in one's place} \\ &= (4 + 7 + 1) \cdot 10 + 3, \text{ distributive property} \\ &= 12 \cdot 10 + 3, \text{ sum digits in ten's place} \\ &= (1 \cdot 10 + 2) \cdot 10 + 3, \text{ compose groups of ten in ten's place} \\ &= 1 \cdot 10 \cdot 10 + 2 \cdot 10 + 3, \text{ distributive property} \\ &= 123, \text{ numeration} \end{aligned}$$

The algorithm begins by summing the digits in the one's places,

$$\begin{array}{r} \overset{1}{4}\,8 \\ +\ \underline{7\,5} \\ 3 \end{array}$$

which shows the result of the sum $8 + 5 = 13$ in the one's place followed by the composition of 1 ten, carried to the ten's place, with 3 remaining in the one's place below the line. In the next step, add the digits in the ten's place, $1 + 4 + 7 = 12$, compose this into 1 ten, to be carried to the next place to the left, with the remaining 2 recorded below the line in the ten's place:

$$\begin{array}{r} 1\,\overset{1}{4}\,8 \\ +\ \underline{7\,5} \\ 2\,3 \end{array}$$

3.2. Computing Sums

The process then continues in the next place to the left. In this example, there is only 1 in the hundred's place, so this is recorded below the line in the hundred's place and the final answer for the sum is 123. □

Example 3.22. A demonstration with soda straws throws additional light on this algorithm. Suppose we have a set of 48 straws which we want to combine with another set of 75 straws. How many straws will be in this union?

Solution: Imagine that we have counted the first set of straws into groups of ten, so that we have 8 straws in the one's cup and 4 groups of ten in the ten's cup. In the same way, the second set of straws is counted into groups of ten giving us 5 straws in its one's cup and 7 groups of ten in its ten's cup.

The union of these two sets can be taken by putting all the groups of ten into a single ten's cup and all the single straws together in a one's cup. It makes no difference whether we form the union of the ten's cup first or of the one's cup first. Now we have $4 + 7 = 11$ groups of ten in the ten's cup and $8 + 5 = 13$ straws in the one's cup.

By the usual numeration process, we would begin by counting the 13 straws in the one's cup into groups of ten, getting 1 such group, which we *carry* to the ten's cup, leaving 3 straws in the one's cup. We now have $11 + 1 = 12$ groups of ten in the ten's cup. Count these into 1 group of ten, which we place in a hundred's cup, leaving 2 groups of ten in the ten's cup. We have 123 straws altogether.

The process shows us that we shouldn't bother counting the elements in the ten's cup until after we have counted those in the one's cup and carried any resulting groups of ten over to the ten's cup.

Oral exercise 3.23. (1) Use soda straws to demonstrate the addition algorithm for finding the sum $28 + 37$.

(2) Use soda straws to demonstrate the addition algorithm for finding the sum in base five numerals of $24_5 + 32_5$.

3.2.2. Scratch Addition. A sum of three or more digits can take us beyond the basic addition facts. For example, if we do the sum $5 + 9 + 4$ from left to right, then after the first step we have $14 + 4$, which is not a basic addition fact. The process of scratch addition gives us a technique for adding more than two numbers in such a way that each sum involves only basic addition facts.

In adding a column of numbers, decide to go down the column, or up the column. When the sum with a digit exceeds 9, draw a diagonal line through the digit and write the excess over 10 as a subscript to the digit. Continue by finding the sum of this excess with the next digit. The number of diagonal lines, called the scratch marks, gives the amount to carry to the next place, that is, the number of groups of 10 of the given place to be carried to the next place to the left. Consider the example $275 + 739 + 524 = 1538$ in which addition in each column is done from top to bottom. Starting at the top of the one's place, we find $5 + 9 = 14 = 1 \cdot 10 + 4$. The 1 group of ten is indicated with a scratch through the 9 and the 4 remaining is indicated with a subscript 4 on the 9. The sum in this column continues with this 4, to give $4 + 4 = 8$, which is recorded under the line. The single scratch in the one's column is carried to the ten's column, and the process is repeated in that column, and the next, until all places are done.

$$
\begin{array}{ccc}
1 & 1 & 1 \\
2 & 7 & 5 \\
\not{7}_0 & \not{3}_1 & \not{9}_4 \\
5 & 2 & 4 \\
\hline
1\ 5 & 3 & 8
\end{array}
$$

Oral exercise 3.24. When you use the addition algorithm, do you add from top to bottom or bottom to top? Does the result depend on the order? Why?

3.2.3. Addition algorithm in other bases. The addition algorithm is the same in our numeration system in any base $b > 1$. Doing some examples in other bases provides a good test of your understanding of the algorithm, because a base other than ten makes the territory unfamiliar enough that each step must come from understanding rather than rote memory.

In base five, the basic addition facts are the sums of the digits $0, 1, 2, 3, 4$. The sums which are not digits are $1 + 4 = 10_5$, $2 + 3 = 10_5$, $2 + 4 = 11_5$, $3 + 3 = 11_5$, $3 + 4 = 12_5$, and $4 + 4 = 13_5$. Remember that in base five the numeral for five is 10_5, the numeral for twenty-five is 100_5, for one-hundred-twenty-five is 1000_5, and so on. The addition algorithm applied to $243_5 + 402_5$ looks like the following, where we have omitted the subscript 5 on the numerals.

$$
\begin{array}{r}
111 \\
2\ 4\ 3 \\
+4\ 0\ 2 \\
\hline
1\ 2\ 0\ 0
\end{array}
$$

The formal explanation is

$$
\begin{aligned}
243_5 + 402_5 &= 2 \cdot 100_5 + 4 \cdot 10_5 + 3 + 4 \cdot 100_5 + 2, \text{ base five numeration} \\
&= 2 \cdot 100_5 + 4 \cdot 100_5 + 4 \cdot 10_5 + 10_5, \text{ add digits in one's place} \\
&= 2 \cdot 100_5 + 4 \cdot 100_5 + (4+1) \cdot 10_5, \text{ distributive property} \\
&= 2 \cdot 100_5 + 4 \cdot 100_5 + 10_5 \cdot 10_5, \text{ add digits in five's place} \\
&= (2+4+1) \cdot 100_5, \text{ distributive property} \\
&= 12_5 \cdot 100_5, \text{ add digits in } 100_5\text{'s place} \\
&= (1 \cdot 10_5 + 2) \cdot 100_5, \text{ compose groups of five} \\
&= 1 \cdot 10_5 \cdot 100_5 + 2 \cdot 100_5, \text{ distributive property} \\
&= 1 \cdot 1000_5 + 2 \cdot 100_5 = 1200_5, \text{ base five numeration}
\end{aligned}
$$

Scratch addition also works, without change, for any base. Additional examples in base five and base two are in the exercises.

§3.2 Exercises

(1) Give a formal explanation and then describe the standard addition algorithm for calculating $264 + 789$.

(2) Explain the scratch addition algorithm for calculating $264 + 859 + 488$.

(3) Explain how properties of addition can be used to simplify the following sums by finding combinations which compose into tens.
(a) $13 + 8 + 7$. (b) $24 + 26$. (c) $27 + 36$. (d) $14 + 8 + 6 + 2 + 7$.

(4) Give a formal explanation to describe how the addition algorithm works to find the sum $237 + 585$.

(5) What are the basic addition facts in base five? What are the corresponding subtraction facts? Apply the addition algorithm, with formal explanation, to find the following sums of numbers written in base five:
(a). $21_5 + 12_5$. (b). $32_5 + 44_5$. (c). $243_5 + 402_5$.

(6) Explain the use of the scratch addition algorithm to find the sum $231_5 + 423_5 + 142_5$. What do scratches represent now?

(7) What are the basic addition facts in base two? Apply the addition algorithm, with formal explanation, to find the following sums of numbers written in base two: (a). $11_2 + 1$. (b). $101101_2 + 11101_2$.

(8) Explain what is being done here. Is it correct?

$$\begin{array}{r} 25 \\ +47 \\ \hline 6(12) \\ 6(10+2) \\ (6+1)2 \\ 72 \end{array}$$

(9) Sue took 1 hour and 20 minutes to do her homework. She started at 8:50 p.m. When did she finish doing her homework? Explain how the calculation accounts for changing 60 minutes into 1 hour.

(10) Explain how to add $3.45 and $2.65 in the following two ways.
  (a) Use the addition algorithm to find $345 + 265$ and then rewrite it in money notation.
  (b) Add 3 dollars and 45 cents plus 2 dollars and 65 cents by adding the dollars then adding the cents and then exchanging each 100 cents for 1 dollar.

(11) Harlan has 357 baseball cards. Shonnie has 85 cards more than Harlan has. How many baseball cards does Shonnie have? How many do they have altogether?

3.3. Computing Differences

Begin by learning the differences of numbers less than 19. These are the ones learned with the basic addition facts. For example,

$$7 = 4 + 3 \text{ means } 7 - 4 = 3 \text{ and } 7 - 3 = 4$$

and

$$13 = 5 + 8 \text{ means } 13 - 5 = 8 \text{ and } 13 - 8 = 5$$

Next, consider the differences of digit multiples of 10. For example, $70 - 40$ is 7 tens take away 4 tens, which is $7 - 4 = 3$ tens, which is 30. The formal explanation uses the distributive property over subtraction, which reduces the problem to basic subtraction facts. For example,

$$\begin{aligned} 70 - 40 &= 7 \cdot 10 - 4 \cdot 10 \\ &= (7-4) \cdot 10, \text{ by distributive property} \\ &= 3 \cdot 10 = 30 \end{aligned}$$

We are now ready to try $47-23$. The algorithm is to subtract place by place. To gain insight into this, let's see how it works with soda straws. Begin with 47 soda straws arranged into 4 groups of ten put into the ten's cup and 7 individual straws put into the one's cup. By take-away subtraction, the difference $47 - 23$ is the number of straws remaining after we remove 23

3.3. Computing Differences

straws. Since 23 straws is 2 groups of ten and 3 individual straws, we can remove these by taking 3 straws from the one's cup, leaving $7-3=4$ straws there, and taking 2 groups of ten from the ten's cup, leaving $4-2=2$ groups of ten there. The number remaining is 2 groups of ten in the ten's cup and 4 straws in the one's cup, which is 24 straws.

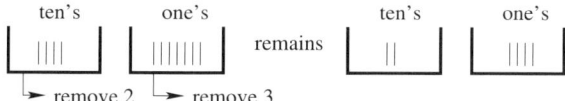

Here is a formal explanation in which we write digit multiples of ten, such as $4 \cdot 10$ as 40, etc.

$$
\begin{aligned}
47 - 23 &= (40 + 7) - (20 + 3) \\
&= (40 - 20) + (7 - 3), \text{ Example 2.40} \\
&= 20 + 4 = 24, \text{ by base ten numeration}
\end{aligned}
$$

We did the subtraction in the one's place and in the ten's place independently to arrive at the answer. Aligned in columns this algorithm looks like this:

(3.3)
$$
\begin{array}{r}
47 \\
-23 \\
\hline
24
\end{array}
$$

In words, the subtractions are carried out in each place where only basic subtraction facts are used.

3.3.1. Decomposing. It may happen that the subtraction in a given place is undefined. For example, subtraction in the one's place is undefined in the problem $43 - 27$, since $3 - 7$ is undefined in whole numbers. The way to get around this difficulty is to use knowledge of the base ten numeration system to rewrite the minuend as $43 = 40 + 3 = 30 + 10 + 3 = 30 + 13$ by decomposition of an element in the ten's place back into a group of ten elements in the one's place, so that the 13 represents 13 elements in the one's place. Then the subtraction in the one's place becomes $13 - 7 = 6$ and the subtraction in the ten's place is now $3 - 2 = 1$, so that the answer is 16.

Carried out with soda straws, the algorithm would look like this:

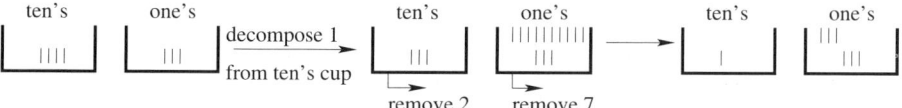

A formal explanation for this computation of 43 − 27 is

$$43 - 27 = (40 + 3) - (20 + 7), \quad \text{numeration}$$
$$= (30 + 10 + 3) - (20 + 7), \quad \text{decompose element of ten's place}$$
$$= (30 - 20) + (13 - 7), \quad \text{Example 2.40}$$
$$= 10 + 6 = 16, \quad \text{numeration}$$

Aligned in columns, this subtraction algorithm for 43 − 27 goes as follows. Begin with the subtraction in the one's place, 3 − 7. As this cannot be done in whole numbers, go to the ten's place in the minuend where there is a 4, and take one of these, leaving 3, and decompose it into a group of ten in the one's place, denoted by a small one before the 3 in the one's place in order to indicate 13 in the one's place of the minuend. Now the subtraction problem in the one's place is 13 − 7 = 6 and in the ten's place is 3 − 2 = 1. This algorithm is written as follows, with the process beginning from the right.

(3.4)
$$\begin{array}{r} \overset{3}{\cancel{4}} \ {}^{1}3 \\ -2 7 \\ \hline 1 6 \end{array}$$

The central concept in this process is the decomposition of an element in a place of the minuend into a group of ten elements of the next place to its right, in order to make possible the subtraction in that place. This process is also called borrowing, a commonly used term, but one which does not indicate the process of decomposing an element in a place into ten elements of the next place to its right.

If 43 − 27 is carried out with soda straws, as illustrated above, one observes that 2 elements may be removed from the ten's cup at the first step, then one may decompose one of the remaining elements there into ten elements to put into the one's cup so that 7 elements can be removed from there.

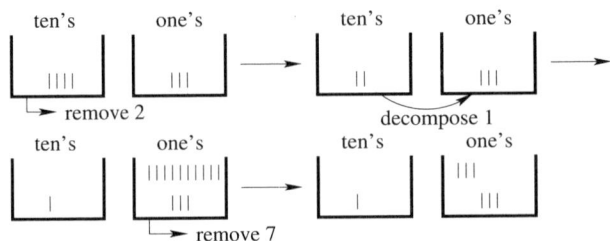

3.3. Computing Differences

This modified subtraction algorithm would be written as follows, now with the process starting from the left.

(3.5)
$$\begin{array}{r} 4{}^1 3 \\ -27 \\ \hline \overset{1}{\cancel{2}}6 \end{array}$$

It is best to teach a single algorithm, but this modification should be kept in mind when a student seems to be making a mistake by starting from the left rather than from the right. If you carry out the process with soda straws and cups, you will see how natural this modification is.

The decomposition process must be carried out more than once in some problems, such as in $734 - 568$.

$$\begin{array}{r} {}^6{}^1 2 \\ 7\cancel{3}{}^1 4 \\ -568 \\ \hline 166 \end{array}$$

Explanation: Begin with the subtraction problem in the one's place. In this example this is $4 - 8$ which is not defined in whole numbers. Go to the 3 in the ten's place of the minuend and decompose one of these into a group of ten elements of the next place to the right, which is the one's place, and add these to the elements already there, $10 + 4 = 14$ in this example. The subtraction problem in the one's place becomes $14 - 8 = 6$.

Now proceed to the subtraction problem in the ten's place, which is $2 - 6$, also not defined in whole numbers. Carry out the decomposition of an element in the next place to the left in the minuend. This reduces the 7 in the hundred's place to a 6 and the other element is decomposed into a group of ten elements of the place to the right and added to the 2 already there in the ten's place to give 12 in the ten's place. The subtraction problem in the ten's place is then $12 - 6 = 6$. The final subtraction problem is $6 - 5 = 1$ in the hundred's place. The formal explanation for this example is

$734 - 568 = (700 + 30 + 4) - (500 + 60 + 8)$, numeration
$= (700 + 20 + (10 + 4)) - (500 + 60 + 8)$, decompose a 10 to 10 ones
$= ((700 + 20) - (500 + 60)) + (14 - 8)$, Example 2.40
$= ((600 + (100 + 20)) - (500 + 60)) + 6$, decompose a 100 to 10 tens
$= (600 - 500) + (120 - 60) + 6$, Example 2.40
$= 100 + (12 - 6) \cdot 10 + 6$, subtract in 100's place; dist. prop.
$= 100 + 60 + 6 = 166$, subtract in 10's place; numeration

Theorem 3.25 (Subtraction algorithm). *To find the difference $a - b$ of whole numbers expressed in the base ten numeration system, proceed as follows.*

Subtract the digit in the one's place of the subtrahend b from the digit in the one's place of the minuend a. If this subtraction is not defined in whole numbers, decompose an element in the next place to the left (the ten's place) in the minuend into ten elements of the one's place, add to this the digit already in the one's place of the minuend, and now carry out the subtraction in the one's place. This subtraction is defined in whole numbers because ten is larger than any digit. The difference is a digit and is the digit in the one's place of the difference $a - b$.

This step required an element in the ten's place of the minuend, that is, the digit in its ten's place is not 0. If it is 0, then proceed to the left to the first place where the digit is not 0. There must be such a place if the original subtraction problem is defined in whole numbers. From this place decompose an element into ten elements of the place to its right and continue decomposing elements until ten elements are put into the ten's place. Now proceed as before.

Repeat this process for the ten's place, and subsequently for each place that appears in the minuend. The difference at each place will be the digit in that place of $a - b$.

Proof. To clarify the statement of the theorem, we shall restate it for the example $602 - 185$.

$$\begin{array}{r} \overset{5}{\cancel{6}} \ \overset{9}{^1\cancel{0}} \ \ ^12 \\ -1 \ \ \ 8 \ \ \ 5 \\ \hline 4 \ \ \ 1 \ \ \ 7 \end{array}$$

Begin with the one's place, where now $2 - 5$ is not defined in whole numbers. The minuend 602 has a 0 in the ten's place, which gives no element to decompose. In such a situation, move to the first place to the left in which there is a nonzero digit. In this case, it is the 6 in the hundred's place. Take one of these and decompose it into a group of ten elements of the place to its right, the ten's place here. Now take one of these 10 elements and decompose it into ten elements of the place to its right, which is the one's place in this instance. Now the subtraction problem in the one's place is $12 - 5 = 7$, the subtraction problem in the ten's place is $9 - 8 = 1$, and the subtraction problem in the hundred's place is $5 - 1 = 4$. These are the digits of the difference: $602 - 185 = 417$. The justification for this process is contained in the formal explanation

$$\begin{aligned}
602 - 185 &= (600 + 2) - (100 + 80 + 5), \text{ numeration} \\
&= (500 + 100 + 2) - (100 + 80 + 5), \text{ decompose a 100 to 10 tens} \\
&= (500 + 90 + (10 + 2)) - (100 + 80 + 5), \text{ decompose a 10 to 10 ones} \\
&= ((500 + 90) - (100 + 80)) + (12 - 5), \text{ Example 2.40} \\
&= (500 - 100) + (90 - 80) + (12 - 5), \text{ Example 2.40} \\
&= 400 + 10 + 7, \text{ subtract in each place} \\
&= 417, \text{ numeration}
\end{aligned}$$

$\square$

In the United States it is common to use the term *borrowing* for the process of decomposing an element of a place into ten elements of the next place to the right and adding these ten elements to that place. Liping Ma in [**Ma99**] convinced me that *decomposing an element of a place into ten elements of the next place to the right* is a more accurate description of the actual process. It emphasizes the process of forming a base ten numeral more closely than the idea of borrowing, which has attached to it the idea of repaying the loan. Elements are not being borrowed, but shifted from one place to another, and this shifting requires the process of decomposing an element into ten elements of the place to the right.

Oral exercise 3.26. Nicholas asks what to do if it should happen that in the last place to the left the subtraction problem is not defined in whole numbers, because then there is no place from which to decompose elements. For example, he says, what if the preceding problem were $602 - 605$, so that the subtraction in the hundred's place would have been $5 - 6$? How would you answer him? Is Theorem 3.16 involved?

3.3.2. The subtraction algorithm in base 5. The subtraction algorithm remains the same for any base b, provided one remembers that now an element in a place is a group of b elements of the next place to its right.

Let us test our understanding of this algorithm by computing the difference $42_5 - 13_5$, where now we use base five numerals. Then $42_5 = 4 \cdot 10_5 + 2 = 3 \cdot 10_5 + (12_5)$ after decomposing a unit from the five's place into five elements of the one's place. From the addition table we read the basic subtraction facts for base 5. These include $12_5 - 3 = 4$ in the one's column and $3 - 1 = 2$ in the five's column. A formal explanation, all in base five numerals, is the following. Remember that the base five numeral for five is 10_5.

$$42_5 - 13_5 = (40_5 + 2) - (10_5 + 3), \text{ base five numeration}$$
$$= (30_5 + (10_5 + 2)) - (10_5 + 3), \text{ decompose a } 10_5 \text{ to five ones}$$
$$= (30_5 - 10_5) + (12_5 - 3), \text{ Example 2.40}$$
$$= 20_5 + 4, \text{ subtract in each place}$$
$$= 24_5, \text{ base five numeration}$$

In columns, the subtraction algorithm in base five looks like this:

$$\begin{array}{r} \overset{3}{\cancel{4}}{}^1 2_5 \\ -1\ 3_5 \\ \hline 2\ 4_5 \end{array}$$

Oral exercise 3.27. Use soda straws to illustrate the above algorithm for computing $42_5 - 13_5$.

Oral exercise 3.28. Give a justification in terms of subtraction properties of the following mental math computations.

(1) $36 - 29 = (36 + 1) - (29 + 1) = 37 - 30 = 7$

(2) $47 - 28 = (47 + 2) - (28 + 2) = 49 - 30 = 19$

(3) $31 - 22 = (31 - 1) - (22 - 1) = 30 - 21 = 9$

(4) $35 - 18$ You supply the path.

§3.3 Exercises

(1) Explain the algorithm for the computation $83 - 56$. Explain how to use soda straws to demonstrate how the algorithm works. Give a formal explanation for why it works.

(2) Explain the modified algorithm (start at the left) for the computation $46 - 18$. Explain how to use soda straws to demonstrate this.

(3) Explain the algorithm for the computation $725 - 438$. Explain how to use soda straws to demonstrate how the algorithm works. Give a formal explanation for why it works.

(4) Explain the algorithm for the computation $204 - 57$. Explain how to use soda straws to demonstrate how the algorithm works. Give a formal explanation for why it works.

(5) What are the basic subtraction facts in base five numeration? List those which involve more than just digits. Explain the algorithm for the

computation $232_5 - 134_5$, where these are base five numerals. Explain how to use soda straws to demonstrate how the algorithm works.

(6) Explain the algorithm for the computation $203_5 - 14_5$, where these are base five numerals.

(7) Explain the algorithm for the computation $1101_2 - 110_2$, where these are base two numerals. Explain how to use soda straws to demonstrate how the algorithm works.

(8) Explain the algorithm for the computation $101001_2 - 11011_2$, where these are base two numerals.

(9) Cite the subtraction properties which justify the mental math computations: (a) $37 - 28 = 40 - 31 = 9$. (b) $37 - 28 = (40 - 3) - (30 - 2) = 10 - 3 + 2 = 9$.

(10) A box of cookies costs $4.95. A box of chocolates costs $9.50. How much cheaper is the box of cookies? Explain what calculation must be made and how to make it. Explain how decomposing 1 dollar into 100 cents enters the process.

(11) Sue started her homework at 8:50 p.m. She finished her homework at 10:10 p.m. How long did she spend doing homework? Explain what calculation must be made and how to make it. Explain how decomposing 1 hour into 60 minutes enters the process.

(12) Addison bought 3 gallons and 2 quarts of milk. He used 1 gallon, 3 quarts and 1 pint of it. How much milk remains? Explain how decomposing gallons into quarts and quarts into pints enters into the calculation.

3.4. Computing Products

Products of the digits must be learned as basic multiplication facts. Learning these facts can be aided by the following special exercises. Division facts can be learned at the same time by emphasizing, for example, that $3 \cdot 5 = 15$ means that $15/3 = 5$ and $15/5 = 3$, and so on for the rest of the basic multiplication facts.

3.4.0.1. *Doubling.* $2 \cdot a = a + a$, for any whole number a. While learning basic addition facts, special attention should be given to sums of the form $a + a$, for any digit a.

3.4.0.2. *Counting by 2's, 3's, 4's, and 5's.* Multiplication of 2, 3, 4, and 5 can be related to counting by 2's, 3's, 4's, or 5's, respectively. For example, $7 \cdot 3 = 21$ is arrived at by counting by 3's from 3 seven times: 3, 6, 9, 12, 15, 18, 21. Counting by 2's, 3's, 4's, etc. involves the idea of multiples, defined in Definition 2.44. The even numbers are the multiples of 2. Counting by 3 is to cite the multiples of 3. Multiples of 10 are 10, 20, 30, and so on.

3.4.0.3. *Commutative property.* $7 \cdot 3 = 21$ and $3 \cdot 7 = 21$. $8 \cdot 5 = 40$ and $5 \cdot 8 = 40$, and so on.

3.4.0.4. *Squares.* Emphasize squares, $2 \cdot 2 = 2^2 = 4$, $3 \cdot 3 = 9$, etc. Relate these to the area of a square whose side length is the given digit. Relate to the number of elements in a square array.

3.4.0.5. *Multiplication by 10.* In the Hindu-Arabic numeration system, multiplication by the base simply annexes a zero onto the right end of the numeral of the multiplicand. Multiplication by the base shifts each place one place to the left. The following examples in base ten illustrate the principle.

Multiplying a digit, say 6, by 10 gives $10 \cdot 6 = 6 \cdot 10 = 60$, by the definition of the numeration system, and this is 6 with a zero annexed on the right. The 6 was shifted from the one's place to the ten's place because each element is multiplied by ten, so it has become a group of ten. Next, $10 \cdot 10 = 100$ and $10 \cdot 100 = 1000$, for the same reason. Now

$$\begin{aligned} 10 \cdot 28 &= 10 \cdot (2 \cdot 10 + 8), \text{ numeration} \\ &= 2 \cdot 10 \cdot 10 + 8 \cdot 10, \text{ distributive property} \\ &= 280, \text{ numeration} \end{aligned} \quad (3.6)$$

which is 28 with a zero annexed on the right.

This can be demonstrated with soda straws as follows. Take a set of 28 soda straws and form 2 groups of ten, with 8 single straws remaining. Multiplication of a number a by 10 means take 10 groups of a, so multiplication of the 2 groups of ten straws by 10 means take 10 groups of these. But 10 groups of one of these groups of ten is then an element of the next place to the left, the hundred's place in this case. Thus, multiplication of the 2 groups of ten straws by 10 gives us 2 elements of the hundred's place. In the same way, multiplication of the 8 single straws by 10 forms 8 groups of ten straws, which is 8 elements of the ten's place. In summary, multiplication by 10 transforms the elements of any place into the same number of elements of the next place to the left and this is accomplished by annexing a 0 onto the right end of the numeral.

As mentioned above, annexing a zero on the right is the rule for multiplying by the base, no matter what the base is. For example, in base five, the numeral for five is 10_5 and multiplication of 34_5 by five is the base five numeral 340_5. In fact, in base five numeration

$$\begin{aligned} 10_5 \cdot 34_5 &= 10_5 \cdot (3 \cdot 10_5 + 4), \text{ base five numeration} \\ &= 3 \cdot (10_5 \cdot 10_5) + 4 \cdot 10_5, \text{ distributive property} \\ &= 340_5, \text{ base five numeration} \end{aligned} \quad (3.7)$$

By the commutative property, multiplication of 10 follows the same rule as multiplication by 10. The same comment applies to the numeration system of any base.

3.4. Computing Products

Back to base ten, since $100 = 10 \cdot 10$, it follows that multiplication by 100 is multiplication by 10 then multiplication by 10 again. Therefore, multiplication by 100 annexes two zeros on the right. It shifts each place two places to the left.

Oral exercise 3.29. What is the rule for multiplication by 1000? by 10000? Explain.

3.4.0.6. *Multiplication by 9.* To help remember the basic facts involving multiplication with 9, use the fact that $9 = 10 - 1$. Then $9 \cdot 5 = (10 - 1) \cdot 5 = 50 - 5 = 45$, and $9 \cdot 8 = 80 - 8 = 72$, etc. The rule is,

$$9 \cdot a = 10 \cdot a - a$$

3.4.0.7. *What remains.* The remaining cases are $6 \cdot 8 = 48$, $7 \cdot 8 = 56$, and $6 \cdot 7 = 42$. These must be memorized. A check on memory is provided by

$$6 \cdot 7 = (5 + 1) \cdot 7 = 5 \cdot 7 + 7 = 35 + 7 = 42$$
$$6 \cdot 8 = (5 + 1) \cdot 8 = 5 \cdot 8 + 8 = 40 + 8 = 48$$
$$7 \cdot 8 = (6 + 1) \cdot 8 = 6 \cdot 8 + 8 = 48 + 8 = 56$$

Next consider multiplication of a digit and a multiple of 10. For example,

(3.8)
$$\begin{aligned} 3 \cdot 20 &= 3 \cdot 2 \cdot 10, \text{ numeration} \\ &= 6 \cdot 10, \text{ multiplication facts} \\ &= 60, \text{ numeration} \end{aligned}$$

Following the same principles, we can compute the product of multiples of 10. For example,

$$\begin{aligned} 30 \cdot 50 &= (3 \cdot 10) \cdot (5 \cdot 10), \text{ numeration} \\ &= 3 \cdot 5 \cdot 10 \cdot 10, \text{ Comm. and Assoc.} \\ &= 15 \cdot 10 \cdot 10 \\ &= 1500, \text{ two multiplications of 10} \end{aligned}$$

From this we see that the rule for computing $30 \cdot 50$ is $3 \cdot 5$ with as many zeros annexed on the right of this as the total number of zeros appearing in the given factors. This follows from the rule for multiplication by 10, as each such multiplication annexes a 0 on the right.

Proceed now to multiplication of a two digit number by a digit, as in

$$\begin{aligned} 4 \cdot 27 &= 4 \cdot (20 + 7), \text{ numeration} \\ &= 4 \cdot 20 + 4 \cdot 7, \text{ distributive property} \\ &= 80 + 28, \text{ basic multiplication facts} \\ &= 108, \text{ addition algorithm} \end{aligned}$$

In columns, this goes as follows, in what we'll call the formal algorithm with margin notes.

(3.9)
$$\begin{array}{r} 27 \\ \times 4 \\ \hline 28 \\ +80 \\ \hline 108 \end{array} \quad \begin{array}{l} 4 \cdot 7 \\ 4 \cdot 20 \\ 28 + 80 \end{array}$$

The algorithm begins with multiplication of the digits in the one's place, which is $4 \cdot 7 = 28$ in this example. This is called a *partial product* and is recorded under the horizontal line in such a way that the places of 28 line up with their respective places in the multiplicand at the top and the multiplier under it. The next step is to multiply 4 times the digit in the ten's place of the multiplicand, the 2 in this example. In order to carry out this algorithm correctly, we must understand that at this point we are calculating $4 \cdot 20 = 80$, even though we need only think about calculating $4 \cdot 2 = 8$. As a consequence, we put this partial product on the next line with the 8 in the ten's place, since it is really 80. As the above formal explanation shows us, the last step is to sum the two partial products, $28 + 80 = 108$ in this example.

In practice, we shorten this algorithm by carrying out the sum of the partial products during the process of carrying out the two multiplications. In the following, the 8 of $28 = 4 \cdot 7$ is recorded in the one's place below the line, but the 2 is recorded as a small superscript in the same way we indicated groups of ten being carried to the next place to the left in the addition algorithm. This 2 is then added to the 8 in the ten's place of the partial product $80 = 4 \cdot 20$, and the sum, $2 + 8 = 10$, is recorded in the ten's place, which by the rules of the base ten numeration, leads to a 0 in the ten's place and a 1 in the next place to the left, which is the hundred's place. Compare the following contracted form with the expanded algorithm (3.9).

$$\begin{array}{r} 27 \\ \times 4 \\ \hline 1\overset{2}{0}8 \end{array} \quad 4 \cdot 7 = 28, \text{ record } 8, \text{ then } 2 + 4 \cdot 2 = 10 \text{ in ten's place}$$

Theorem 3.30 (Multiplication algorithm). *If two whole numbers are expressed in the base ten numeration system, then their product is found as follows: Multiply each digit of the multiplier times each digit of the multiplicand, using basic multiplication facts and rules for multiples of ten. Add together all of these partial products.*

Proof. The essence of the proof is contained in the following formal explanation of $35 \cdot 89$. In the second and third lines we are expanding out a

3.4. Computing Products

product of two sums $(30 + 5) \cdot (80 + 9)$ as in equation (2.6).

$$\begin{aligned}
35 \cdot 89 &= (30 + 5) \cdot (80 + 9), \text{ numeration} \\
&= (30 + 5) \cdot 80 + (30 + 5) \cdot 9, \text{ distributive property} \\
&= 30 \cdot 80 + 5 \cdot 80 + 30 \cdot 9 + 5 \cdot 9, \text{ distributive property twice} \\
&= 2400 + 400 + 270 + 45, \text{ multiplication facts and by 10} \\
&= 3115, \text{ addition algorithm}
\end{aligned}$$

In columns with margin notes the expanded algorithm records this as

$$\begin{array}{rl}
89 & \\
\times 35 & \\
\hline
45 & 5 \cdot 9 \\
400 & 5 \cdot 80 \\
270 & 30 \cdot 9 \\
+2400 & 30 \cdot 80 \\
\hline
3115 &
\end{array}$$

This calculation amounts to finding $5 \cdot 89$ and $30 \cdot 89$ and adding the results, a process justified by the distributive property. As in the case of a 1-digit multiplier, each of these multiplications is contracted as follows.

$$\begin{array}{rcccl}
& & 8 & 9 & \\
\times & & 3 & 5 & \\
\hline
& & {}^4 & & \\
& 4 & 0 & 5 & \quad 5 \cdot 9 + 5 \cdot 80 = 45 + 400 \\
& {}^2 & & & \\
2 & 4 & 7 & 0 & \quad 30 \cdot 9 + 30 \cdot 80 = 270 + 2400 \\
\hline
{}^1 & {}^1 & & & \\
3 & 1 & 1 & 5 & \quad \text{addition algorithm}
\end{array}$$

The small superscripted digit in each partial product is the amount carried from the product in the preceding place. The small superscripted digits in the answer are the groups of ten carried into that place from the addition of the two partial products. The contracted multiplication algorithm requires quite a few mental calculations. □

3.4.1. Multiplication algorithm in other bases. As with the algorithms of addition and subtraction, the multiplication algorithm is essentially the same in all bases. The only real difference is the set of basic multiplication facts needed. The smaller the base, the smaller is this set.

In base five, the basic multiplication facts consist of the products of the digits $0, 1, 2, 3, 4$. Products of these which are not again digits include $2 \cdot 3 = 11_5$, $2 \cdot 4 = 13_5$, $3 \cdot 4 = 22_5$, $4 \cdot 4 = 31_5$, and so on. The rule for multiplying by the base is exactly the same as it is in base ten. In fact, in base five, the numeral for five is 10_5, and multiplication by 10_5 is done by annexing a zero onto the right of the base five numeral of the multiplicand.

For example, $10_5 \cdot 23_5 = 230_5$, because

$$\begin{aligned}10_5 \cdot 23_5 &= 10_5 \cdot (2 \cdot 10_5 + 3), \text{ base five numeration} \\ &= 2 \cdot 10_5 \cdot 10_5 + 3 \cdot 10_5, \text{ dist., assoc., comm. properties} \\ &= 230_5 \text{ base five numeration}\end{aligned}$$

In base five, products of multiples of five follow the same rules as products of multiples of ten in base ten. For example, $30_5 \cdot 40_5 = 3 \cdot 4 \cdot 100_5 = 2200_5$, and so on, for exactly the same reasons as given in the base ten case.

A formal explanation of $23_5 \cdot 43_5 = 2144_5$ is

$$\begin{aligned}23_5 \cdot 43_5 &= (20_5 + 3) \cdot (40_5 + 3), \text{ numeration} \\ &= (20_5 + 3) \cdot 40_5 + (20_5 + 3) \cdot 3, \text{ distributive property} \\ &= 20_5 \cdot 40_5 + 3 \cdot 40_5 + 20_3 \cdot 3 + 3 \cdot 3, \text{ distributive property} \\ &= 1300_5 + 220_5 + 110_5 + 14_5, \text{ basic multiplication facts} \\ &= 2144_5, \text{ addition algorithm}\end{aligned}$$

Writing this in columns, we arrive at the expanded multiplication algorithm in base five, with margin notes:

$$\begin{array}{rl} 43_5 & \\ \times\ 23_5 & \\ \hline 14_5 & 3 \cdot 3 \\ 220_5 & 3 \cdot 40_5 \\ 110_5 & 20_5 \cdot 3 \\ 1300_5 & 20_5 \cdot 40_5 \\ \hline 2144_5 & \text{addition algorithm} \end{array}$$

As with the base ten case, this algorithm can be contracted further, but we omit that step because the point of studying the algorithm in bases other than ten is to use the unfamiliar territory to see just how the algorithm works.

3.4.2. Mental Math. Apply the distributive property twice to derive an expansion of a product of a sum with a difference:

(3.10) $$(a+b) \cdot (a-b) = a^2 - b^2$$

for any whole numbers a and b. Compare this to the expansion of a product of sums in Equation (2.6). It can be applied to computations such as

$$21 \cdot 19 = (20+1) \cdot (20-1) = 20^2 - 1^2 = 400 - 1 = 399$$

Oral exercise 3.31. Use this technique to find: (a) $32 \cdot 28$. (b) $101 \cdot 99$. (c) $61 \cdot 59$. (d) $88 \cdot 92$. (e) $33 \cdot 27$. (f) $22 \cdot 18$. (g) $23 \cdot 17$.

3.4. Computing Products

3.4.3. Counting subsets. Some problems involve counting the number of subsets of a given set. For example, the number of different bridge hands is the number of 13-element subsets (the hands) of a set of 52 elements (the deck of cards). The method of counting is similar to the method used to count the number of permutations of a set.

How many 2-element subsets in a set of 3 elements $\{a, b, c\}$? We count the subsets by forming them in a systematic way. To form a 2-element subset, we choose a first element. There are 3 possible choices for this first element:

$$a, \quad b, \quad c$$

Once the first element is chosen, there remain 2 choices for the second element.

$$\{a\,b\}, \{a\,c\} \quad \{b\,a\}, \{b\,c\} \quad \{c\,a\}, \{c\,b\}$$

The total number of choices is 3 groups (the number of choices for first element) of 2 (the number of choices for second element), which is $3 \cdot 2$ choices.

But, each subset has been formed twice by this method. For example, if we choose a first, then b second, we get the subset $\{a\,b\}$. If we choose b first, then a second, we get the subset $\{b\,a\}$, which is the same subset, but with the elements permuted. In fact, we have chosen each subset as many times as the elements of these subsets can be permuted. A set of 2 elements has $2! = 2$ permutations. We conclude that the number of choices is

$$3 \cdot 2 = (\text{number of 2-elements subsets}) \cdot 2!$$

so the number of 2-element subsets of $\{a, b, c\}$ is

$$(3 \cdot 2) \div 2! = 3$$

which we can verify from the list above.

Theorem 3.32 (Number of Subsets). *If a set A has n elements and if $k \leq n$, then the number of k-element subsets of A is*

$$(n \cdot (n-1) \cdots (n - (k-1))) \div k!$$

The method of proof of this formula is what should be learned here, not the formula itself, which is easily forgotten.

Proof. We shall illustrate the principles of the proof by counting the number of 3-element subsets of a set A of 5 elements. Suppose $A = \{a\,b\,c\,d\,e\}$. To form a subset, there are 5 choices for first element.

$$a \quad b \quad c \quad d \quad e$$

For each such choice, there remain 4 choices for second element, so 5 groups of 4 choices for the first two elements. This is $5 \cdot 4$ choices, the number of elements in the following array.

$$\begin{array}{llll} \{a\,b\} & \{a\,c\} & \{a\,d\} & \{a\,e\} \\ \{b\,a\} & \{b\,c\} & \{b\,d\} & \{b\,e\} \\ \{c\,a\} & \{c\,b\} & \{c\,d\} & \{c\,e\} \\ \{d\,a\} & \{d\,b\} & \{d\,c\} & \{d\,e\} \\ \{e\,a\} & \{e\,b\} & \{e\,c\} & \{e\,d\} \end{array}$$

For each choice of first and second element, there remain 3 choices for the third element of the subset, so there are $5 \cdot 4$ groups of 3 such choices. That comes to a total of $5 \cdot 4 \cdot 3$ choices, which is $5 \cdot (5-1) \cdot (5-(3-1))$ choices.

This method of choosing elements has produced every possible 3-element subset. But each subset occurs in every possible permutation of its elements. There are 3! permutations of a set with 3 elements, so our method produced each subset 3! times, and

$$5 \cdot 4 \cdot 3 = (\text{number of 3-element subsets}) \cdot 3!$$

The number of 3-element subsets is therefore

$$(5 \cdot 4 \cdot 3) \div 3! = (5 \cdot (5-1) \cdot (5-(3-1))) \div 3!$$

which is the formula to be proved. $\square$

Example 3.33. The number of distinct bridge hands is the number of 13-element subsets of a set of 52 elements, which is

$$52 \cdot 51 \cdot 50 \cdot 49 \cdot 48 \cdot 47 \cdot 46 \cdot 45 \cdot 44 \cdot 43 \cdot 42 \cdot 41 \cdot 40 \div 13!$$
$$= 635,013,559,600$$

by Theorem 3.32, since $52 - (13-1) = 40$. The first associative property of Theorem 2.89 simplifies this computation enough that it can be done by hand (with some patience).

Example 3.34. A bag contains four white marbles and 6 black marbles. Reach into the bag, without looking inside, and take out two marbles at the same time. Call the resulting two marbles an outcome.

(1) How many different outcomes are there? This would be the number of 2-element subsets in a set of 10 elements. There are 10 choices for the first element and for each choice there are 9 choices for the second element, for a total of $10 \cdot 9 = 90$ choices. The same subset chosen in a different order has counted as a separate choice. We have counted each outcome as many times as there are permutations of a set of 2 elements. There are $(10 \cdot 9)/2 = 45$ outcomes.

(2) How many different outcomes are there for which both marbles are white? Now each 2-element subset must come from the set of four white marbles. There are $(4 \cdot 3)/2 = 6$ outcomes for which both marbles are white.

(3) How many different outcomes are there for which one marble is white and the other is black? To form such a subset, there are 4 possible choices for the white marble, and for each of these there are 6 possible choices for the black marble. This makes 4 groups of 6 choices, a total of $4 \cdot 6 = 24$ outcomes for which 1 marble is white and the other is black.

§3.4 Exercises

Use line segment diagrams to explain your solution to the word problems.

(1) Compute $28 \cdot 43$, written as the expanded algorithm with margin notes, then as the contracted algorithm.

(2) Give a formal explanation of $96 \cdot 87$. Follow that by writing the computation in the expanded algorithm with margin notes.

(3) Compute $492 \cdot 988$, written in expanded algorithm with margin notes. Give a formal explanation.

(4) Use the distributive property and Theorem 2.33 to derive $(a+b) \cdot (a-b) = a \cdot a - b \cdot b = a^2 - b^2$, for any whole numbers a and b.

(5) Using base five numerals, give a formal explanation of $34_5 \cdot 42_5$, and then exhibit the expanded multiplication algorithm with margin notes.

(6) What are the basic multiplication facts needed in the base two numeration system? Using base two numerals, give a formal explanation of $1011_2 \cdot 1101_2$ and then exhibit the expanded multiplication algorithm with margin notes.

(7) A drover bought 120 sheep at 6 dollars each, 27 cows at 47 dollars apiece, 65 pigs at 7 dollars apiece, and 24 horses at 134 dollars each: how much did all cost? *Ans.* 5660 dollars.

(8) If there are 640 acres in a square mile, how many acres in 36 square miles?

(9) With the $800 she received for her birthday, Shelby bought a saddle and boots. The boots cost $128.99 and the saddle cost 3 times as much. How much money did she have left?

(10) Danielle earns $2437 a month. If she spends $1830 each month and saves the rest, how much will she save in 12 months?

(11) Newell bought 27 calves at $487 each. Then he bought 13 bulls at $2078 each. How much did he spend altogether?

(12) Sigrid has a collection of rag dolls and porcelain dolls. She has 127 porcelain dolls. She has 4 times as many rag dolls as porcelain dolls. How many dolls altogether in her collection?

Hint: There are more rag dolls than porcelain dolls.

    Porcelain ⌊127⌋
    Rag   ⌊ | | | ⌋

1 unit = how many dolls?
Sigrid's doll collection comprises how many units?

(13) There are 12 eggs in a carton. Suella bought 2 cartons of brown eggs and 3 times as many cartons of white eggs.
  (a) How many eggs did she buy? Make a line segment diagram, where each unit represents 1 carton.
  (b) How many more white eggs than brown eggs did she buy? (How many more cartons of white eggs than cartons of brown eggs did she buy?)

(14) During the past year Austin's salary was $2076 a month. During this year she saved $4972. How much did she spend during this year?

(15) During January Kyle's bakery sold 243 dozen cookies. He sold 3 times as many in February. In March he sold 2 times as many as he sold in February. How many dozen cookies did he sell in these 3 months?

(16) Estimate 39 · 28 by approximating 39 and 28 by numbers which you can multiply in your head. Without calculating the actual value, explain how you know your estimate is too big or too small. Calculate the actual product.

(17) Estimate 72 · 37 by 70 · 40. Explain why you think the estimate is too big or too small. Use Equation (3.10) to calculate the actual product. Was your first answer correct? If not, explain your mistake.

(18) Find the area of the shaded triangle. Each square of the grid has side length of 3 inches.

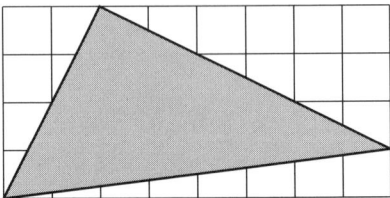

(19) Derive the number of subsets with 2 elements in a set of 6 elements.

(20) Six people sit around the dinner table. Someone proposes a toast to the cook and everyone clinks their glass with everyone else. How many

clinks are there? *Hint:* Each person makes how many clinks? Are there any duplications?

(21) A bag contains three white marbles and 7 black marbles. Reach into the bag, without looking inside, and take out two marbles at the same time. Call the resulting two marbles an outcome.
 (a) How many different outcomes are there? *Hint:* How many 2 element subsets in a set of 10 elements?
 (b) (i) How many different outcomes are there for which both marbles are white? (ii) For which both marbles are black?
 (c) How many different outcomes are there for which one marble is white and the other is black?

(22) How many different 5-card hands can be dealt from a deck of 52 cards? Use the first associative property of Theorem 2.89 to simplify the calculation. *Ans:* 2,598,960.

3.5. Computing Quotients

The object of this section is to describe two common algorithms for computing the quotient q and remainder r in a division problem $a \div b$, for any whole numbers a and $b \neq 0$. The test of whether answers q and r are correct is always to see if $a = q \cdot b + r$.

The table of basic multiplication facts provides also the basic division facts. For example, $8 \cdot 9 = 72$ means that $72 \div 8 = 9$ and that $72 \div 9 = 8$. The basic division facts means those division facts coming from the multiplication table for the digits.

Long division, Theorem 2.97, was interpreted in terms of measurement division, where $a \div b$ is the number of b's in a, which is the same as the number of times b can be subtracted from a until the remainder is less than the divisor b. The basic division facts can be used to estimate this number. Measurement division is an exercise in estimating the number of these subtractions possible, in which multiplication is used to speed up the computation of repeated additions.

This section introduces two division algorithms, a partitive division algorithm which is the standard algorithm, and a measurement division algorithm.

3.5.1. A Partitive Division Algorithm. Consider the partitive division problem $56 \div 3$, which asks: 3 groups of what make 56? The following solution serves an an introduction to the partitive division algorithm. The demonstration with soda straws should be done in the classroom.

Imagine that we have 56 soda straws organized according to the base ten numeration system: 5 groups of ten straws (tied in bundles of ten) in the ten's cup and 6 individual straws in the one's cup. We want to divide these equally among 3 people. This algorithm begins with the highest place and works from left to right.

Begin by dividing the 5 bundles of ten straws equally among 3 people. This is the partitive division $5 \div 3$. By basic division facts

$$5 = 3 \cdot 1 + 2$$

so give 1 bundle to each person and have 2 left over. The 2 remaining bundles cannot be divided equally among 3 people, so for the next step we decompose (untie) each bundle into its ten individual straws and put them into the one's cup. Of course, this is the process of decomposing 2 tens into 2 groups of ten ones, which is 20 ones. Add these to the 6 straws we already have.

We now divide 26 straws equally among 3 people. This is the partitive division $26 \div 3$. By basic division facts

$$26 = 3 \cdot 8 + 2$$

so give 8 straws to each person and have 2 left over. The 2 remaining straws cannot be divided equally among 3 people. As long as we leave each straw whole, we cannot divide further.

In conclusion, each person receives 1 bundle of ten and 8 individual straws, which is 18 straws. There are 2 straws remaining. Write this as

$$56 \div 3 = 18, \quad r = 2$$

which means that 56 is 3 groups of 18 plus 2 more

$$56 = 3 \cdot 18 + 2$$

Here is how we write the partitive division algorithm for $56 \div 3$.

(1) Consider the partitive division problem: 3 of how many tens is 5 tens? The answer is $5 \div 3 = 1, r = 2$, which means 1 ten in each part with $r = 2$ tens remaining. Write

$$\begin{array}{r} 1 \\ 3 \overline{\smash{)}56} \\ -3 \\ \hline 2 \end{array} \quad \begin{array}{l} 5 \text{ tens} \div 3 = 1 \text{ ten, remainder 2 tens} \\ \\ -3 \cdot 1 \text{ tens} \\ \text{remainder of 2 tens} \end{array}$$

(2) Decompose the remainder of 2 tens into 2 groups of ten ones, which is 20 ones, and add the 6 ones contained in the original dividend. This is accomplished by bringing down the 6 and putting it beside the 2 of the 2 tens, to give $20 + 6 = 26$ ones. Now repeat Step 1 for the problem 26

ones divided equally among the 3 parts. From basic division facts, this is $26 \div 3 = 8$ ones, with remainder 2 ones. The whole process is then written in the following standard form, with margin notes.

$$
\begin{array}{r|l}
18 & 5 \div 3 = 1 \text{ ten}, r = 2 \text{ tens}; 26 \div 3 = 8, r = 2 \\
3\,\overline{)\,56} & 56 = 5 \text{ tens } + 6 \text{ ones. Find } 5 \text{ tens } \div 3 \\
-3 & -3 \cdot 1 \text{ tens} \\ \hline
26 & 2 \text{ tens decomposed into } 20 \text{ ones } + 6 \text{ ones. Find } 26 \div 3 \\
-24 & -(3 \cdot 8 = 24 \text{ ones}) \\ \hline
2 & \text{remainder}
\end{array}
$$

To check the correctness of this result, verify that $3 \cdot 18 + 2 = 56$.

A *formal explanation* of this process goes as follows. By the base ten numeration
$$56 = 5 \cdot 10 + 6$$
Look at $5 \div 3$ to get $5 = 3 \cdot 1 + 2$ and substitute this into the preceding equation to get
$$56 = (3 \cdot 1 + 2) \cdot 10 + 6$$
$$= 3 \cdot 1 \cdot 10 + 2 \cdot 10 + 6, \text{ distributive property}$$
$$= 3 \cdot 1 \cdot 10 + 26, \text{ by } 2 \cdot 10 + 6 = 20 + 6 = 26$$
Next consider $26 \div 3$, which by basic division facts is $26 = 3 \cdot 8 + 2$. Substitute this into the preceding equation to get
$$56 = 3 \cdot 1 \cdot 10 + 3 \cdot 8 + 2$$
$$= 3 \cdot (1 \cdot 10 + 8) + 2, \text{ distributive property}$$
$$= 3 \cdot 18 + 2, \text{ numeration}$$
which agrees with the answer given by the partitive division algorithm.

This procedure extends easily to cases where the dividend has more than two digits.

Example 3.35. Show the division algorithm with margin notes for $526 \div 4$. Then give a formal explanation of it.

Algorithm with margin notes to compute $526 \div 4$.

$$
\begin{array}{r|l}
131 & \text{quotient} \\
4\,\overline{)\,526} & 4 \text{ into } 5 \text{ hundreds } = 1 \text{ hundred} \\
-4 & 4 \cdot 1 \text{ hundred } = 4 \text{ hundreds} \\ \hline
12 & 1 \text{ hundred } = 10 \text{ tens. Add } 2 \text{ tens. } 4 \text{ into } 12 \text{ tens} \\
-12 & 4 \cdot 3 \text{ tens } = 12 \text{ tens} \\ \hline
06 & 0 \text{ tens } +6 \text{ ones } = 6 \text{ ones. } 4 \text{ into } 6 \text{ ones} \\
-4 & 4 \cdot 1 \text{ one } = 4 \text{ ones} \\ \hline
2 & \text{remainder}
\end{array}
$$

To check this answer, compute $4 \cdot 131 + 2 = 526$, as desired.

Formal explanation: The goal is to write 526 as the product of 4 times the quotient plus a remainder smaller than 4.

$$526 = 5 \cdot 100 + 2 \cdot 10 + 6, \text{ numeration}$$
$$= (4 \cdot 1 + 1) \cdot 100 + 2 \cdot 10 + 6, \text{ find } 5 \div 4$$
$$= 4 \cdot 1 \cdot 100 + 1 \cdot 100 + 2 \cdot 10 + 6, \text{ distributive prop.}$$
$$= 4 \cdot 100 + 10 \cdot 10 + 2 \cdot 10 + 6, \text{ decompose 100 to } 10 \cdot 10$$
$$= 4 \cdot 100 + (10 + 2) \cdot 10 + 6, \text{ distributive prop.}$$
$$= 4 \cdot 100 + (4 \cdot 3) \cdot 10 + 6, \text{ find } 12 \div 4$$
$$= 4 \cdot 100 + 4 \cdot 30 + (4 \cdot 1 + 2), \text{ find } 6 \div 4$$
$$= 4 \cdot (100 + 30 + 1) + 2, \text{ distributive property}$$
$$= 4 \cdot 131 + 2, \text{ numeration}$$

Sometimes the divisor is larger than the first digit, in which case we decompose that digit into 10 times itself in the place to its right.

Example 3.36. Show the partitive division algorithm with margin notes for $594 \div 7$. Then give a formal explanation of it.

Algorithm with margin notes to compute $594 \div 7$. Now 5 hundreds cannot be divided equally among 7, so we decompose the 5 hundreds into 50 tens, and to this add the 9 tens already there.

```
       84
  7 | 594    divide 59 tens into 7 equal parts. Find 59 ÷ 7
     -56     -(7 · 8 = 56 tens)
       34    3 tens decomposed into 30 ones + 4 ones. Find 34 ÷ 7
      -28    -(7 · 4 = 28 ones)
        6    remainder
```

Check the answer by calculating $7 \cdot 84 + 6$, which is 594 as required.

Formal explanation of $594 \div 7$.

$$594 = 59 \cdot 10 + 4, \text{ numeration}$$
$$= (7 \cdot 8 + 3) \cdot 10 + 4, \text{ find } 59 \div 7$$
$$= 7 \cdot 8 \cdot 10 + 3 \cdot 10 + 4, \text{ distributive property}$$
$$= 7 \cdot 80 + 34, \text{ numeration}$$
$$= 7 \cdot 80 + (7 \cdot 4 + 6), \text{ find } 34 \div 7$$
$$= 7 \cdot (80 + 4) + 6, \text{ distributive property}$$
$$= 7 \cdot 84 + 6, \text{ numeration}$$

Theorem 3.37 (Partitive division algorithm). *If a and b are whole numbers expressed in base ten numerals and $b \neq 0$, then compute $a \div b$ as follows:*

3.5. Computing Quotients

Begin with the left most digit, a_1, of a. By basic division facts, or by computing multiples of b, carry out the partitive division problem $a_1 \div b$ to get $a_1 = b \cdot q_1 + r_1$, where q_1 is a digit, possibly 0, and $r_1 < b$. Place a_1 above the q_1, to indicate that it is a digit of the quotient occupying the same place as a_1. There remains r_1 in this place, but $r_1 < b$, so r_1 cannot be divided equally among b. Decompose each of these r_1 elements into groups of ten elements of the next place to the right, and add to this the number of elements already in this place in the dividend. Repeat the above process with this place. Continue with each place of the dividend through to the one's place.

Proof. We shall illustrate the essence of the proof with the example, $735 \div 8$. The first digit of the dividend is 7, which is in the hundred's place. Consider the partitive division problem of dividing 7 hundreds into 8 parts. From basic division facts, $7 = 0 \cdot 8 + 7$, so the answer is 0 hundreds, and 0 is placed above the 7 to indicate that the quotient has 0 in the hundred's place.

Decompose the 7 hundreds into 70 tens, add the 3 tens already present and consider the problem of dividing 73 tens equally into 8 parts. From basic division facts, $73 = 8 \cdot 9 + 1$, so that 9 is the digit in the ten's place of the quotient. We write these two steps as follows.

$$\begin{array}{r} 09 \\ 8 \overline{\smash{)}735} \\ -72 \\ \hline 1 \end{array} \quad \begin{array}{l} \text{0 in hundred's place, 9 in ten's place} \\ \text{Divide 73 tens equally among 8} \\ -9 \cdot 8 \text{ tens} \\ \text{1 ten remainder in } 73 = 8 \cdot 9 + 1 \end{array}$$

Next, decompose the remaining 1 ten into 10 groups of ones and add these to the 5 in the one's place to get 15 ones remaining. Repeat the process by considering the problem of dividing 15 equally among 8. By basic division facts we have $15 = 8 \cdot 1 + 7$, so 1 is the digit in the one's place of the quotient, and the final remainder is 7. Write the whole process, with margin notes, as follows. In the base ten numeration system it is not required to write the leading 0 in the hundred's place of the quotient.

$$\begin{array}{r} 91 \\ 8 \overline{\smash{)}735} \\ -72 \\ \hline 15 \\ -8 \\ \hline 7 \end{array} \quad \begin{array}{l} \text{9 in ten's place, 1 in one's place} \\ \text{Divide 73 tens equally among 8, } 73 = 8 \cdot 9 + 1 \\ -8 \cdot 9 \text{ tens} \\ \text{1 ten, is 10 ones, add the 5 ones, } 15 = 8 \cdot 1 + 7 \\ -8 \cdot 1 \\ \text{remainder of 7 ones} \end{array}$$

Check by verifying that $8 \cdot 91 + 7 = 735$. Additional features, and complications, of this algorithm and its proof will be illustrated in the next subsection. $\square$

3.5.2. Divisor of Two Digits. Consider the problem $571 \div 23$. Computation of quotients when the divisor is an arbitrary two-digit number poses a substantial challenge to our ability to estimate mentally. Consequently, such problems provide an excellent source of mental math exercises. The partitive division algorithm for this calculation goes as follows.

In base ten numeration, 571 is 5 hundreds plus 7 tens and 1 one. Beginning with the left-most digit of 571, we consider the partitive division problem $5 \div 23$, whose solution we see easily is $5 = 23 \cdot 0 + 5$, since 5 hundreds cannot be divided equally into 23 parts. Thus, 0 is the digit in the hundred's place of the quotient.

We decompose the 5 hundreds into 50 tens, add the 7 tens already present and consider the partitive division problem of dividing 57 tens into 23 equal parts. This problem is beyond basic division facts.

We must try to estimate $57 \div 23$. One method to estimate $57 \div 23$ is discussed below. For now, we shall consider a very elementary method to estimate this quotient. We do this by calculating the multiples of 23 off to the side. These are

$$23 \cdot 2 = 46, \quad 23 \cdot 3 = 69, \quad 23 \cdot 4 = 92, \quad 23 \cdot 5 = 115$$

and so on. From these multiples we see that the largest multiple of 23 which is no larger than 57 is $23 \cdot 2 = 46$, from which we find $57 = 23 \cdot 2 + 11$, since $57 - 46 = 11$. Thus, 2 is the digit in the ten's place of the quotient, and there are 11 ten's remaining. We write the process so far as follows.

```
          02     0 hundreds and 2 tens
    23 ⟌ 571    5 = 23·0 + 5 and then 5·100 = 50·10, plus 7 ten's
         −0     (5 − 23·0)·100 = 5·100
         ───
          57    5·100 + 7·10 = (50+7)·10,  57 = 23·2 + 11
         −46    57 − 23·2 tens
         ───
          11    11 tens
```

Decompose the 11 remaining tens into 110 ones and add these to the 1 already in the one's place. In the notation of the algorithm, this is accomplished by bringing the 1 in the one's place of the dividend down to the one's place in the present last line.

Repeat the process on the one's place, which is now the partitive division problem $111 \div 23$. If we think of dividing 111 equally among 23 people by first handing each person one element, then making another round handing each person one element, then each such round uses 23 elements. The number of elements each person receives is the same as the number of rounds made. Since each round uses 23 elements, the number of rounds is the number of 23's in 111. Lo and behold, we are now considering the measurement division problem of how many 23's are in 111. This is the largest multiple of 23 which is no larger than 111.

From our table of multiples of 23, we see that this multiple is $4 \cdot 23$, and therefore $111 = 23 \cdot 4 + 19$, since $19 = 111 - 23 \cdot 4 = 111 - 92$. Hence, 4 is the digit in the one's place of the quotient, and 19 is the final remainder. Write the whole process as follows, where we omit the leading 0 in the quotient and the step involving $5 - 23 \cdot 0$.

(3.11)
$$\begin{array}{r} 24 \quad \text{quotient} \\ 23 \overline{\smash{\big)}\ 571} \quad 57 = 23 \cdot 2 + 11 \text{ tens} \\ -46 \quad 57 - 23 \cdot 2 \text{ tens} \\ \hline 111 \quad 11 \cdot 10 + 1 = 110 + 1 = 111 \text{ ones}, 111 = 23 \cdot 4 + 19 \\ -92 \quad 111 - 23 \cdot 4 = 19 \\ \hline 19 \quad \text{remainder} \end{array}$$

3.5.3. Estimating quotients. Another method to estimate $57 \div 23$ is to change the divisor to a nearby multiple of 5 or 10. The effect is to allow us to do mentally the side computation of multiples of the divisor. It is generally better to change the divisor upward, so that resulting estimates of the partial quotients are never too large.

In the preceding example of $571 \div 23$, replace 23 by 25 and consider the problem $57 \div 25$. Knowing that the multiples of 25 are 25, 50, 75, etc., we see that the quotient of $57 \div 25$ is 2. In addition, because $23 < 25$ the second order property of division in Theorem 2.89 tells us that the quotient of $57/25$ is less than or equal to the quotient of $57/23$, so we know that the estimate of 2 for the quotient of $57 \div 23$ is possibly too small, but not too large. Checking this estimate, we find that $57 - 23 \cdot 2 = 11 < 23$, which shows that the quotient of $57/23$ is 2.

Overestimates at any step require repeating that step. Underestimates do not require repeating a step, although they can require more than the minimum number of steps. In general, it is better to underestimate than overestimate, as we shall illustrate in the next example.

Example 3.38. Consider $973 \div 18$. The first step is to find the quotient of $97 \div 18$. How many 18's in 97? Ask instead how many 20's in 97. Knowing the multiples of 20, we know that the quotient of $97 \div 20$ is 4, so we use 4 as our estimate of the quotient of $97 \div 18$. The second order property of Theorem 2.89 tells us that it is possible that the quotient of $97 \div 18$ is greater than 4, but we use 4 as our first estimate in the partitive division algorithm. We begin by handing out 4 tens to each of 18. As shown in the calculation below, $97 - 18 \cdot 4 = 25 > 18$, which means that the quotient of $97/18$ is indeed greater than 4. The algorithm permits us to continue with the problem $25 \div 18$, whose quotient we put above the 4, because we are still working in the ten's place. The algorithm allows us to make another round, giving 1 to each of 18 on this second round.

$$
\begin{array}{r}
11 \\
43 \\
18 \overline{\smash{\big)}\,973} \\
-72 \\
\hline
25 \\
-18 \\
\hline
73 \\
-54 \\
\hline
19 \\
-18 \\
\hline
1
\end{array}
\quad
\begin{array}{l}
25 \text{ tens} \div 18 = 1 \text{ ten. } 19 \text{ ones} \div 18 = 1 \text{ one} \\
97 \text{ tens} \div 20 = 4 \text{ tens. } 73/20 = 3 \text{ ones} \\
\\
-18 \cdot 4 \\
\\
97 - 72 = 25 \text{ tens. } 25 > 18, \text{ so find } 25 \div 18 \\
-18 \cdot 1 \\
\\
7 \text{ tens} = 70 \text{ ones, plus bring down the 3 ones} \\
18 \cdot 3 \\
\\
19 > 18, \text{ so find } 19 \div 18 \\
-18 \cdot 1 \\
\\
\text{remainder}
\end{array}
$$

The quotient is $11 + 43 = 54$, with remainder 1. This example illustrates how to proceed when the estimate of the quotient at any step is too small. In the first step we estimated 97 tens $\div 18$ by replacing this division problem with the problem of finding the quotient of $97/20$, which is 4. Since $20 > 18$, we know that the quotient of $97/18$ is greater than or equal to the quotient of $97/20$. We discovered that the quotient of $97/18$ is greater than 4, but that we could continue with the algorithm in the normal fashion.

At the third step, the problem $73 \div 18$ was estimated by the quotient of $73 \div 20$, which is 3. Again, according to the second order property of division, we know that the quotient of $73 \div 18$ is greater than or equal to the quotient of $73 \div 20$, which is 3. As shown in the displayed algorithm, this quotient is too small, because $73 - 18 \cdot 3 = 19 > 18$. Nevertheless, we were able to continue, by finding the quotient of $19 \div 18$, which is 1, which we recorded above the 3, because we are now working in the one's place. The quotient of $973 \div 18$ is then $43 + 11 = 54$, and the remainder is 1, as can be checked by verifying that $54 \cdot 18 + 1 = 973$.

One way to picture how this algorithm works is to think of the problem of dividing 973 dollars among 18 people. Suppose that the money is in the form of 9 hundred dollar bills, 7 ten dollar bills, and 3 ones. Since we cannot divide 9 hundreds among 18 people, we change each hundred dollar bill into 10 tens, so that in place of our 9 hundreds we now have $9 \cdot 10 = 90$ tens, plus the 7 tens we already had, for a total of 97 tens. To divide these evenly among the 18 people, we start with the estimate of the quotient of $97/20$, which is 4, and give 4 tens to each person. This uses a total of $18 \cdot 4 = 72$ tens, leaving us with 25 tens. Since $25 > 18$, we can divide these tens further. That the quotient of $25 \div 18$ is 1 means that we can give each person 1 more ten, and we will have $25 - 18 = 7$ tens left over. Now each person has $4 + 1 = 5$ tens. Next, we change the 7 remaining tens into 70 ones, so

3.5. Computing Quotients

that with the 3 ones we already had we now have 73 ones to divide equally among 18 people. Again we estimate the quotient of 73/18 by the quotient of 73/20, which is 3, and give each person 3 ones, leaving $73 - 18 \cdot 3 = 19$ ones left over. We can give each person the quotient of 19/18 more, which is one more dollar bill, and then there will be $19 - 1 \cdot 18 = 1$ one dollar bill remaining. Tallying what each person has received, we see that each has $4 + 1 = 5$ tens and $3 + 1 = 4$ ones, which comes to 54 dollars.

Notice how important it was to underestimate the quotient at each step (accomplished by changing the divisor upward). If at any step we had overestimated the quotient, then we would have had to take back the money handed out on that step and to start over. Some of those folks might not want to give back the money, causing all kinds of trouble.

Oral exercise 3.39. State the order property for multiplication. How do you derive from this that since $20 \cdot 3 = 60 > 57$ it must be that $23 \cdot 3 > 57$?

3.5.4. Parititive division algorithm in other bases. This algorithm remains essentially unchanged in our numeration system for any base.

3.5.4.1. *Base five.*

Oral exercise 3.40. What are the basic division facts in base five?

Oral exercise 3.41. Use soda straws to demonstrate how to divide 42_5 straws equally among 3 people. How many straws will each person receive? How many straws are left over?

Example 3.42. Find $202_5 \div 3$ in base five numeration.

Algorithm with margin notes. In the first step, we cannot divide 2 five-of-fives equally among 3, so we decompose the 2 five-of-fives into 20_5 fives. The first step appearing in the algorithm is to find the quotient of $20_5 \div 3$, which is found by basic division facts.

$$
\begin{array}{r}
32_5 \\
3 \overline{\smash{)}202_5} \\
-14_5 \\
\hline
12_5 \\
-11_5 \\
\hline
1
\end{array}
\quad
\begin{array}{l}
\text{quotient} \\
20_5 \text{ fives } \div 3 = 3 \text{ fives} \\
-(3 \cdot 3 = 14_5) \text{ fives} \\
1 \text{ five } +2 = 12_5 \text{ ones. Find } 12_5 \div 3 \\
-(3 \cdot 2 = 11_5) \text{ ones} \\
\text{remainder}
\end{array}
$$

Check by showing that $3 \cdot 32_5 + 1 = 202_5$.

Formal explanation for $202_5 \div 3$.

$$202_5 = 20_5 \cdot 10_5 + 2, \text{ numeration}$$
$$= (3 \cdot 3 + 1) \cdot 10_5 + 1, \text{ find } 20_5 \div 3$$
$$= 3 \cdot 3 \cdot 10_5 + 1 \cdot 10_5 + 2, \text{ distributive property}$$
$$= 3 \cdot 30_5 + 12_5, \text{ numeration}$$
$$= 3 \cdot 30_5 + (3 \cdot 2 + 1), \text{ find } 12_5 \div 3$$
$$= 3 \cdot (30_5 + 2) + 1, \text{ distributive property}$$
$$= 3 \cdot 32_5 + 1, \text{ numeration}$$

which shows that $202_2 \div 3$ has quotient 32_5 and remainder 1.

Oral exercise 3.43. Suppose that you have 2 quarters and 2 pennies. Explain the above partitive division algorithm for finding $202_5 \div 3$ in terms of dividing this money equally among 3 people. You may change any quarter into five nickels and any nickel into five pennies.

Oral exercise 3.44. Use soda straws to demonstrate the partitive division algorithm for finding $202_5 \div 3$.

3.5.4.2. *Base two.* The division algorithm in base two is especially easy, because at each step the quotient is either 0 or 1, depending on whether the divisor is greater than the dividend or not at that step.

Example 3.45. Show the partitive division algorithm for $110111_2 \div 101_2$ with margin notes.

$$
\begin{array}{r|l}
 1011_2 & \text{quotient, leading zeros omitted} \\
101_2\ \overline{)\ 110111_2} & \text{first nonzero quotient is } 110_2 \div 101_2 \\
\underline{-101_2} & 101_2 \cdot 1 = 101_2 \\
111_2 & 110_2 - 101_2 = 1 \text{ eights, plus } 11_2 \text{ fours} \\
\underline{-101_2} & 101_2 \cdot 1 = 101_2 \\
101_2 & 111_2 - 101_2 = 10_2 \text{ twos, plus } 1 \\
\underline{-101_2} & 101_2 \cdot 1 \\
0 & \text{remainder is } 0 \\
\end{array}
$$

This can be checked by verifying that $101_2 \cdot 1011_2 = 110111_2$, which can be done with the multiplication algorithm.

Oral exercise 3.46. Physical demonstration of the partitive division algorithm for multidigit numbers in base ten is impractical because the numbers would be so large. In base two, however, it can be done. Start with fifty-five soda straws and put them into groups of two, tying each group with a string or rubber band. Now put these groups of two into groups of two, tying them with a string or rubber band. Continue until no more groups of two can be

3.5. Computing Quotients

made. Explain how the result illustrates the base two numeral for fifty-five, which is 110111_2, the dividend of the preceding example.

Illustrate the above partitive division algorithm in base two, by going through the process of dividing these soda straws evenly among 101_2 people. How many people is that? Notice that at each step you can give each person exactly 1 bundle, or none. There is no problem with estimating the partial quotient at any step.

3.5.5. A Measurement Division Algorithm. The following algorithm is based on measurement division, interpreted as repeated subtraction. Consider the measurement division $972 \div 38$. The answer is the number of 38's contained in 972, which is also the number of times 38 can be subtracted from 972. As it would be quite inefficient simply to subtract 38 repeatedly, we must use the fact that the result of subtracting 38 a certain number of times, say 10 times, is the same as subtracting $10 \cdot 38 = 380$. This algorithm, as well as the standard one, requires that we know, say by calculating to the side, the multiples of the divisor, by digits and by multiples of 10. For the given example, we calculate to the side

$$2 \cdot 38 = 76, \quad 3 \cdot 38 = 114, \quad 4 \cdot 38 = 152, \quad 5 \cdot 38 = 190, \quad 6 \cdot 38 = 228, \ldots$$

and, of course we know that $10 \cdot 38 = 380$. In particular, we see that $20 \cdot 38 = 760$, which is less than 972, so after subtracting 38 from 972 twenty times, we have left

$$972 - 20 \cdot 38 = 972 - 760 = 212$$

Now we are faced with the measurement division $212 \div 38$. From the above table of multiples of 38 we conclude that 38 can be subtracted from 212 only 5 times. Then

$$212 - 5 \cdot 38 = 212 - 190 = 22$$

which is smaller than 38, and therefore is the remainder. We conclude that 38 can be subtracted from 972 a total of $20 + 5 = 25$ times, with a remainder of 22. The following is a scheme, with margin notes, for carrying out this algorithm.

```
    38 | 972  | 20        first partial quotient
        -760             -20 · 38
        ----
         212    5         second partial quotient
        -190             -5 · 38
        ----
          22              remainder
               ----
                25        quotient is sum of partial quotients
```

The measurement division algorithm seems conceptually simpler than the partitive division algorithm, for the same reason that measurement division is conceptually simpler than partitive division. Notice that the measurement division algorithm makes no use of the base 10 numeration system. It is really very primitive. It can be illustrated easily, even for numbers as large as the preceding example. Start with 972 soda straws, and start making groups of 38 from them. After creating 25 such groups, there will be only 22 straws remaining, from which you conclude that the quotient of 972/38 is 25, and the remainder is 22.

Example 3.47 (Limits to the method). With multidigit divisors, the partitive division algorithm can become the same as the measurement division algorithm. Consider $1134 \div 252$. In the partitive division algorithm, the first division occurs after every place is decomposed to the one's place, and then the division problem is $1134 \div 252$, which is the original problem. Calculate the multiples of 252 to find that the quotient is 4 and the remainder is 126.

3.5.6. Averages.

Definition 3.48 (Averages). The *average* of a set of numbers is the sum of these numbers divided by the number of elements in the set.

For example, the average of the set of numbers $\{21, 33, 17, 42, 22\}$ is $(21 + 33 + 17 + 42 + 22)/5 = 135/5 = 27$. If the sum, which is 135, were divided equally among the number of elements of the set, there would be $135/5 = 27$ in each part. The average is the amount in each part if the total were divided equally among the number of elements of the set.

One sort of application just applies the definition. For example, Anthony is 33 inches tall and Addison is 27 inches tall. The average height of these two boys is $(33 + 27)/2 = 30$ inches.

Another sort of application gives the average and asks for the sum. For example, if seven truck loads of coal average 30 tons per load, then they total $7 \cdot 30 = 210$ tons of coal.

Definition 3.49. *Average speed* is the distance travelled divided by the time of travel.

If Cameron drove 120 miles in 2 hours, then her average speed was $120/2 = 60$ mph. If she drove at an average speed of 70 mph for 4 hours, then she travelled a distance of $4 \cdot 70 = 280$ miles, since the distance travelled divided by the time must be the average speed of 70.

Example 3.50. Sophie drove from St. Louis to Jefferson City in 2 hours at an average speed of 60 mph. She drove back in a rain storm for 3 hours at an average speed of 40 mph. What was her average speed for the round trip?

Solution: At 60 mph for 2 hours, she travelled $2 \cdot 60 = 120$ miles.

3.5. Computing Quotients

Total distance travelled was $120 + 120 = 240$ miles.

Total time travelled was $2 + 3 = 5$ hours.

Average speed for round trip was $240/5 = 48$ mph.

As long as we work only in the set of whole numbers, the average is defined only when the division involved is defined in whole numbers. Once we have defined fractions, we will find that the average is always defined.

3.5.7. Word problem solutions. When faced with a word problem, the first step is to solve it. After finding the solution, look back on what you did with the view to articulating your method. Pay special attention to explaining how you decide at each step which arithmetic operation to use. Some explanations are best done in words, others are conveyed eloquently by a picture. This text emphasizes the use of line segment diagrams for this purpose. Avoid the use of variables and other methods of algebra, because these techniques assume the conceptual understanding that we are trying to teach.

Develop a style which concisely explains the essential details without leaving the reader stranded on some gap in the exposition. In this chapter we are concerned with the process of making the calculations, so these should be included now. In subsequent chapters it will be sufficient to say something like "using scratch addition, we find the sum to be" and then record the answer. Strive for a style of explanation which is simple and correct.

Example 3.51. A miller bought 5 loads of wheat, the first weighing 6224 pounds; the second, 6420 pounds; the third, 6582 pounds; the fourth, 5857 pounds; and the last, 6172 pounds. What was the average weight per load?

Solution: The average weight per load is the total weight delivered divided equally among the 5 loads. The total weight delivered is the weight of the five loads combined, which is the sum of the weights of the loads, as can be seen from the line segment diagram

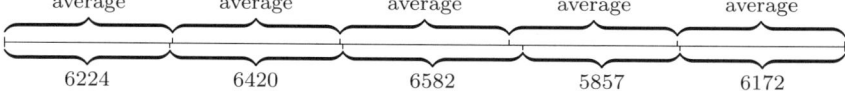

Compute the sum of the five weights using the scratch method.

$$
\begin{array}{r}
3\,\overset{2}{6}\overset{2}{2}\overset{1}{2}\,4 \\
\cancel{6}_4\,4\,2\,0 \\
\cancel{6}_0\cancel{5}_3\cancel{8}_3\,2 \\
5\,\cancel{8}_1\,5\,\cancel{7}_3 \\
\cancel{6}_1\,1\,\cancel{7}_5\,2 \\
\hline
3\,1\,2\,5\,5
\end{array}
\quad
\begin{array}{l}
\text{first load} \\
\text{second load} \\
\text{third load} \\
\text{fourth load} \\
\text{fifth load} \\
\text{total weight}
\end{array}
$$

The average weight per load is then $31255 \div 5 = 6251$ pounds, computed by the standard algorithm as follows.

```
         6251    quotient = average weight per load
     5 ⟌31255
        30       6 · 5
        ──
        12       1 thousand plus 2 hundreds = 12 hundreds
        10       2 · 5
        ──
         25      2 hundreds plus 5 tens = 25 tens
         25      5 · 5
         ──
          5      0 tens plus 5 ones
          5      1 · 5
          ─
          0      remainder
```

Example 3.52. Driving at a constant speed, Walter took 19 hours to drive 1254 miles. At the same speed, how far would he drive in 8 hours?

Solution: Unitary method. How far did Walter drive in 1 hour?

miles 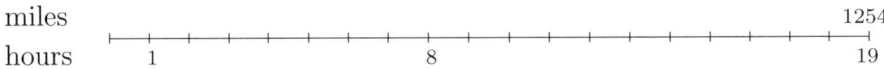 1254

hours   1   8   19

19 hours = 1254 miles.

1 hour = $1254/19 = 66$ miles, so speed is 66 miles per hour.

8 hours = $8 \cdot 66 = 528$ miles

Calculations: In the division, estimate the quotients at each step by using the divisor 20, which is the multiple of 5 next bigger than 19.

```
                  ¹
                6 5  = 66              6 6
         19 ⟌ 1254              ×       8
                                      ─────
            20    114                  ⁴
                  ──                 4 8 8
                  114                5 2 8
                   95
                   ──
                   19
                   19
                   ──
```

Use the standard multiplication algorithm to find $8 \cdot 66$.

Example 3.53. Dona, Samantha, and Andrea have 823 dolls altogether in their doll collections. Dona has 32 more dolls than Samantha and 2 times as many dolls as Andrea. How many dolls has Andrea?

Solution: The given information is displayed in the line segment diagram

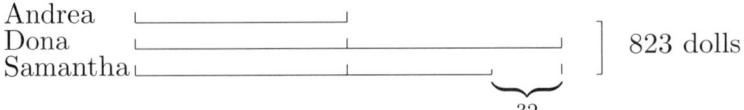

3.5. Computing Quotients

5 units = 823 + 32 dolls.

1 unit = (823 + 32)/5 = 855/5 = 171 dolls.

Andrea has 1 unit = 171 dolls.

3.5.8. Longitude and Time. The sun appears to us to go once around the earth in 1 day, moving from east to west (because the earth is spinning west to east). This is 360° of arc in 24 hours. The meridian lines on the earth can be used to keep track of this motion. In the United States, the meridian lines are labeled by the number of degrees of arc they lie west of Greenwich, England. Looking down on the north pole:

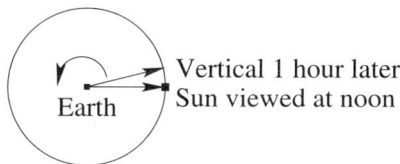

An angle of 1° is divided into 60 equal parts, each called a minute of arc, written 1' of arc. Each minute of arc is divided into 60 equal parts, each called a second of arc, written 1" of arc.

Oral exercise 3.54. (1) Explain why the sun appears to move through 15° of arc in 1 hour?

(2) Explain why it takes the sun 4 minutes to pass through 1° of arc?

(3) Explain why it takes the sun 4 seconds to pass through 1' of arc?

Example 3.55. Honolulu airport is at longitude 157°55' W. The airport at Miles City, Montana, is at longitude 105°52' W. How long does it take the sun to pass from directly over the meridian through Miles City to directly over the meridian through Honolulu?

Solution: Difference in longitude is 157° 55' − 105°52' = 52° 03'.

15° = 1 hour.

How many 15's in 52? Quotient of 52/15 is 3.

3 · 15 = 45° takes 3 hours. There remain 52°03' − 45° = 7°03'.

1° takes 4 minutes.

7° takes 7 · 4 = 28 minutes.

1' takes 4 seconds.

03' takes 3 · 4 = 12 seconds.

45° + 7° + 03' takes 3 hours +28 minutes +12 seconds.

The total time it takes the sun to go through the whole arc is then 3 hours 28 minutes 12 seconds, which can be write in compact form as 3 : 28 :: 12.

§3.5 Exercises

(1) Compute 73 ÷ 4 by the standard algorithm with margin notes. Use the money metaphor with one and ten dollar bills to describe the process. Then write a formal explanation.

(2) Compute 749 ÷ 5 by the standard algorithm with margin notes. Use the money metaphor with one, ten, and hundred dollar bills to describe the process. Then write a formal explanation.

(3) Compute 381 ÷ 4 by the standard algorithm with margin notes. Describe an illustration of this process using soda straws. Then write a formal explanation.

(4) Compute 477 ÷ 32 by the standard algorithm with margin notes, then write a formal explanation. Describe how you estimate the partial quotients.

(5) Compute 286 ÷ 39 by the standard algorithm with margin notes, then write a formal explanation. Describe how you estimate the partial quotients.

(6) Compute 1678 ÷ 127 by the standard algorithm with margin notes, then write a formal explanation. Describe how you estimate the partial quotients.

(7) Compute 187,242 ÷ 413 by the standard algorithm with margin notes, then write a formal explanation. Describe how you estimate the partial quotients.

(8) Compute 7,242 ÷ 413 by the measurement model algorithm, with margin notes. Describe how you decide which multiple of 413 to subtract at each step. This multiple is arbitrary, so describe how you chose the multiples you used.

(9) Kevin desires to trade 123 acres of land at 117 dollars an acre for woodland at 39 dollars an acre: how many acres of woodland will he get? *Ans:* 369 acres.

(10) Morris owns 2000 acres of pasture land which presently sells for $43 an acre. He wants to sell some of it and use the money to buy 237 acres of irrigated farm land, which costs 3 times as much as pasture land. How many acres of pasture land must he sell?

(11) According to the postal scale, first class postage for a package is $2.21. How many 37 cent stamps must be put on this package to cover its postage?

§3.5 Exercises   151

(12) Morris is 28 years older than Joshua. This year he is 3 times older than Joshua. How old is Josh? *Hint:*

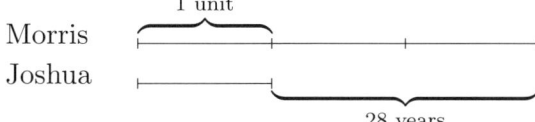

(13) In base five numeration compute $341_5 \div 4$ by the standard algorithm with margin notes, then write a formal explanation. Use the money metaphor with pennies, nickels, and quarters to describe the process.

(14) In base two numeration compute $10,001,110,101_2 \div 1101_2$, by the standard algorithm with margin notes.

(15) Suppose that the snowfall during each of the last five years was 16, 21, 40, 23, and 60 inches, respectively. What was the average annual snowfall during these five years?

(16) If there are 640 acres in a square mile, how many square miles in 17280 acres?

(17) If 17 horses cost 1802 dollars, how much will 9 horses cost?

(18) If 15 people can husk 1095 bushels of corn in a day, how many bushels can 27 people husk?

(19) At the end of the year, Nels had an extra $13011. He spent $1869 to repair his front porch, and divided the rest equally among his grandchildren. If each grandchild received $1238, how many grandchildren has he?

(20) A turtle can crawl 6 blocks in 2 hours. How long does it take the turtle to crawl 8 blocks at the same speed? Express the answer in hours and minutes.

(21) A school of 411 students is going on a field trip in buses each of which holds 47 students. If each bus is loaded to capacity until the last, how many students are in the last bus?

(22) Caitlin mixed 14 quarts of syrup with 210 quarts of water to make soft drinks. She then poured the whole mixture into 28 equal sized jugs, filling all 28 of them. How many quarts of soft drink are in each jug? Explain which arithmetic operations to use and show the calculations.

(23) Sigrid, Sue, and Jen have 678 dolls altogether in their doll collections. Sigrid has 13 more dolls than Sue and Sue has 2 times as many dolls as Jen. How many dolls has Jen? *Think:* The 678 dolls are divided among them as

```
Jen    ⊢────┘
Sue    ⊢────┴────┘
Sigrid ⊢────┴────┴──┘
                  13
```

(24) Addison's candy machine business made a profit last year of $4488. What was his average monthly profit?

(25) The following table gives the scores on a math test.

Boys		Girls	
Anthony	75	Bethany	76
Joseph	66	Caitlin	80
Joshua	80	Danielle	85
Nicholas	71	Shelby	71

Find the average score of the boys. *Ans:* 73.

(26) The average weight of Niels, Leah, and Ragna is 177 pounds. Niels is 30 lbs light of being twice as heavy as Leah. Ragna weighs 41 lbs. more than Leah. Find Niels's weight. *Hint:*

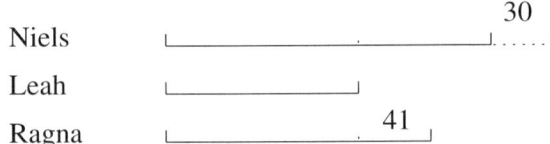

(27) George drove for 2 hours at an average speed of 60 mph. Traffic thinned for the final 3 hours of his trip, so that for the whole 5 hour trip he averaged 66 mph. What was his average speed during the final 3 hours of the trip? *Hint:* Answer the question marks in the given order.

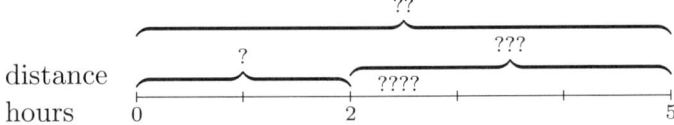

(28) Joshua has three times as many baseball cards as Nicholas. Anthony has 30 more baseball cards than Nicholas. They have 400 baseball cards altogether. How many baseball cards does Joshua have? Note: Use no algebra. Use line segment diagrams to explain your solution.

(29) The longitude at St. Louis airport is 90°23′ W and the longitude of San Francisco is 122°26′ W. How long does it take the sun to pass from directly over the meridian through St. Louis to directly over the meridian through San Francisco? *Ans:* 2 hours 8 minutes 12 seconds.

(30) When the sun is directly over the meridian through San Francisco (longitude 122°26′ W), Patti set her very accurate watch to noon. Then she set sail from the bay out through the Golden Gate heading for Hawaii. A week later her watch says 2:15 p.m. when the sun is directly over head. What is her longitude at this time? *Ans:* 156°11′.

Chapter 4

Number Theory

The arithmetic of whole numbers endows them with many properties, some interesting in their own right, and others important for later applications, such as the arithmetic of fractions. This chapter is concerned mostly with natural numbers, which are the whole numbers except 0. Recall that we denote the set of natural numbers by **N**.

The goal of the chapter is the Unique Prime Factorization Theorem (Fundamental Theorem of Arithmetic), which states that any number has a unique prime factorization. The existence of a prime factorization is easy, and will be presented first. The uniqueness of the prime factorization is much harder and requires the use of Euclid's algorithm for finding the greatest common factor (gcf) of two numbers.

The chapter begins with a discussion of factors, factorizations, and primes and then presents the Factorization Theorem, which states that every natural number greater than 1 has a prime factorization. The first section also includes Euclid's proof that there are an infinite number of primes.

As an aid to finding factorizations, the section presents some divisibility rules, which depend on the base ten numeration system. This is also an appropriate place to review some elementary logic, such as the notions of theorem, converse, and contrapositive.

The next section presents Euclid's algorithm for finding the greatest common factor of two numbers. Euclid's algorithm is the heart of this chapter.

The final section presents the Unique Prime Factorization Theorem followed by applications to finding the gcf, the least common multiple (lcm), and the set of all factors of a number.

All of the material in this chapter, with the exception of the divisibility rules, comes from Euclid's *Elements*. Precise references are cited for many of the results.

4.1. Factors and Primes

Definition 4.1. A *factor* of a number is any number that will exactly divide it (with no remainder). Thus, 1 and the number itself are always factors of a number. A *factorization* of a number is any way of expressing the number as a product of numbers larger than 1. The number by itself is regarded as a factorization of it.

The numbers in a factorization of a number are factors of the number.

Example 4.2. The numbers 1, 2, 3, 4, 6, and 12 are the factors of 12. Factorizations of 12 are

$$12, \quad 2 \cdot 6, \quad 3 \cdot 4, \quad 2 \cdot 2 \cdot 3$$

This is the whole set of factorizations of 12, except for variations in the order of the factors. We count $2 \cdot 6$ and $6 \cdot 2$ as the same factorization, for example.

In subsequent discussions we shall use the expression *d is a factor of a* interchangeably with the expression *d divides a* (meaning exactly in whole numbers, with no remainder). If d does not divide a, or d is not a factor of a, then this means that $a = q \cdot d + r$, where $0 < r < d$. For example, 3 divides 9 because $9 = 3 \cdot 3$, but 4 is not a factor of 9 because $9 = 4 \cdot 2 + 1$.

Recall Definition 2.44 where we defined *multiple* of a number a to be any product of a with a natural number. The notions of multiple and factor are related. If d is a factor of a, then a is a multiple of d, because $a = d \cdot q$, for some natural number q. Conversely, if a is a multiple of d, then d is a factor of a. For example, 4 is a factor of 12 and 12 is a multiple of 4.

Keep in mind that if d is a factor of a, then $d \leq a$.

Oral exercise 4.3. (1) Is 6 a factor of 96? Is 7 a factor of 96? Is 96 a multiple of 6? Is 98 a multiple of 7?

(2) Is 4 a factor of 24? Is 4 a factor of 72? Is 4 a *common factor* of 24 and 72, meaning is 4 a factor of both 24 and 72?

(3) Is 27 a multiple of 3? Is 42 a multiple of 3? Are 27 and 42 *common multiples* of 3, meaning are 27 and 42 both multiples of 3?

Remark 4.4. A factor of a factor is a factor. This vaguely poetic line means the following. If d is a factor of a and if a is a factor of m, then d is a factor of m. For example, 2 is a factor of 6 and 6 is a factor of 30, so 2 is a factor

4.1. Factors and Primes

of 30. In detail, 2 is a factor of 6 because $6 = 2 \cdot 3$, and 6 is a factor of 30 because $30 = 6 \cdot 5$, so
$$30 = 6 \cdot 5 = 2 \cdot 3 \cdot 5$$
shows that 2 is a factor of 30.

Oral exercise 4.5. Is 7 a factor of 56? Is 56 a factor of 168? Is 7 a factor of 168? Find another factor of 56. Is it a common factor of 56 and 168?

Definition 4.6. A *prime* number is a number greater than 1 which has no factors other than 1 and itself. A *composite* number is one that has at least one factor other than 1 and itself. It is agreed that the number 1 is neither prime nor composite.

The idea of prime number p can be illustrated by attempting to put the elements of a set of p elements into a rectangular array for which there are at least two rows and at least two columns. For example, take a set of 6 tokens and experiment with putting the tokens into a rectangular array. The following arrays are found.

The arrays indicate that we can put the 6 tokens into 3 groups of 2 (each group is a row) or into 2 groups of 3. These arrays correspond to the factorizations $6 = 3 \cdot 2$ and $6 = 2 \cdot 3$, respectively, and therefore we see that 6 is a composite number. On the other hand, the same exercise done with a set of 7 tokens leads to the following attempted arrays.

We conclude that 7 tokens cannot be put into a rectangular array except for the case of only 1 row or only 1 column. That is, the only factorization of 7 is by itself. Therefore, 7 is a prime number.

4.1.1. Sieve of Eratosthenes. The prime numbers less than 20 are 2, 3, 5, 7, 11, 13, 17, and 19. This can be found systematically by a Sieve of Eratosthenes. List the numbers from 2 through 20 (or through whatever range you desire).

$$2\ 3\ 4\ 5\ 6\ 7\ 8\ 9\ 10\ 11\ 12\ 13\ 14\ 15\ 16\ 17\ 18\ 19\ 20$$

Begin with 2, which is prime and then strike out all larger multiples of 2, which are composite because 2 is a factor.

$$2\ 3\ \not{4}\ 5\ \not{6}\ 7\ \not{8}\ 9\ \not{10}\ 11\ \not{12}\ 13\ \not{14}\ 15\ \not{16}\ 17\ \not{18}\ 19\ \not{20}$$

Go to the next number which hasn't been crossed out, 3 in this case. It is prime, because its only possible factor other than 1 and 3 is 2, which is not a factor of 3 because we have just determined that 3 is not a multiple of 2. Strike out all larger multiples of 3.

$$2\ 3\ \cancel{4}\ 5\ \cancel{6}\ 7\ \cancel{8}\ \cancel{9}\ \cancel{10}\ 11\ \cancel{12}\ 13\ \cancel{14}\ \cancel{15}\ \cancel{16}\ 17\ \cancel{18}\ 19\ \cancel{20}$$

The next uncrossed out number on the list, 5, must be prime because 2 and 3 are not factors of it (since 5 is not a multiple of either) and the only other possible factor, other than 1 and 5, is 4. Now 4 has been crossed out, because it is a multiple of an earlier number, a multiple of 2 in this case. Since 2 is a factor of 4, if 4 were a factor of 5, then 2 would have to be a factor of 5 also, and this isn't the case since 5 wasn't crossed out. Therefore, 5 is a prime number. Next, cross out all multiples of 5 beyond 5 on the list. For our list, this has already been done. The next uncrossed out number, which is 7, is a prime number by the same argument we used to see that 5 is prime. Cross out the multiples of 7 larger than 7. On our list, this has already been done. Continue in this fashion to the end of the list. The prime numbers in the list are precisely those which have not been crossed out.

Example 4.7. The composite numbers less than 21 are 4, 6, 8, 9, 10, 12, 14, 15, 16, 18, and 20. In the Sieve of Eratosthenes from 1 to 20, these are exactly the numbers that were crossed out.

Example 4.8. If $\mathbf{P}$ denotes the set of all prime numbers and if $\mathbf{C}$ denotes the set of all composite numbers, then $\mathbf{N} = \mathbf{P} \cup \mathbf{C} \cup \{1\}$, because every natural number is either prime, composite, or 1; and $\mathbf{P} \cap \mathbf{C} = \emptyset$, because no natural number is both prime and composite.

Example 4.9 (Even numbers). Numbers for which 2 is a factor are called even numbers. The even numbers are 2, 4, 6, 8, etc., which are the multiples of 2: $2 \cdot 1, 2 \cdot 2, 2 \cdot 3, 2 \cdot 4$, etc., which are all divisible by 2. Any even number is of the form $2 \cdot n$, where n is some natural number.

Example 4.10 (Odd numbers). The odd numbers are all the natural numbers which are not divisible by 2. If n is an odd number, then 2 is not a factor of it, which means that the division $n/2$ must have a nonzero remainder r. But $r < 2$, and therefore, $r = 1$ and the number $n = 2 \cdot q + 1$, where q is the quotient of $n/2$. In summary, odd numbers have the form

$$2 \cdot q + 1$$

where q is any whole number (the case $q = 0$ says 1 is an odd number).

Example 4.11. The sum of an odd number and an even number is always odd. To prove this, suppose a is any odd number and that b is any even

4.1. Factors and Primes   157

number. We have determined that the odd number must have the form $a = 2 \cdot q + 1$ for some whole number q, while the even number must have the form $b = 2 \cdot m$ for some natural number m. Now

$$a + b = (2 \cdot q + 1) + (2 \cdot m) = 2 \cdot (q + m) + 1, \text{ distributive property}$$

which is an odd number, as claimed. For example, $15 = 2 \cdot 7 + 1$ is odd and $38 = 2 \cdot 19$ is even, and

$$15 + 38 = (2 \cdot 7 + 1) + (2 \cdot 19) = 2 \cdot (7 + 19) + 1 = 2 \cdot 26 + 1$$

is an odd number.

Definition 4.12. A *prime factorization* of a number is a factorization for which all factors are prime.

Prime factorizations of 12 and 33 are $12 = 2 \cdot 2 \cdot 3$ and $33 = 3 \cdot 11$. A prime factorization of 7 is 7 by itself, since 7 is a prime number.

Theorem 4.13 (Prime Factorization). *If a is a natural number greater than 1, then a has a prime factorization.*

Proof. If a is a prime number, then a by itself is a prime factorization of it. If a is not a prime number, then it is a composite number. By definition, a composite number a has a factorization for which each factor is smaller than a and larger than 1. If some of the factors are still composite, then they have factorizations for which each factor is smaller than that number. Since the factors keep getting smaller, the factorizations must eventually end, with all factors being prime. Since any factor of a factor is also a factor of the original number, the prime factorizations of the factors give a prime factorization of the original number.

Here is how this proof goes for the number 6370. The possible factors of this number are any natural number from 2 through 6370. A systematic search for its factors begins by checking to see if 2 divides it exactly. If it doesn't, then try 3, and so on through the list until a factor is found. If there were no factor less than 6370, then the number would be prime. In this case, 2 is a factor and we have

$$6370 = 2 \cdot 3185$$

and the other factor, 3185, is smaller than 6370. We know that 2 is prime, so we look for factors of 3185 by the same process, but now there are fewer cases to check. We find that 2 is not a factor of 3185, but 3 is and we have

$$6370 = 2 \cdot 3 \cdot 637$$

We know that 2 and 3 are prime and $637 < 3185$, so its list of potential factors is smaller than the list for 3185. Repeating the process of searching for a factor of 637, we find that 7 is a factor and we have

$$6370 = 2 \cdot 3 \cdot 7 \cdot 91$$

We know that 7 is prime, so we are left with finding the factors of 91, which is smaller than 637. At each step we are faced with finding the factors of a number smaller than the step before. Repeating the process of searching for a factor of 91, we find that 7 is a factor and we have

$$6370 = 2 \cdot 3 \cdot 7 \cdot 7 \cdot 13$$

Now each factor is a prime number, and therefore this is a prime factorization of 6370. The point to observe about this process is that it must terminate, no matter what number we began with, because at each step the numbers to be factored are smaller than those of the preceding step. □

Remark 4.14. This theorem tells us that any number has a prime factorization. This means that in order to find a prime factorization, we need to search only among the prime numbers which are no greater than the number. Actually, we need only to try prime numbers whose square is no greater than the number, as we show in the next proposition.

Proposition 4.15. *Suppose that n is a natural number greater than 1. If n is composite, then n has a prime factor p such that $p^2 \leq n$.*

It could also have a prime factor whose square is larger than n. For example, $119 = 7 \cdot 17$, and 7 is a prime factor with $7^2 < 119$, but the prime factor 17 has square larger than 119.

Proof. If n is not prime, then it must have at least two prime factors, say p and q, where $p \leq q$. Being factors of n, they must satisfy $p \cdot q \leq n$ (the product could be less than n because there could be additional prime factors of n). Then the order property of multiplication says that

$$p \cdot p \leq p \cdot q \leq n$$

Therefore, p is a prime factor of n such that $p^2 \leq n$. □

The contrapositive of Proposition 4.15 is: If n has no prime factor p such that $p^2 \leq n$, then n is prime. The contrapositive is logically equivalent to the original form.

One consequence of this Proposition is that in a Sieve of Eratosthenes for the numbers from 2 through n, we need go through the striking process only for the primes in the set of all prime numbers whose square is no greater than n. For example, in the Sieve made in Example 4.1.1, we could stop after striking the multiples of 2 and 3, because the next prime is 5 whose square is greater than 20.

4.1. Factors and Primes

Example 4.16. Find a prime factorization of 83. The set of primes whose square is no greater than 83 is $A = \{2, 3, 5, 7\}$. We check that 2 is not a factor, 3 is not a factor, 5 is not a factor, and 7 is not a factor. Therefore, 83 is prime according to the contrapositive statement of Proposition 4.15.

Oral exercise 4.17. (1) What is the set of all primes whose square is no greater than the number $n = 123$? Are any of the primes in this set factors of n?

(2) Same questions for $n = 127$.

Oral exercise 4.18. Proposition 4.15 is in the form: "If P, then Q". What is the statement P? What is the statement Q?

The contrapositive form of Proposition 4.15 is "If not Q, then not P". What is the statement not Q (the negation of Q)? What is the statement not P?

In August 2002, Agrawal et al. in [**AKS02**] found the AKS algorithm to determine whether a number is composite or prime. Used on a computer, this algorithm will return a definite answer in a reasonably short time. This result was sufficiently important that it was reported in the *New York Times* of August 8, 2002. The AKS algorithm will tell us that a number is composite, but with no hint as to what the prime factorization is.

The search described in the above proof of Theorem 4.13, modified according to the Remark following it, is an algorithm for finding a prime factorization of a number. It is inefficient, however, to such a degree that even a computer will take months to find the prime factorization of a really big number, say a number whose base ten numeral consists of 300 digits.

Various sophisticated methods for finding a prime factor of a number have been developed in the field of number theory. To date, even the most sophisticated methods need several months to find the prime factorization of a really big number.

Example 4.19. Find the prime factorization of 2773. According to the contrapositive statement of Proposition 4.15, one way to do this is to find the set A of all primes whose square is not greater than 2773. Starting with 2 and working up through the primes, we find

$$A = \{2, 3, 5, 7, 11, 13, 17, 19, 23, 29, 31, 37, 41, 43, 47\}$$

Then we must test which of the primes in A are factors of 2773. We find that only the last one is a factor and that $2773 = 47 \cdot 59$ is the prime factorization. Important applications, such as RSA encryption, are based on the fact that the length of time needed to factor a number grows quickly as the size of the number grows.

4.1.2. Divisibility. The next theorem restates several parts of Theorem 2.89 and adds some slight variations that are useful in this chapter.

Theorem 4.20. *Suppose a, b, and c are whole numbers.*

(1) If d is a factor of a and d is a factor of b, then d is a factor of $a + b$.

(2) If d is a factor of a and if d is not a factor of b, then d is not a factor of $a + b$.

(3) If d is a factor of a and if d is a factor of b and if $a > b$, then d is a factor of $a - b$.

(4) If d is a factor of one of $a > b$, but not the other, then d is not a factor of $a - b$.

(5) If d is a factor of a or d is a factor of b, then d is a factor of $a \cdot b$.

Proof. (1) is a restatement of the first distributive property of division in Theorem 2.89, where it is proved. An example is: 4 is a factor of 12 and 4 is a factor of 16, so 4 is a factor of $12 + 16$.

(2) The statements "d is a factor of a" and "d is not a factor of b" mean that
$$a = q \cdot d \quad \text{and} \quad b = p \cdot d + r$$
for some whole numbers q, p, and r, where $0 < r < d$. Using associative and distributive properties, we see that
$$a + b = q \cdot d + p \cdot d + r = (q + p) \cdot d + r$$
where the remainder satisfying $d > r > 0$ shows that d is not a factor of $a + b$. An example is: 3 is a factor of 12 and 3 is not a factor of 16, so 3 is not a factor of $12 + 16$.

(3) This is the same as the second distributive property of division in Theorem 2.89, where it is proved. An example is: 3 is a factor of 15 and 3 is a factor of 9, so 3 is a factor of $15 - 9$.

(4) This is proved in the same way as (2) above. An example is: 5 is a factor of 20, but not of 12, so 5 is not a factor of $20 - 12$.

(5) This is the same as the first associative property in Theorem 2.89, where it is proved. An example is: 4 is a factor of 12, so 4 is a factor of $12 \cdot 17$. □

Oral exercise 4.21. Six is a factor of 18 and of 24, because
$$18 = 3 \cdot 6 \quad \text{and} \quad 24 = 4 \cdot 6$$
By the distributive property,
$$18 + 24 = 3 \cdot 6 + 4 \cdot 6 = (3 + 4) \cdot 6$$
which shows that 6 is a factor of $18 + 24$. Use similar reasoning in the following problems.

4.1. Factors and Primes

(1) Three is a factor of 9 and 12. Use the distributive property of multiplication over addition to explain why 3 is a factor of $9 + 12$.

(2) Use the distributive property to explain why 3 is not a factor of $10 + 12$.

(3) Use the distributive property to explain why 3 is not a factor of $13 - 9$.

Divisibility has nothing to do with the numeration system being used. For example, 2 divides 6 is a property of the numbers 2 and 6 and of the operation of multiplication. It is true no matter whether one writes these numbers in Hindu-Arabic numeration or Roman numerals. The next theorem gives divisibility tests which depend on the Hindu-Arabic numeration system to base ten. That is, the tests depend on the numeration, even on the base.

Theorem 4.22. *Divisibility tests in Hindu-Arabic numeration base ten.*

(1) A number is divisible by 2 if and only if the digit in its one's place is divisible by 2 (that is, the last digit is 0, 2, 4, 6, or 8).

(2) A number is divisible by 5 if and only if its last digit on the right is 0 or 5.

(3) A number is divisible by 10 if and only if the last digit on the right is 0.

(4) A number is divisible by 4 if and only if the number formed by its last two digits on the right is divisible by 4.

(5) A number is divisible by 3 if and only if the sum of its digits is divisible by 3.

(6) A number is divisible by 9 if and only if the sum of its digits is divisible by 9.

Application of these tests often requires using the *contrapositive* statements. The contrapositive of "If P then Q" is "If not Q then not P", which is logically equivalent. For example, a statement of (1) is: If a number is even, then its last digit is divisible by 2. The contrapositive to this is: If the last digit of a number is not divisible by 2, then the number is not even.

Proof. Suppose a is a natural number. From the base ten numeration system, we know that

$$a = q \cdot 10 + r$$

where $0 \leq r < 10$ and r is the digit of a in the one's place. For example, if $a = 5286$, then $5286 = 528 \cdot 10 + 6$ and if $a = 5287$, then $5287 = 528 \cdot 10 + 7$.

(1) We know that 2 divides 10 and therefore 2 divides any multiple, $q \cdot 10$ of 10. If 2 divides $a = q \cdot 10 + r$, then 2 divides $r = a - q \cdot 10$, by (3) of Theorem 4.20.

Conversely, if 2 divides r, then 2 divides $q \cdot 10 + r$, by (1) of Theorem 4.20.

(2) Since 5 divides 10, the same reasoning as above shows that 5 divides a if and only if 5 divides r. The only digits divisible by 5 are 0 and 5.

(3) Since 10 divides 10, the same reasoning as above shows that 10 divides a if and only if 10 divides r. The only digit divisible by 10 is 0.

(4) The explanations of the first three parts won't work for 4 because 4 does not divide 10. But 4 divides 100 and we can write

$$a = q \cdot 100 + s \cdot 10 + r$$

where s is the digit in the ten's place of a and r is the digit in the one's place of a. As in the above explanations, 4 divides a if and only if 4 divides $s \cdot 10 + r$, which is statement (4).

(5) To reduce the level of abstraction, consider how the argument goes for the number $a = 5286$. We want to see why 3 divides 5286 if and only if 3 divides $5 + 2 + 8 + 6$. The proof uses the fact that if any power of 10 is divided by 3 then the remainder is 1. For example,

(4.1)
$$10 = 3 \cdot 3 + 1$$
$$100 = 33 \cdot 3 + 1$$
$$1000 = 333 \cdot 3 + 1$$

and so on. By base ten numeration we have

$$5286 = 5 \cdot 1000 + 2 \cdot 100 + 8 \cdot 10 + 6$$

Substitute equations (4.1) into this, apply the distributive property, and then apply it again to factor out a 3, to get

$$5286 = 5 \cdot (333 \cdot 3 + 1) + 2 \cdot (33 \cdot 3 + 1) + 8 \cdot (3 \cdot 3 + 1) + 6$$
$$= 5 \cdot 333 \cdot 3 + 5 \cdot 1 + 2 \cdot 33 \cdot 3 + 2 \cdot 1 + 8 \cdot 3 \cdot 3 + 8 \cdot 1 + 6$$
$$= (5 \cdot 333 + 2 \cdot 33 + 8 \cdot 3) \cdot 3 + (5 + 2 + 8 + 6)$$

As in the preceding arguments, we now apply Theorem 4.20 to conclude that 3 is a factor of 5286 if and only if 3 is a factor of $5 + 2 + 8 + 6$.

(6) The argument here is the same as for the preceding part because if any power of 10 is divided by 9 then the remainder is 1. For example,

$$10 = 1 \cdot 9 + 1$$
$$100 = 11 \cdot 9 + 1$$
$$1000 = 111 \cdot 9 + 1$$

□

4.1. Factors and Primes

Example 4.23. Test 81342 for divisibility by 2, 3, 4, 5, 9, and 10. It is divisible by 2 because the digit in the one's place is even. It is not divisible by 4 because 4 is not a factor of 42. The sum of its digits is $8+1+3+4+2 = 18$, which is divisible by 3 and by 9. Therefore, the number is divisible by both 3 and 9. The last digit is not 0 or 5, so the number is neither divisible by 5 nor by 10.

Example 4.24 (Casting out 3's or 9's)**.** In determining whether the sum of the digits is divisible by 3, each partial sum which is a multiple of 3 can be discarded and the sum restarted with the next digit. The end result will be divisible by 3 if and only if the whole sum is divisible by 3. For example, for the number 819251, the process of casting out 3's goes like this. Starting from the left (one could also start from the right) we have: $8+1 = 9$ divisible by 3, so cast it out and begin the sum anew with the next digit, which is 9, already divisible by 3, so cast it out and begin anew at the next digit. $2+5+1 = 8$, which is not divisible by 3. Conclusion: 819251 is not divisible by 3. The same procedure can be used for casting out 9's to determine divisibility by 9.

Oral exercise 4.25. Use casting out nines to test 34293654612 for divisibility by 9.

Example 4.26. Find a prime factorization of 11250. Because the number ends in 0, we know it is divisible by 10 and the other factor is 1125, which we know is divisible by 5. In addition, we know a prime factorization of 10 is $10 = 2 \cdot 5$. The process continues as follows.

$$\begin{aligned} 11250 &= 10 \cdot 1125 = 2 \cdot 5 \cdot 5 \cdot 225 \\ &= 2 \cdot 5 \cdot 5 \cdot 5 \cdot 45 = 2 \cdot 5 \cdot 5 \cdot 5 \cdot 5 \cdot 9 \\ &= 2 \cdot 5 \cdot 5 \cdot 5 \cdot 5 \cdot 3 \cdot 3 \end{aligned}$$

which is a prime factorization of 11250.

Example 4.27. Find a prime factorization of 3191941. Our divisibility tests show that this number is not divisible by 2, 3, or 5. We must check divisibility by each prime less than about 1800. This is hard work if the first prime factor is fairly large. In this case, a prime factorization is $449 \cdot 7109$, and so indeed it would be hard work to find this by hand calculations. Elementary calculators, such as the Explorer Plus, will factor fairly large numbers. Computer software will find prime factorizations of very large numbers. This problem was done using *Find the Prime Factors of a Number* located on the web at http://www.math.wustl.edu/primes.html.

4.1.2.1. *Divisibility tests in base five.* The base five numeral of six is 11_5, the last digit of which is not even. Remember that six is even, no matter what numeration system is being used. For example, six written in Roman numerals is VI, and still six is even. The above *test* for evenness, however, depends on the base ten numeration system. This test is not correct in base five, as the example 11_5 shows. This is not surprising when we remember that the proof of the test used the fact that 2 divides the base, 10. Since 2 does not divide 5, the same argument would not work for base five.

The correct test for evenness in base five is: *A number is even if and only if the sum of the digits of its base five numeral is even.* Notice that the sum of the digits of 11_5 is $1 + 1 = 2$, which is even. The proof of this rule is exactly the same as that used above for the divisibility rules for 3 and 9 in base ten, where the argument used the fact that if any power of ten is divided by 3 or 9 the remainder is 1. In the case at hand, if any power of 5 is divided by 2 then the remainder is 1, because the product of odd numbers is always odd.

4.1.3. The Set of Primes. The Prime Factorization Theorem 4.13 shows that the prime numbers are the building blocks of the natural numbers. Are there an infinite number of prime numbers? In a Sieve of Eratosthenes, every second number past 2 is eliminated, then every third number past 3 is eliminated, and so on. Only the numbers remaining after all this are prime. It seems possible that beyond some point the Sieve will have eliminated every number. Euclid proved that this is not the case.

Theorem 4.28 (Infinite number of primes). *The set of all prime numbers is infinite.*

This is Proposition 20, Book IX of Euclid's *Elements*, see [**Euc56**, Vol. 2, pp. 412-413]. The proof, which applies the Prime Factorization Theorem 4.13, involves some subtle, but elementary, mathematical reasoning, well within the scope of this book. The following proof is the one given by Euclid. The argument represents an intellectual tradition of over 2000 years and continues to intrigue each new generation of students.

Proof. Recall that definition 1.16 defines a set to be infinite if it is not finite, and finite means that the elements can be listed. We will prove that no finite set of prime numbers contains all of the prime numbers. That proves that the set of prime numbers is infinite.

Suppose that **P** is some finite set of prime numbers. To keep the argument concrete, suppose that

$$\mathbf{P} = \{2, 3, 5, 7, 11, 13\}$$

We want to prove, by an argument which illustrates the general case, that **P** does not contain all prime numbers. To do this, consider the number n

obtained by taking the product of all of the numbers in **P** and then adding 1.
$$n = 2 \cdot 3 \cdot 5 \cdot 7 \cdot 11 \cdot 13 + 1$$
None of the numbers in **P** can divide n, by (3) of Theorem 4.20. For example, 3 is in **P**, so 3 divides the product $2 \cdot 3 \cdot 5 \cdot 7 \cdot 11 \cdot 13$. If 3 divides n, then it must divide
$$n - 2 \cdot 3 \cdot 5 \cdot 7 \cdot 11 \cdot 13 = 1$$
but 3 does not divide 1. Therefore, 3 does not divide n. This argument shows that none of the numbers in **P** can divide n. The Prime Factorization Theorem 4.13 says that n must have a prime factor. This prime factor cannot be in the set **P**, because none of the numbers in **P** divides n. Therefore, there is a prime number not in the set **P**. □

Oral exercise 4.29. Does the above proof imply that $2 \cdot 3 \cdot 5 \cdot 7 \cdot 11 \cdot 13 + 1$ a prime number?

Study carefully the way that the Prime Factorization Theorem is used in the above proof. Notice that we have **not** claimed that n is a prime number, although this is a possibility. In fact, for the example chosen to illustrate the proof, the number $n = 30031 = 59 \cdot 509$ is not prime. The proof shows that the prime factors of n are not elements of **P**.

§4.1 Exercises

(1) Use long division to decide if 37 is a factor of 1961. Is 1961 a multiple of 37?

(2) Use long division to decide if 221 is a factor of 10829. What is the largest number not greater than 10829 which is a multiple of 221?

(3) Use a Sieve of Eratosthenes to find all prime numbers less than 100. How many such primes are there? *Ans:* 2, 3, 5, 7, 11, 13, 17, 19, 23, 29, 31, 37, 41, 43, 47, 53, 59, 61, 67, 71, 73, 79, 83, 89, 97. There are 25.

(4) Use a Sieve of Eratosthenes to find all prime numbers between 100 and 200. How many are there? Notice that you need only write down the numbers from 100 to 200 (might as well omit the even ones) and begin by striking out the multiples of the primes less than 100 whose square is less than 200. *Ans:* There are 21.

(5) Find a prime factorization for each of the following numbers n. Find the set A of all prime numbers whose square is less than or equal to n.
  (a) 160 *Ans:* $160 = 2 \cdot 2 \cdot 2 \cdot 2 \cdot 2 \cdot 5$, $A = \{2, 3, 5, 7, 11\}$.
  (b) 330
  (c) 145 *Ans:* $145 = 5 \cdot 29$, $A = \{2, 3, 5, 7, 11\}$.

(d) 91

(e) 97 *Ans:* $97 = 97$, $A = \{2, 3, 5, 7\}$.

(f) 127

(6) Find a prime factorization of 84. Find all factors of 84. *Ans:* $84 = 2 \cdot 2 \cdot 3 \cdot 7$ is a prime factorization. The set of all factors of 84 is

$$\{1, 2, 3, 4, 6, 7, 12, 14, 21, 28, 42, 84\}$$

(7) Find a prime factorization of 90. Find all factors of 90.

(8) Locker Door Problem (from [**KR70**, p. 140]). Imagine a long hallway with 100 lockers in a line along the wall. The door of each locker is closed and the lockers are numbered 1 through 100. At one end of the hall there are 100 people in a line. The first person goes down the hall and opens every locker door. The second person goes down the hall and closes every even numbered door. The third person goes down the hall and changes the status (closes it if it is open, opens it if it is closed) of every door whose number is divisible by three. And so it goes for all 100 people, with the hundredth person changing the status of the door on locker 100. After the hundredth person has passed down the hall, which locker doors are open and which are closed? How would you characterize the door numbers of the open doors? From your answer, explain how you would predict the status of door 121 if there had been 200 doors and 200 people.

(9) Prove: The sum of any two even numbers is even. *Hint:* Use the method of example 4.11. (Prop. 21, Book IX [**Euc56**, p.414, v.2]).

(10) Prove: The difference of any two even numbers is even. That is, if a and b are even numbers and if $a \geq b$, then $a - b$ is even. (Prop. 24, Book IX [**Euc56**, p.414, v.2]).

(11) Prove: The sum of any two odd numbers is even. (Prop. 23, Book IX [**Euc56**, p.414, v.2]).

(12) Prove: The difference of any two odd numbers is even. (Prop. 26, Book IX [**Euc56**, p.415, v.2]).

(13) Prove: The product of any two odd numbers is odd. (Prop. 29, Book IX [**Euc56**, p.416, v.2]).

(14) Prove: If an odd number is subtracted from an even number, then the difference is odd. (Prop. 25, Book IX [**Euc56**, p.415, v.2]).

(15) Prove: The product of an even number by an odd number is even. (Prop. 28, Book IX [**Euc56**, p.416, v.2]).

(16) Prove that the sum, difference, and product of any two multiples of 12 are again multiples of 12. Explain why the same proof works if 12 is replaced by any natural number n. *Hint:* See example 4.11.

§4.1 Exercises

(17) The converse of the theorem "If P then Q" is the theorem "If Q then P", where P and Q are statements of some kind. For example, (1) of Theorem 4.20 is "If d is a factor of a and d is a factor of b, then d is a factor of $a+b$." Here P is the statement "d is a factor of a and d is a factor of b" while Q is the statement "d is a factor of $a+b$". The converse of (1) is the theorem: If d is a factor of $a+b$, then d is a factor of a and d is a factor of b. The converse of (1) is false because there are numbers a, b, and d for which it is false. A set of numbers for which a theorem is false is called a *counterexample* of the theorem. To say that a theorem is false means that there is a counterexample to it.

Find numbers a, b, and d such that d is a factor of $a+b$, but d is not a factor of a or d is not a factor of b.

(18) State the converse to the remaining parts of Theorem 4.20. Is the converse to any of these statements true? Find a counterexample if you think a converse is false; prove it if you think it is true.

(19) Each part of Theorem 4.22 has the form "P if and only if Q". Such a statement is shorthand for the two theorems: "If P then Q" and "If Q then P". Each of these theorems is the converse of the other. State each of the six parts of Theorem 4.22 in the form "If P then Q" and then in the form "If Q then P".

(20) State the contrapositive of each part of Theorem 4.22. Remember, from the preceding exercise, each part consists of two theorems, each of which has a contrapositive form.

(21) If a number is divisible by 9, is it necessarily divisible by 3? Explain. If it is divisible by 3, is it necessarily divisible by 9? Explain.

(22) Test the following numbers for divisibility by 2, 3, 4, 5, 9, and 10.
  (a) 746988. (b) 81342. (c) 15810. (d) 4201012.

(23) Explain why it is true that in base five numeration a number is divisible by 4 if and only if the sum of its digits is divisible by 4. Hint: the argument follows the same lines as that used to prove that in base ten a number is divisible by 9 if and only if the sum of its digits is divisible by 9. The argument is based on the fact that the remainder is 1 when five, or any power of five, is divided by 4. You should include a proof of this last assertion.

(24) Follow the proof of Theorem 4.28 to prove that the set $Q = \{2, 3, 5, 7\}$ does not contain all of the prime numbers. What are the prime factors of $2 \cdot 3 \cdot 5 \cdot 7 + 1$?

(25) Follow the proof of Theorem 4.28 to prove that the set

$$Q = \{2, 3, 5, 7, 11, 13, 17\}$$

does not contain all of the prime numbers. What are the prime factors of 1 plus the product of the numbers in Q?

(26) Explain why there is no largest prime number. That is, prove that if p is a prime number, then there is a prime number larger than p.

4.2. Euclid's Algorithm

This algorithm is the heart of number theory. It is an application of long division.

4.2.1. Greatest Common Factor.

Definition 4.30. A *common factor* of a set of numbers is a number which is a factor of each number in the set.

For example, 3 is a common factor of $\{18, 24, 60\}$. Also, 2 and 6 are each common factors of $\{18, 24, 60\}$. Let's determine all common factors of $\{18, 24, 60\}$. The set of factors of 18 must be a subset of the numbers 1 through 18. Testing each, we find that the set of factors of 18 is

$$\{1, 2, 3, 6, 9, 18\} = \text{all factors of 18}$$

In the same way, we find that the set of all factors of 24 is

$$\{1, 2, 3, 4, 6, 8, 12, 24\} = \text{all factors of 24}$$

and the set of all factors of 60 is

$$\{1, 2, 3, 4, 5, 6, 10, 12, 15, 20, 30, 60\} = \text{all factors of 60}$$

The set of all common factors of $\{18, 24, 60\}$ is the intersection of these three sets, which we determine to be

$$\{1, 2, 3, 6\} = \text{all common factors of 18, 24, and 60}$$

Definition 4.31. The *greatest common factor* of a set of numbers is the greatest number which is a factor of each element of the set. If the greatest common factor is 1, then the numbers of the set are called *relatively prime*. The notation $\gcf(a, b, c)$ means the greatest common factor of $\{a, b, c\}$.

For example, from our description of the set of all common factors of $\{18, 24, 60\}$ we see that

$$\gcf(18, 24, 60) = \max(\{1, 2, 3, 6\}) = 6$$

Oral exercise 4.32. (1) Find a common factor of 27 and 24? Find their gcf.

(2) Find a common factor of 36 and 48? Find their gcf.

(3) Find a common factor of 72 and 96? Find their gcf.

4.2. Euclid's Algorithm

Proposition 2 of Book VII of Euclid's *Elements*, [**Euc56**, Vol. 2, pp. 298-300], gives a method for finding the greatest common factor of any two numbers. This method is called Euclid's Algorithm.

Theorem 4.33 (Euclid's Algorithm). *To find the greatest common factor of two numbers, divide the larger number by the smaller. Divide the previous divisor by the remainder and continue dividing each successive remainder into the previous divisor, until the remainder becomes 0. This last divisor is the greatest common factor of the numbers.*

Proof. The following proof is Euclid's. The way to understand why these procedures work is to study the principles of the proof in several examples. As a representative example, consider how the method goes to find the greatest common factor of 91 and 52.

Step 1. Divide the larger by the smaller, 91 by 52 in this case, to get a quotient of 1 and remainder 39. Write the result as

$$(4.2) \qquad 91 = 1 \cdot 52 + 39$$

Step 2. Divide the divisor of the first step by the remainder. In this case, divide 52 by the remainder 39, to get a quotient of 1 and remainder 13. Write the result as

$$(4.3) \qquad 52 = 1 \cdot 39 + 13$$

Step 3. Divide the divisor of the preceding step by the remainder. In this case, divide 39 by the remainder 13, to get a quotient of 3 and remainder 0. Write the result as

$$(4.4) \qquad 39 = 3 \cdot 13$$

Because the remainder is 0, the greatest common factor of 91 and 52 is 13, the divisor in the last step. Let's see why. By equation (4.4), 13 is a factor of 39. Then equation (4.3) shows that 13 is a factor of 52. Then equation (4.2) shows that 13 is a factor of 91. Thus 13 is a common factor of 52 and 91.

To show that 13 is the greatest common factor of 52 and 91, we shall see that any other common factor divides 13, and therefore is no larger than 13. For this purpose, suppose d is any common factor of 52 and 91. Then equation (4.2) shows that d is a factor of 39. Then equation (4.3) shows that d is a factor of 13. Hence, $d \leq 13$, and 13 is the greatest common factor of 52 and 91. $\square$

Example 4.34. The above proof also shows that there are multiples of 52 and of 91 whose difference is their gcf. More precisely, it shows that there exist whole numbers x and y such that

$$(4.5) \qquad x \cdot 52 - y \cdot 91 = \gcd(52, 91)$$

In fact, solve for the gcf, which is 13, in equation (4.3) to get
$$13 = 52 - 1 \cdot 39 = 52 - 39$$
Into this substitute the value of 39 obtained from equation (4.2) to get
$$13 = 52 - (91 - 52) = (52 + 52) - 91 = 2 \cdot 52 - 1 \cdot 91$$
by the distributive property of multiplication over subtraction and then the third subtraction associative property from (4) of Theorem 2.33. This is the desired result of Equation (4.5) with $x = 2$ and $y = 1$.

Example 4.35. Here is another illustration of why Euclid's Algorithm finds the gcf. Find the gcf(336, 812). The results of the divisions can be written as

$$(4.6) \quad \begin{aligned} 812 \div 336 &\Rightarrow 812 = 2 \cdot 336 + 140 \\ 336 \div 140 &\Rightarrow 336 = 2 \cdot 140 + 56 \\ 140 \div 56 &\Rightarrow 140 = 2 \cdot 56 + 28 \\ 56 \div 28 &\Rightarrow 56 = 2 \cdot 28 \end{aligned}$$

Since division by 28 left no remainder, we conclude that $28 = \text{gcf}(336, 812)$, which we verify as follows. Reading these equations from the bottom to the top, we see that 28 divides 56 and therefore 28 divides 140, by (1) of Theorem 4.20. The next equation up shows that 28 must then divide 336, and then the top equation shows that 28 also divides 812. Therefore, 28 is a common factor of 336 and 812. If d is any common factor of 336 and 812, then the top equation shows that d divides 140, and then the next equation shows that d divides 56 and the next equation shows that d divides 28. Therefore, $d \leq 28$ and we conclude that 28 is the greatest common factor of 336 and 812.

As in the preceding example, we can also see that the difference of some multiples of 336 and 812 is equal to 28. In fact, solve for 28 in the third equation in (4.6) and then use the second and first equations as follows.

$$\begin{aligned} 28 &= 140 - 2 \cdot 56 = 140 - 2 \cdot (336 - 2 \cdot 140) = 140 - (2 \cdot 336 - 2 \cdot 2 \cdot 140) \\ &= 140 - (2 \cdot 336 - 4 \cdot 140) = (140 + 4 \cdot 140) - 2 \cdot 336 \\ &= 5 \cdot 140 - 2 \cdot 336 = 5 \cdot (812 - 2 \cdot 336) - 2 \cdot 336 \\ &= (5 \cdot 812 - 10 \cdot 336) - 2 \cdot 336 = 5 \cdot 812 - (10 + 2) \cdot 336 \\ &= 5 \cdot 812 - 12 \cdot 336 \end{aligned}$$

where subtraction associative properties (4) of Theorem 2.33 are used at each step. Therefore, $28 = 5 \cdot 812 - 12 \cdot 336$.

Corollary 4.36. *If n is a common factor of the natural numbers a and b, then n is a factor of gcf(a, b).*

4.2. Euclid's Algorithm

Proof. The proof of this corollary is contained in the proof of Euclid's Algorithm, where we demonstrated that any common factor of a and b is a factor of the gcf found by the algorithm. $\square$

Corollary 4.37. *To find the greatest common factor of more than two numbers, first find the greatest common factor of two of them, then of this factor and one of the remaining numbers, and so on to the last. The last greatest common factor is the greatest common factor of all the numbers.*

Proof. The general case is illustrated in the case of three numbers $\{a, b, c\}$. If $d = \gcf(a, b)$, then any common factor of a and b is a factor of d, by Corollary 4.36. Therefore, any common factor of $\{a, b, c\}$ is a common factor of $\{d, c\}$. Of course, any factor of d is a common factor of a and b. Therefore, any common factor of d and c is a common factor of $\{a, b, c\}$. It follows that the set of common factors of $\{a, b, c\}$ equals the set of common factors of $\{d, c\}$, so the greatest number in this set is the $\gcf(a, b, c)$ and the $\gcf(d, c)$. Hence, $\gcf(a, b, c) = \gcf(d, c)$, which is what we wished to prove. $\square$

Corollary 4.38. *If a and b are natural numbers and if $d = \gcf(a, b)$, then there are natural numbers x and y such that*

$$(4.7) \qquad x \cdot a - y \cdot b = d \text{ or } x \cdot b - y \cdot a = d$$

Proof. The proof has already been illustrated in the two preceding examples. Here is another example worked out to exhibit the general proof. The divisions of Euclid's Algorithm in finding the $\gcf(407, 1067) = 11$ are written as follows.

$$1067 = 2 \cdot 407 + 253$$
$$407 = 1 \cdot 253 + 154$$
$$253 = 1 \cdot 154 + 99$$
$$154 = 1 \cdot 99 + 55$$
$$99 = 1 \cdot 55 + 44$$
$$55 = 1 \cdot 44 + 11$$
$$44 = 4 \cdot 11$$

Solve for 11 in the second from the last equation and work upward, using subtraction associative properties (4) Theorem 2.33, as follows.

$$11 = 55 - 44 = 55 - (99 - 55) = (55 + 55) - 99 = 2 \cdot 55 - 99$$
$$= 2 \cdot (154 - 99) - 99 = (2 \cdot 154 - 2 \cdot 99) - 99 = 2 \cdot 154 - 3 \cdot 99$$
$$= 2 \cdot 154 - 3 \cdot (253 - 154) = 5 \cdot 154 - 3 \cdot 253$$
$$= 5 \cdot (407 - 253) - 3 \cdot 253 = 5 \cdot 407 - 8 \cdot 253$$
$$= 5 \cdot 407 - 8 \cdot (1067 - 2 \cdot 407) = 21 \cdot 407 - 8 \cdot 1067$$

Therefore, $11 = x \cdot 407 - y \cdot 1067$ where $x = 21$ and $y = 8$. $\square$

Definition 4.39. If gcf$(a, b) = 1$, then we say that the natural numbers a and b are *relatively prime*.

Theorem 4.40. *If a and b are natural numbers whose greatest common factor is d, then the quotients a/d and b/d are relatively prime.*

Proof. The general idea of the proof can be illustrated in the example of $a = 273$ and $b = 357$. Their greatest common factor can be found by Euclid's Algorithm to be $d = 21$. Then the quotients $273/21 = 13$ and $357/21 = 17$. We want to show that gcf$(13, 17) = 1$, in a way that illustrates the general argument.

By Corollary 4.38, either $21 = x \cdot 273 - y \cdot 357$ or $21 = x \cdot 357 - y \cdot 273$, for some natural numbers x and y. Suppose the first case occurs. If the second case occurs, a minor modification of the following argument still applies. Dividing both sides of $21 = x \cdot 273 - y \cdot 357$ by 21, and using the first division associative property and the second division distributive property, in Theorem 2.89, we get

$$1 = 21/21 = (x \cdot 273)/21 - (y \cdot 357)/21 = x \cdot 13 - y \cdot 17$$

Hence, if p is a common factor of $13 = 273/21$ and $17 = 357/21$, then it must divide

$$x \cdot 13 - y \cdot 17 = 1$$

by the distributive property over subtraction (5) of Theorem 2.89. Therefore, $p = 1$ and we conclude that gcf$(13, 17) = 1$, as was to be proved. □

4.2.2. Least Common Multiple. Review the definition of multiple of a number given in Definition 2.44.

The multiples of 5 are 5, 10, 15, 20, etc. The multiples of 5 are listed by counting by 5. The set of all multiples of 5 is

$$\{5, 10, 15, \ldots\} = \{5 \cdot n : n \in \mathbf{N}\} = \text{set of all multiples of 5}$$

The set of all multiples of 5 is the set of all numbers for which the remainder is 0 when they are divided by 5.

The set of all multiples of a natural number is always infinite, whereas the set of all factors of a natural number is always finite.

Oral exercise 4.41. Explain why the sum, difference, or product of two multiples of 3 is again a multiple of 3. Is this true for multiples of 7 or of 8?

Oral exercise 4.42. Explain why even numbers are the multiples of 2. Explain why odd numbers are whole numbers which are 1 less than some even number.

4.2. Euclid's Algorithm

Oral exercise 4.43. (1) How many numbers are between two consecutive multiples of 3?

(2) Let A be the set of all natural numbers for which the remainder is 1 when they are divided by 3. Let B be the set of all natural numbers for which the remainder is 2 when they are divided by 3. List a few elements of A. List a few elements of B.

(3) Explain why the sum of two elements of A is an element of B.

(4) Is the sum of two elements of B an element of A? Explain.

Theorem 4.44. *If d is a factor of the natural number a, then d is a factor of any multiple of a.*

Proof. This follows from Remark 4.4, because a is a factor of any of its multiples. $\square$

Definition 4.45. A *common multiple* of a set of natural numbers is a number that is a multiple of each element of the set.

A common multiple of a set of natural numbers is any number for which each element of the set is a factor. Notice that the product of all the numbers in the set is a common multiple of the set.

A common multiple of $\{36, 18, 12\}$ is 72, because each element in the set is a factor of 72. In fact, $72 = 2 \cdot 36$, $72 = 4 \cdot 18$, and $72 = 6 \cdot 12$. The product $36 \cdot 18 \cdot 12 = 7776$ is also a common multiple of this set.

Corollary 4.46. *A common factor of a set of numbers is a factor of any common multiple of the set of numbers.*

Proof. A common factor of the set of numbers is a factor of each number in the set, thus a factor of any multiple, by Theorem 4.44. In particular, it is a factor of any common multiple. $\square$

In the above example, 6 is a common factor of the set of numbers $\{36, 18, 12\}$ and it is a factor of the common multiple $72 = 6 \cdot 12$.

Definition 4.47. The *least common multiple* of a set of natural numbers is the smallest of the common multiples of the set of numbers. Denote the least common multiple of a set of numbers, say of $\{a, b, c\}$, by $\operatorname{lcm}(a, b, c)$.

The product of a set of numbers is a common multiple of these numbers. Therefore, the least common multiple of a set of numbers must be no greater than their product.

Let us determine the least common multiple of $\{6, 10\}$, that is, determine the $\operatorname{lcm}(6, 10)$. The set of multiples of 6 is

$$\{6, 12, 18, 24, 30, 36, 42, 48, 54, 60, \dots\} = \{6 \cdot n : n \in \mathbf{N}\}$$

and the set of multiples of 10 is
$$\{10, 20, 30, 40, 50, 60, 70, 80, 90, \dots\} = \{10 \cdot n : n \in \mathbf{N}\}$$
The common multiples of $\{6, 10\}$ are those numbers which appear in both sets, which we observe to be
$$\{30, 60, \dots\} = \{30 \cdot n : n \in \mathbf{N}\}$$
which is the intersection of the sets of multiples of each number in $\{6, 10\}$. The least common multiple of $\{6, 10\}$ is the smallest number in the intersection, which is 30. This intersection can be pictured on the number line as follows.

```
          10        20        30        40        50        60
|++++++++|++++++++|++++++++|++++++++|++++++++|++++++++|
0    6    12    18    24   30    36   42    48   54   60
```

This picture illustrates the result of equation (4.7) in Corollary 4.38, once we calculate the gcf$(6, 10) = 2$. In the picture we can see where multiples of 6 and 10 differ by their gcf and therefore give solutions x and y to Equations (4.7). For example
$$2 = 2 \cdot 6 - 1 \cdot 10$$
$$2 = 2 \cdot 10 - 3 \cdot 6$$
$$2 = 7 \cdot 6 - 4 \cdot 10$$
$$2 = 5 \cdot 10 - 8 \cdot 6$$
The first of these equations would be the one found by using Euclid's Algorithm as in the proof of Corollary 4.38.

Oral exercise 4.48. List some multiples of 3? of 10? of 12? Describe these sets of multiples in set language.

Oral exercise 4.49. List some common multiples of 3 and 6? Of 6 and 8? Of 5 and 7?

Oral exercise 4.50. What is the set of common multiples of 3 and 6? Of 6 and 8? Of 5 and 7? What is the least common multiple of 3 and 6? Of 6 and 8? Of 5 and 7?

We may interpret Corollary 4.38 as saying that outside of the common multiples, the minimum distance between the multiples of a and the multiples of b is the gcf(a, b). For example, the gcf$(6, 10) = 2$ and an equation predicted by the corollary is $2 \cdot 6 - 1 \cdot 10 = 2$. Write multiples of 6 and 10, respectively, as $m \cdot 6$ and $n \cdot 10$, for some natural numbers m and n and suppose that $m \cdot 6 > n \cdot 10$. Then 2 is a factor of both multiples, so 2 is a factor of the difference $m \cdot 6 - n \cdot 10$, which means that 2 is less than or equal to this difference, as claimed. On the above number line picture of

4.2. Euclid's Algorithm

some multiples of 6 and 10, observe that outside of the common multiples, they are never closer together than 2.

The following theorem gives us a method for finding the lcm, because we already have a method for finding the gcf.

Theorem 4.51. *If a and b are any whole numbers, then*
$$a \cdot b = \text{gcf}(a,b) \cdot \text{lcm}(a,b)$$

Proof. The proof is an application of Euclid's Algorithm. What may seem to be a simpler proof is given as Corollary 4.60 in the next section.

Let $g = \text{gcf}(a,b)$ and let $l = \text{lcm}(a,b)$. Then
$$a = g \cdot m \quad \text{and} \quad b = g \cdot n$$
for some natural numbers m and n and
$$a \cdot p = l = b \cdot q$$
for some natural numbers p and q. Combining these,
$$g \cdot m \cdot p = a \cdot p = l = b \cdot q = g \cdot n \cdot q$$
implies that g can be factored from each end to give
$$m \cdot p = n \cdot q$$
Since $a \cdot n$ is a multiple of a and $b \cdot m$ is a multiple of b,
$$a \cdot n = g \cdot m \cdot n = b \cdot m$$
is a common multiple of a and b. If we show that $a \cdot n$ is a factor of l, then $a \cdot n$ must equal l, because $l = \text{lcm}(a,b)$.

Now $\text{gcf}(m,n) = 1$ by Theorem 4.40, so
$$1 = x \cdot m - y \cdot n \quad \text{or} \quad 1 = x \cdot n - y \cdot m$$
for some natural numbers x and y, by Corollary 4.38. For the sake of argument, suppose the first case occurs and multiply this equation by p to get
$$p = p \cdot 1 = x \cdot p \cdot m - y \cdot p \cdot n \ = x \cdot n \cdot q - y \cdot p \cdot n = (x \cdot q - y \cdot p) \cdot n$$
Multiply this by a to get
$$l = a \cdot p = (x \cdot q - y \cdot p) \cdot (a \cdot n)$$
which shows that $a \cdot n$ is a factor of l and thus $a \cdot n = l$, as mentioned above. Hence,
$$a \cdot b = a \cdot g \cdot n = g \cdot a \cdot n = g \cdot l$$
which is what we wanted to prove, since $g = \text{gcf}(a,b)$ and $l = \text{lcm}(a,b)$. □

One consequence of this theorem is that the lcm of two numbers can be found by first finding the gcf, for example by using Euclid's Algorithm, and then finding the lcm as the product of the two numbers divided by the gcf,

$$\text{lcm}(a,b) = \frac{a \cdot b}{\gcf(a,b)}$$

For example, from Euclid's Algorithm, $\gcf(308, 630) = 14$, and therefore

$$\text{lcm}(308, 630) = \frac{308 \cdot 630}{14} = 308 \cdot (630/14) = 308 \cdot 45 = 13860$$

§4.2 Exercises

(1) Use Euclid's Algorithm to find the greatest common factor of the given set of numbers. In each case, follow the proof of Euclid's Algorithm to demonstrate that the number you have found is a common factor and that any common factor divides it.
  (a) 308 and 630.
  (b) 252 and 1134.
  (c) 112, 140, and 560.
  (d) 4718, 6951, and 8876.

(2) Newell has three fields, containing 24 acres, 18 acres, and 42 acres. He wishes to cut them into smaller fields of an equal number of acres each. How large can the fields be? Explain your answer and how you arrived at it. *Ans.* 6 acres.

(3) Nicholas, Joshua, and Caitlin living on a new street own land fronting as follows: Nicholas 600 ft.; Joshua 720 ft.; Caitlin 900 ft. They wish to cut their land into lots of an equal front width. How wide can the lots be and how many will each have? Explain your solution to this problem and explain how you do the calculations. *Ans.* 60 ft. Nicholas 10 lots; Joshua 12 lots; Caitlin 15 lots.

(4) Morris has 4 tracts of land containing 420 acres, 350 acres, 490 acres, and 560 acres, respectively. He wishes to divide them into farms of an equal number of acres each. How large can the farms be, and how many of these will there be? Explain your answer and how you arrived at it. *Ans.* 70 acres; 26 farms.

(5) Ragna has a lot with the following dimensions: 450 ft. long on the north side, 540 ft. on the south side, 420 ft. on the east side, and 495 ft. on the west side. What is the greatest length of her panels of fence if they are equal, and how many rails and posts will be required to fence it if the fence is 5 rails high? Note: there is a post at each end of a panel

4.3. Unique Prime Factorization

and a panel consists of horizontal boards called rails. *Ans.* Panels 15 ft., 635 rails and 127 posts.

(6) Consider the set of numbers which are 1 more than a multiple of 3. List a few elements of this set. Is the product of two numbers from this set again an element of this set? Explain why your answer is correct. Consider the set of numbers which are 2 more than a multiple of 3. What is true about the product of two numbers from this set? Prove your claim.

(7) Use Euclid's Algorithm to find the gcf(12, 30). Use Theorem 4.51 to find the lcm(12, 30). *Ans:* gcf = 6, lcm = $\frac{12 \cdot 30}{6} = 12 \cdot 5 = 60$.

(8) Use Euclid's Algorithm to find the gcf(190, 798). Use Theorem 4.51 to find the lcm(190, 798).

(9) (a) Use Euclid's Algorithm to find the gcf of 2805 and 6783.
  (b) Use Theorem 4.51 to find the lcm of these two numbers.
  (c) Find a multiple of 2805 and a multiple of 6783 whose difference is the gcf(2805, 6783).

(10) On a number line from 0 to 36, mark the multiples of 6 in one color and the multiples of 9 in another color. Find the gcf and lcm of $\{6, 9\}$. Corollary 4.38 says that there exist natural numbers x and y such that $x \cdot 9 - y \cdot 6 = \gcd(6, 9)$, but there are many solutions. From your number line picture, find three different pairs of numbers x, y for which this is true.

(11) Find the gcf and lcm of $\{33, 35\}$. Find a multiple of 33 and a multiple of 35 which differ by the gcf.

(12) Cameron and Stephanie meet on a Monday afternoon and enjoy themselves so much that they decide to meet every 3 days. How many days until they meet again on a Monday?

(13) Suppose that a certain variety of pin oak has an extra abundant crop of acorns every 7 years and that a squirrel population which feeds on these acorns peaks every 5 years. If the extra abundant crop and the population peak both occurred in the year 2000, what is the next year when they will both occur? What are two successive years for which the extra abundant crop occurs in one of the years and the population peak occurs in the other? Relate this last question to Corollary 4.38.

4.3. Unique Prime Factorization

It was easy to prove the existence of a prime factorization, but to prove the uniqueness of such a factorization another idea is needed. Actually, a few examples might cast doubt on its truth. For example, $44 = 2 \cdot 2 \cdot 11$ is close to $45 = 3 \cdot 3 \cdot 5$. A small adjustment in the factorization of 44 obtained by increasing a 2 to 3 and decreasing the 11 to 7 gives $2 \cdot 3 \cdot 7 = 42$, which doesn't quite equal 44, but perhaps a bit more persistence in this vein would find a number with two different prime factorizations.

Proof of the uniqueness of the prime factorization requires the following application of Euclid's Algorithm.

Theorem 4.52. *If p is a prime number and if a and b are natural numbers such that p is a factor of $a \cdot b$, then p is a factor of a or p is a factor of b. Consequently, if p divides a product of primes, then it must equal one of those primes.*

The truth of this result requires that the factor p be prime. A composite factor of a product need not be a factor of either term of the product. For example, 6 is a factor of $3 \cdot 4$, but 6 is neither a factor of 3 nor of 4.

Proof. Suppose that p is not a factor of a. Then $\gcd(p, a) = 1$, since 1 and p are the only factors of the prime number p, and thus 1 is the only common factor of $\{p, a\}$. Then

$$x \cdot p - y \cdot a = 1 \quad \text{or} \quad x \cdot a - y \cdot p = 1$$

for some natural numbers x and y, by Corollary 4.38. For the sake of argument suppose that the first case occurs. The argument is similar for the other case. Multiply the equation by b to get

$$x \cdot p \cdot b - y \cdot a \cdot b = b$$

Then p is a factor of $x \cdot p \cdot b$ and p is a factor of $y \cdot a \cdot b$ (since p is a factor of $a \cdot b$) and therefore p is a factor of b, by (3) of Theorem 4.20.

By the same kind of argument, if p is not a factor of b, then we can conclude that p is a factor of a.

Suppose that the prime number p is a factor of a product of primes. We want to prove that p must be one of these primes. We illustrate the argument with the case where the product of primes is $2 \cdot 3 \cdot 5$. By the associative property of multiplication, we can group this product as $2 \cdot (3 \cdot 5)$. We have already proved that if p is a factor of $2 \cdot (3 \cdot 5)$, then p is a factor of 2 or p is a factor of $3 \cdot 5$. If p is a factor of 2, then $p = 2$, since the prime $p \neq 1$ and the only factors of 2 are 1 and 2. If p is a factor of $3 \cdot 5$, then we have already proved that p is a factor of 3 or p is a factor of 5. If p is a factor of 3, then $p = 3$, because the only factors of the prime number 3 are 1 and 3. If p is a factor of 5, then $p = 5$, because the only factors of 5 are 1 and 5. Therefore, p must be one of the primes in $\{2, 3, 5\}$. □

The following theorem is sometimes called the Fundamental Theorem of Arithmetic.

4.3. Unique Prime Factorization

Theorem 4.53 (Unique Prime Factorization). *If n is a natural number greater than 1, then it has exactly one prime factorization, up to the order of the factors.*

The uniqueness part of this theorem is Proposition 14 of Book IX of Euclid's *Elements*; see [**Euc56**, p.402, v.2].

Proof. In Theorem 4.13 we proved that n has a prime factorization. Suppose a number n has two prime factorizations, which we refer to as the first factorization and the second factorization. To keep the argument concrete, let's go through it with the number 6370. A prime factorization of 6370 is $6370 = 2 \cdot 5 \cdot 7 \cdot 7 \cdot 13$. We are supposing that 6370 has another prime factorization which we shall refer to as the second factorization.

The prime number 2 occurs in the first factorization, so 2 must divide our number 6370, which means that it divides the second factorization. By Theorem 4.52, 2 must equal one of the primes in the second factorization. Cancel this prime out of both factorizations. The resulting products are still equal, where the first product is now $5 \cdot 7 \cdot 7 \cdot 13$. Continue to the next prime, 5, in the first factorization. Since 5 divides the product it is in, it must also divide the product remaining of the second factorization. Cancel 5 from both products, so that the first becomes $7 \cdot 7 \cdot 13$. Continue this argument with each prime in the first factorization until eventually all that is left of the first factorization is 1 (after 13 is cancelled from both). But then, the product of the remaining primes in the second factorization must be 1, which means that there cannot be any primes left in the second factorization. Therefore, the second factorization contained exactly the primes of the first factorization and no more. Up to order of the factors, the two factorizations were the same. □

Definition 4.54. The *set of prime factors* of a number is the set of its prime factors, each occuring as many times as it occurs in the prime factorization of the number. The number of times a prime occurs is called its *multiplicity*.

We agree that the set of prime factors of a prime number consists of just the number itself. It has multiplicity one.

Notice how uniqueness of the prime factorization allows us to say *the* set of prime factors rather than *a* set of prime factors.

Example 4.55. A prime factorization of 11250 was found in Example 4.26. The prime factors of 11250 are 2 (once), 3 (two times), and 5 (four times). The set of prime factors is $\{2, 3, 3, 5, 5, 5, 5\}$. The prime factor 2 has multiplicity 1, 3 has multiplicity 2, and 5 has multiplicity 4.

Theorem 4.56. *If $d > 1$ is a factor of a, then the set of prime factors of d is a subset of the set of prime factors of a.*

If any subset of the set of prime factors of a be taken, then the product of the elements of this subset is a factor of a.

For example, 12 is a factor of 60 and the set of prime factors of 12 is $\{2, 2, 3\}$, which is a subset of $\{2, 2, 3, 5\}$, which is the set of prime factors of 60.

The product of the primes in any subset of $\{2, 2, 3, 5\}$ is a factor of 60. In fact, this product times the product of the elements of the complement of the subset is 60. For example, if the subset is $\{2, 5\}$, then $2 \cdot 5 = 10$ is a factor of 60 and the complement of this set is $\{2, 3\}$, the product of whose elements is $2 \cdot 3 = 6$ and we have $10 \cdot 6 = 2 \cdot 5 \cdot 2 \cdot 3 = 60$, as claimed.

Proof. If $d > 1$ is a factor of a, then $a = q \cdot d$, for some natural number q. If p is a prime factor of d, then $d = r \cdot p$, for some natural number r, and therefore $a = q \cdot r \cdot p$, which shows that p is a prime factor of a. This shows that any prime factor of d is a prime factor of a, that is, the set of prime factors of d is a subset of the set of prime factors of a.

The second statement is true because if d is the product of a subset of prime factors of a, then the quotient $a \div d$ is just the product of the remaining prime factors of a. That is, d is a divisor of a. $\square$

4.3.1. Using prime factors to find the gcf.

Theorem 4.57. *The product of the common prime factors of two or more numbers is the greatest common factor of those numbers. If they have no common prime factors, then they are relatively prime.*

Proof. The first statement is true because any factor, other than 1, of a number must be a product of some of the prime factors of the number, by uniqueness of the number's prime factorization. For example, consider the greatest common factor of 198 and 42. The prime factorizations are $198 = 2 \cdot 3 \cdot 3 \cdot 11$ and $42 = 2 \cdot 3 \cdot 7$. Any factor of 198 must be a product of some subset of the primes in $A = \{2, 3, 3, 11\}$, the set of prime factors of 198. Any factor of 42 must be a product of some subset of the primes in $B = \{2, 3, 7\}$, the set of prime factors of 42. Therefore, the greatest common factor of 198 and 42 must be the product of the primes which lie in both A and B, namely in $A \cap B = \{2, 3\}$, and this number is $2 \cdot 3 = 6$. $\square$

Example 4.58. Find the greatest common factor of 30, 40, and 60. To do this, first find the prime factorization of each number.

$$30 = 2 \cdot 3 \cdot 5, \quad 40 = 2 \cdot 2 \cdot 2 \cdot 5, \quad 60 = 2 \cdot 2 \cdot 3 \cdot 5$$

The common prime factors are thus 2 and 5. The greatest common factor is the product of the common prime factors, namely $2 \cdot 5 = 10$. The only common factors, besides 1, are 2, 5 and 10, all of which divide 10.

4.3. Unique Prime Factorization

In the following theorem, notice that the converse of each statement is also true, and obvious.

4.3.2. Using prime factors to find the lcm.

Theorem 4.59. *1. If m is a multiple of a number, then the prime factorization of m contains every prime factor of the number. For example, 12 is a multiple of 4 and the prime factorization of 12 is $2 \cdot 2 \cdot 3$, which contains $2 \cdot 2$, the prime factorization of 4.*

2. If m is a common multiple of two or more numbers, then the prime factorization of m contains all the prime factors of each of those numbers.

3. If m is the least common multiple of two or more numbers, then the prime factorization of m contains all the prime factors of each of those numbers, and no other prime factors.

Proof. The proof of all three statements is based on an application of the uniqueness of the prime factorization of a number. A few examples should illustrate how the arguments go.

1. The number $420 = 35 \cdot 12$ is a multiple of the number 12. The prime factorization of 420 is $2 \cdot 2 \cdot 3 \cdot 5 \cdot 7$, which contains all the prime factors of 12, whose prime factorization is $12 = 2 \cdot 2 \cdot 3$. In fact, notice that the prime factorization of 420 is just the prime factorization of 35 times the prime factorization of 12, so of course it contains the prime factors of 12.

2. Consider the numbers 12 and 33. Their prime factorizations are $12 = 2 \cdot 2 \cdot 3$ and $33 = 3 \cdot 11$. Any multiple of 12 must contain the prime factors of 12, which are 2, 2, and 3, by Part 1. For the same reason, any multiple of 33 must contain the prime factors of 33, which are 3 and 11. Therefore, any common multiple of 12 and 33 must contain the prime factors of both numbers, therefore must contain the primes 2, 2, 3, and 11. The least common multiple of 12 and 33 will contain only these prime factors, and therefore must be $2 \cdot 2 \cdot 3 \cdot 11 = 132$.

3. Consider the numbers 16, 21, and 24. Their prime factorizations are

(4.8) $\qquad 16 = 2 \cdot 2 \cdot 2 \cdot 2, \qquad 21 = 3 \cdot 7, \qquad 24 = 2 \cdot 2 \cdot 2 \cdot 3$

Any multiple of 16 contains its prime factors, which are 2, 2, 2, and 2. Any multiple of 21 contains its prime factors, which are 3 and 7. Any multiple of 24 contains its prime factors, which are 2, 2, 2, and 3. Any common multiple of 16, 21, and 24 must contain the prime factors of each of these numbers, therefore must contain 2, 2, 2, 2, 3, and 7. The least common multiple must contain only these prime factors, therefore is $2 \cdot 2 \cdot 2 \cdot 2 \cdot 3 \cdot 7 = 336$. □

The next corollary is a restatement of Theorem 4.51, but with a simpler proof based on the uniqueness of prime factorizations.

Corollary 4.60. *If a and b are natural numbers, then*

$$\text{lcm}(a,b) = \frac{a \cdot b}{\text{gcf}(a,b)}$$

Proof. By Theorem 4.57, if g is the greatest common factor of a and b, then g is the product of the common prime factors of a and b. Therefore, a/g and b have no common prime factors, so $\frac{a \cdot b}{g}$ contains all the prime factors of a and b, and no other prime factors. This means that $\frac{a \cdot b}{g}$ is the least common multiple of a and b, by statement 3 of Theorem 4.59. □

4.3.3. Simplified Method to find the lcm. Let us review the procedure followed in the above proof of statement 3 of Theorem 4.59 to find the lcm of $\{16, 21, 24\}$. We selected prime factors from those of the three numbers in such a way that the resulting set contains the prime factors of each number and nothing more. To do this, line up the three sets horizontally and select out primes which appear in at least two of the sets, writing these to the left and striking them out from the given sets. The primes thus written on the left together with those remaining in the given sets then constitute the primes needed in the prime factorization of the lcm. We illustrate this process in several steps, each step being the selection of a prime appearing in at least two of the sets. The prime factorizations of 16, 21, and 24 are given in equation (4.8) above.

Common primes	16	21	24	
$\{2\}$	$\{\cancel{2},2,2,2\}$	$\{3,7\}$	$\{\cancel{2},2,2,3\}$	A common 2
$\{2\} \cup \{2\}$	$\{\cancel{2},2,2\}$	$\{3,7\}$	$\{\cancel{2},2,3\}$	A common 2
$\{2,2\} \cup \{2\}$	$\{\cancel{2},2\}$	$\{3,7\}$	$\{\cancel{2},3\}$	A common 2
$\{2,2,2\} \cup \{3\}$	$\{2\}$	$\{\cancel{3},7\}$	$\{\cancel{3}\}$	A common 3
$\{2,2,2,3\}$	$\{2\}$	$\{7\}$	$\emptyset$	No common primes

The set of prime factors of $\text{lcm}(16, 21, 24)$ is then $\{2, 2, 2, 3, 2, 7\}$, which means that the lcm is $2 \cdot 2 \cdot 2 \cdot 3 \cdot 2 \cdot 7 = 336$.

The Simplified Method rewrites this procedure as follows. Write the numbers in a horizontal row. Find a *prime* factor common to at least two of the numbers and write it at the left. Rewrite the horizontal row after cancelling the common prime factor out of those numbers containing it. Continue until no two numbers in the horizontal row contain a common prime factor. The prime factors of the lcm will then be the primes listed in the left column together with the prime factors of the numbers remaining

4.3. Unique Prime Factorization

in the horizontal row. The procedure looks like this.

$$
\begin{array}{c|ccc l}
2 & 16 & 21 & 24 & \text{2 a common prime factor of 16 and 24} \\
2 & 8 & 21 & 12 & \text{16/2 and 24/2. 2 again a common prime factor} \\
2 & 4 & 21 & 6 & \text{8/2 and 12/2. 2 again a common prime factor} \\
3 & 2 & 21 & 3 & \text{4/2 and 6/2. 3 a common prime factor} \\
 & 2 & 7 & 1 & \text{21/3 and 3/3. No more common prime factors}
\end{array}
$$

The lcm is then $2 \cdot 2 \cdot 2 \cdot 3 \cdot 2 \cdot 7 = 336$. Compare this with what was done above.

Example 4.61. Find the lcm of $\{27, 36, 45, 90\}$. Write the numbers in a horizontal line. To the left write a common prime factor of two or more of the numbers. On the next line write the quotients of those numbers divisible by the factor, or just rewrite the number otherwise. Continue until the resulting numbers no longer have any common prime factors.

$$
\begin{array}{c|cccc l}
3 & 27 & 36 & 45 & 90 & \text{3 a common prime factor of all 4 numbers} \\
3 & 9 & 12 & 15 & 30 & \text{Quotients. 3 a common factor of all 4 numbers} \\
5 & 3 & 4 & 5 & 10 & \text{Quotients. 5 a common factor of 5 and 10} \\
2 & 3 & 4 & 1 & 2 & \text{Quotients. 2 a common factor of 4 and 2} \\
 & 3 & 2 & 1 & 1 & \text{Quotients. All pairs relatively prime}
\end{array}
$$

The least common multiple of $\{27, 36, 45, 90\}$ is, therefore, $3 \cdot 3 \cdot 5 \cdot 2 \cdot 3 \cdot 2 = 540$. The reason this works is the same as that shown in the preceding example. To see it in detail, write out the set of prime factors of 27, 36, 45, and 90 and repeat the process illustrated above.

4.3.4. Method to find the gcf and lcm of two numbers. Find the gcf and lcm of 308 and 630. Use the above Simplified Method. At each step find a common prime factor and write it to the left. On the next line write the quotients, which are the numbers with the common factor cancelled out. Continue until the two remaining quotients are relatively prime. The gcf will then be the product of the numbers written in the left column. The lcm will be the product of this with the numbers in the last row.

(4.9)
$$
\begin{array}{c|cc l}
2 & 308 & 630 & \text{2 a common factor} \\
7 & 154 & 315 & \text{Quotients. 7 a common factor} \\
 & 22 & 45 & \text{Quotients. Relatively prime}
\end{array}
$$

The gcf is $2 \cdot 7 = 14$ and the lcm is $2 \cdot 7 \cdot 22 \cdot 45 = 13860$. Here is another view of why the Simplified Method applied to only two numbers gives both the gcf and the lcm. The prime factorizations of our numbers are $308 = 2 \cdot 2 \cdot 7 \cdot 11$ and $630 = 2 \cdot 3 \cdot 3 \cdot 5 \cdot 7$. The set of prime factors of the gcf is

$$\{2, 2, 7, 11\} \cap \{2, 3, 3, 5, 7\} = \{2, 7\}$$

while the set of prime factors of the lcm is the smallest set of primes containing $\{2,2,7,11\}$ and $\{2,3,3,5,7\}$, which is the complement of the intersection in the union of these two sets

$$\{2,2,3,3,5,7,11\}$$

The Simplified Method produces these two sets as follows.

(4.10)
$$\begin{array}{c|ccl}
\{2\} & \{\not{2},2,7,11\} & \{\not{2},3,3,5,7\} & \text{Remove common 2} \\
\{7\} & \{2,\not{7},11\} & \{3,3,5,\not{7}\} & \text{Remove common 7} \\
 & \{2,11\} & \{3,3,5\} & \text{No more common primes}
\end{array}$$

The set of primes in the left column, $\{2,7\}$, is the intersection of the two given sets, and therefore is the set of prime factors of the gcf. The union of this set with the two sets remaining in the bottom row is the set of prime factors of the lcm, because the resulting set contains all the prime factors of 308 and 630 and nothing more. The duplicate primes have been removed and written only once in the left column.

4.3.5. Finding all factors. Theorems 4.13 and 4.56 give us a method for finding all factors of a number, and for counting the number of factors. Consider the problem of finding all factors of 315. Begin by finding the prime factorization of 315, which is $315 = 3 \cdot 3 \cdot 5 \cdot 7$. The set of prime factors of 315 is $\{3, 3, 5, 7\}$. By Theorem 4.56, the set of prime factors of any factor of 315 must be a subset of this set of prime factors. Distinct factors come from distinct subsets, which are formed by choosing the number of 3's, the number of 5's and the number of 7's. There are three possible choices for the number of 3's (none, one, or two), two possible choices for the number of 5's (none or one) and two possible choices for the number of 7's (none or one).

4.3. Unique Prime Factorization

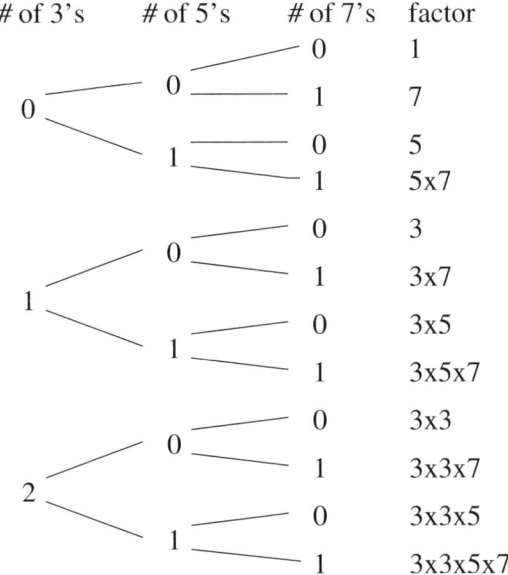

Therefore, the number of distinct subsets of $\{3, 3, 5, 7\}$ is the product of the number of choices, $3 \cdot 2 \cdot 2 = 12$, where the choice of no 3's, no 5's, and no 7's results in the empty set. By convention, we say that the choice of the empty set corresponds to the factor 1.

The total number of factors of 315 is obtained by adding 1 to each multiplicity and then taking the product of the resulting numbers. The 1 must be added to each multiplicity because we allow the choice of 0 of a given prime. This rule is true in general.

4.3.6. Cancellation. Divisions can be carried out by first factoring the divisor and dividend. Repeated application of Theorem 2.89 allows division of common prime factors such as p/p, which is 1 and thus *cancels* out. For example, consider $(6 \cdot 7 \cdot 5 \cdot 3) \div (3 \cdot 6 \cdot 5) = (6/6) \cdot 7 \cdot (5/5) \cdot (3/3) = 7$. To carry this out, write the division as

$$\frac{\not{6} \cdot 7 \cdot \not{5} \cdot \not{3}}{\not{3} \cdot \not{6} \cdot \not{5}} = 7 \tag{4.11}$$

and then strike out any prime appearing both above and below the line, as shown.

Example 4.62. Find $5775 \div 175$. The prime factorizations are $5775 = 3 \cdot 5 \cdot 5 \cdot 7 \cdot 11$ and $175 = 5 \cdot 5 \cdot 7$, so that

$$5775 \div 175 = \frac{3 \cdot 5 \cdot 5 \cdot 7 \cdot 11}{5 \cdot 5 \cdot 7} = 3 \cdot 11 = 33 \tag{4.12}$$

§4.3 Exercises

(1) For each of the following numbers: Find the prime factorization. Use the method described in Subsection 4.3.5 to find all factors. Explain why the number of factors is given by adding 1 to each multiplicity and then taking the product of these numbers.
  (a) 60 *Answer:* 12 factors.
  (b) 330
  (c) 325

(2) Find the prime factorization of your date of birth, written as the number month/day/year. Write the day as a two digit number, possibly beginning with 0. For example, March 19, 1941, would be 3191941, while June 1, 1937, would be 6011937. Do this by hand, or use the Find the Prime Factors of a Number program located at the web site cited above.

(3) Find the greatest common factor of the given numbers by first finding their prime factorizations. Explain why your method works.
  (a) 24 and 35.
  (b) 84, 126, and 210.
  (c) 108, 270, and 432.
  (d) 360, 288, 720, and 648.

(4) Use the Simplified Method to find the lcm and gcf of 600 and 3150.

(5) Find the prime factorizations of 2805 and 6783 and then use the Simplified Method to find their gcf and lcm.

(6) Use cancellation to solve the following.
  (a) Divide $3 \cdot 13 \cdot 14$ by 42.
  (b) Divide $15 \cdot 3 \cdot 84$ by $45 \cdot 12$.
  (c) Divide 2954 by 14.
  (d) How many dozen of eggs at 30 cents a dozen will pay for 6 yards of cashmere at $2.10 a yard?

(7) Write the proof of Theorem 4.52 for the case $p = 7$, $a = 11$, and $b = 14$. What are x and y in this case?

(8) Can a prime number d divide a product of primes all of which are different from d? Explain.

(9) Use the prime factorizations of 8, 16, 21, and 28 to find their lcm. Next find the lcm of these numbers by the Simplified Method. Explain why the method works.

§4.3 Exercises

(10) (a) Explain how the Unique Prime Factorization Theorem 4.53 allows you to conclude, without making any calculations, that the product $2 \cdot 2 \cdot 2 \cdot 3 \cdot 5$ is not equal to the product $2 \cdot 3 \cdot 3 \cdot 7$.
(b) Does the Unique Prime Factorization Theorem 4.53 allow you to conclude that the product $19 \cdot 8383$ is not equal to the product $1577 \cdot 101$? Why?

(11) Use the Simplified Method to find the lcm and gcf of 600 and 3150. Verify that lcm · gcf = $600 \cdot 3150$.

(12) What is the smallest tract of land that may be divided into equal fields of 12 acres each, or of 15 acres each, or of 18 acres each? *Ans.* 180 acres.

(13) The Delmar bus goes by the library every 20 minutes, the Olive bus every 22 minutes, the Lucas bus every 15 minutes, and the Shuttle every 18 minutes. If all start from the library together, how long will it be till they next all arrive at the library at the same time, and how many times will each have gone over its route?

(14) Suppose that there is heavy snow and spring rain at the headwaters of the Missouri River every 12 years, at the headwaters of the Mississippi River every 6 years, in the Missouri River valley between Nebraska and Iowa every 10 years, and in the Mississippi valley between Iowa and Illinois every 8 years. How often would you expect severe flooding in St. Louis?

Chapter 5

Rational Numbers

Fractions were defined in Definition 2.104 as a quantity of some object. A *fraction* is an expression of the form $\frac{p}{q}$, where p and q are whole numbers and $q > 0$. The fraction $\frac{p}{q}$ of an object is the amount obtained by dividing the object into q equal parts and then taking p of these parts.

The goal of this chapter is to interpret fractions as numbers, to be called rational numbers, in such a way that they contain the whole numbers. We want to *extend* the four operations of arithmetic of whole numbers to all rational numbers. In doing this we want to preserve all the properties these operations have in whole number arithmetic. The definition of these operations should be clearly related to the operations in whole numbers. Actual computation should involve only whole number computations.

Since we have adopted the philosophy of the number line, our task is to identify fractions with points on the number line. This is accomplished by identifying a fraction with the amount of number line obtained by taking that fraction of the unit line segment. This simple idea is complicated somewhat by the fact that different fractions represent the same rational number. This issue will be investigated in the first two sections.

Once the rational numbers have been identified with points on the number line, we can use the number line interpretation of addition to define addition of rational numbers. Subtraction is then defined to be the inverse of addition in the same way we defined it for whole numbers. Multiplication turns out to be the same as the process of our Definition 2.104 of fraction, and then division is the inverse of multiplication in the same way as it was defined for whole numbers. Rational numbers are represented by fractions, so we have to understand how to carry out the arithmetic operations with the representing fractions.

5.1. Representation by Fractions

Review Definition 1.2.1 of the number line, and the terminology that goes with it. Recall that the line segment from 0 to 1 is called the unit line segment, and it is the same length as the segment from any whole number to its successor. Any whole number n is the number of elements in the set of unit line segments from 0 to n. Recall also the procedure described in Section 2.5 for dividing a line segment into any given number of equal subsegments.

In Definition 2.104 we defined the fraction $\frac{p}{q}$ of an object to be p of the equal parts when the object is divided into q equal parts. We now want to make fractions a part of the number system. We begin this process by assigning each fraction to a certain point on the number line.

Definition 5.1. The fraction $\frac{p}{q}$ represents the point on the number line arrived at by dividing the unit interval into q equal parts and then going p of these parts to the right from 0. This point is called the *value* of the fraction. A *rational number* is the value of some fraction.

The denominator shows into how many equal parts the unit interval is divided. The numerator shows how many of these equal parts are taken. The denominator denotes the subinterval, which is the *building block* of that fraction. The ordinal number term of the denominator is the name of this subinterval. For example, if the denominator is 5, then the unit interval is divided into 5 equal subintervals, each of which is called a *fifth*. Then $\frac{3}{5}$ means 3 fifths, that is, three of these subintervals called fifths. The value of $\frac{3}{5}$ is the point reached going 3 fifths to the right from 0, as shown on the following number line.

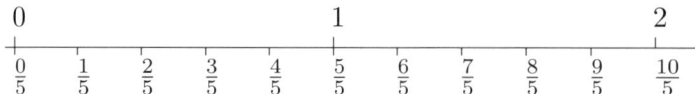

Because a fifth is the result of dividing the unit interval into 5 equal parts, it follows that 5 of these fifths exactly fills up a unit interval. In this way we see that the value of 5 fifths is the whole number 1, as shown on the above number line. As 2 groups of 5 is $2 \cdot 5 = 10$, we see that the value of $\frac{10}{5}$ is 2, and in the same way the value of $\frac{n \cdot 5}{5}$ is n, for any whole number n.

Oral exercise 5.2. (1) How many fifths in 1 unit interval? How many fifths in 2 unit intervals?

5.1. Representation by Fractions

(2) How many halves in 2 unit intervals? How many thirds? How many seventeenths?

(3) How many halves in 3 unit intervals? How many sixths?

(4) How many thirds in 4 unit intervals? In 6 unit intervals?

(5) How many fourths in 5 unit intervals? In 27/4?

Example 5.3. The value of 1/2 is the point arrived at by dividing the unit interval into 2 equal parts, each part called a half, and going one of these halves from 0.

The value of 7/4 is the point attained by dividing the unit interval into 4 equal parts, each part called a fourth, and going 7 of these fourths from 0.

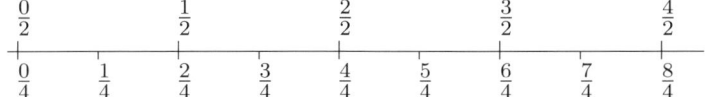

The value of $\frac{2}{4}$ is the point obtained by dividing the unit interval into 4 equal parts, called fourths, and going 2 of them from 0. A look at the above number line indicates that the point arrived at is the same as the value of 1/2. This is true for the following reason. Four fourths make the unit interval, and dividing this set of four fourths into 2 equal parts gives $4 \div 2 = 2$ fourths in each part. Then 2 of these parts is 2 groups of 2, which is 4 fourths, which is the unit interval. Therefore, 2 fourths divides the unit interval into 2 equal parts, which shows that 2 fourths exactly make 1 half. Consequently, $\frac{2}{4}$ and $\frac{1}{2}$ have the same value.

A practical way to draw the unit interval divided into, say, 5 equal parts is to draw the line, mark the 0 point, then mark equal spaces to the right of 0 and let 1 be the point at the end of the fifth such mark. Now each such mark will be one part of dividing the unit into 5 equal parts.

The preceding example shows that different fractions can have the same value. When fractions such as 2/4 and 1/2 have the same value, it is common practice to write $2/4 = 1/2$. This seems natural enough, but we must keep in mind that the equality expressed is of the value, not of the fractions.

This insistence on distinguishing between a fraction and its value serves an important purpose during the beginning stages of learning fractions. Eventually, the distinction between a fraction and its value drops to the background. In several contexts, such as in addition or order relationships, the distinction is important. We add values in addition, but the fractions representing the values must sometimes be replaced by other fractions of the same value in order to carry out the calculations.

Sometimes fractions must be replaced by fractions of the same value in order to plot several fractions on the number line at the same time.

Example 5.4. Suppose we want to plot the values of $\frac{2}{3}$ and $\frac{3}{4}$ on the same number line. To plot $\frac{2}{3}$, we would divide the unit interval into 3 equal parts and go 2 of these thirds from 0. To plot $\frac{3}{4}$, we would divide the unit interval into 4 equal parts and go 3 of these fourths from 0.

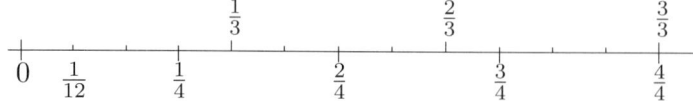

We can relate thirds and fourths by dividing each third into 4 equal parts. There are $3 \cdot 4 = 12$ of these parts in the unit interval, since there are 3 thirds in the unit interval and 3 groups of 4 is $3 \cdot 4$. Thus, each part is a $\frac{1}{12}$ of the unit interval and we have $\frac{4}{12} = \frac{1}{3}$. Divide the set of 12 twelfths into 4 equal parts to get $12 \div 4 = 3$ twelfths in each part. Then 4 of these $\frac{3}{12}$ make the unit interval, which means that $\frac{3}{12} = \frac{1}{4}$.

The examples above show that the whole numbers are also rational numbers. A whole number p is the value of the fraction $\frac{p}{1}$; it is also the value of the fractions $\frac{2 \cdot p}{2}$, $\frac{3 \cdot p}{3}$, etc. The set of whole numbers is a proper subset of the set of rational numbers.

Oral exercise 5.5. (1) Explain why the value of $\frac{3}{1}$ is 3.

(2) Explain why $\frac{2}{3}$ and $\frac{4}{6}$ have the same value. Hint: Follow the argument given in examples 5.3 and 5.4.

(3) Explain why $\frac{2 \cdot 3}{2}$ and $\frac{3}{1}$ have the same value.

5.1.1. Improper and Mixed Fractions. If the numerator is smaller than the denominator, the fraction is called *proper*. Otherwise, it is called *improper*. For example, 3/5 and 27/81 are proper fractions, while 7/4, 85/7, and 3/3 are improper. A *mixed fraction* is a whole number followed by a fraction. For example, $2\frac{3}{5}$ and $5\frac{2}{3}$, even $0\frac{7}{8}$, are mixed fractions. We need to define the value of a mixed fraction.

Definition 5.6. If a is a whole number and r/d is a fraction, then the mixed fraction $a\frac{r}{d}$ represents the point on the number line reached by dividing the unit interval into d equal parts and going r of these parts to the right from a. The point represented by $a\frac{c}{d}$ is called the value of $a\frac{c}{d}$.

Example 5.7. The mixed fraction $2\frac{3}{5}$ represents the point reached by dividing the unit into 5 equal parts, called fifths, and going 3 of these fifths to the right from 2. Since 5 fifths make a unit and 2 units are required to go from 0 to 2, it takes $2 \cdot 5 = 10$ of these fifths to go from 0 to 2 and then another 3 fifths to the right of 2 requires a total of $2 \cdot 5 + 3$ fifths from 0 to the point represented by $2\frac{3}{5}$. Therefore, this point is also represented by $\frac{2 \cdot 5 + 3}{5}$. The value of $2\frac{3}{5}$ is equal to the value of $\frac{2 \cdot 5 + 3}{5}$, as shown below.

5.1. Representation by Fractions

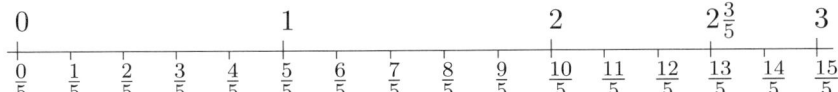

Example 5.8. The fraction $\frac{11}{4}$ represents the point reached by dividing the unit interval into 4 equal parts, called fourths, and going 11 of these fourths to the right from 0. Since 4 fourths make the unit interval, the number of unit intervals in 11 fourths is determined by the long division $11 \div 4$. Writing the quotient and remainder as $11 = 4 \cdot 2 + 3$, we see that 11 fourths is 2 unit intervals with 3 fourths remaining. The point reached is therefore 3 fourths to the right of 2, which is the point represented by the mixed fraction $2\frac{3}{4}$. In summary, the fraction $\frac{11}{4}$ has the same value as the mixed fraction $2\frac{3}{4}$.

The preceding two examples illustrate the proof of the following theorem.

Theorem 5.9. *If $\frac{n}{d}$ is a fraction and if the long division $n \div d$ has quotient q and remainder r so that $n = q \cdot d + r$, then the value of $\frac{n}{d}$ is the same as the value of the mixed fraction $q\frac{r}{d}$. Conversely, if $q\frac{r}{d}$ is a mixed fraction, then it has the same value as the fraction $\frac{q \cdot d + r}{d}$.*

Proof. We illustrate the proof of the first statement with the fraction $\frac{17}{3}$ (that is, $n = 17$ and $d = 3$). Then $17 \div 3$ has quotient 5 and remainder 2, so that $17 = 3 \cdot 5 + 2$. The mixed fraction $5\frac{2}{3}$ represents the point reached by dividing the unit interval into 3 equal subintervals, called thirds, and going 2 of these thirds to the right of 5. On the other hand, there are 3 thirds in a unit interval and 5 unit intervals from 0 to 5, which means that there are $5 \cdot 3$ thirds from 0 to 5, and going 2 more thirds is a total of $5 \cdot 3 + 2 = 17$ thirds from 0. Therefore, $5\frac{2}{3}$ and $\frac{17}{3}$ represent the same point on the number line, which means that they have the same value.

We illustrate the proof of the converse with the mixed fraction $7\frac{3}{5}$, which represents the point reached by dividing the unit interval into 5 equal parts, called fifths, and going 3 fifths to the right from 7. But there are $7 \cdot 5$ fifths from 0 to 7, and $7 \cdot 5 + 3$ fifths from 0 to the point represented by $7\frac{3}{5}$. Therefore, $(7 \cdot 5 + 3)/5$ has the same value as $7\frac{3}{5}$.

The general argument goes in the same way. □

Following our policy of writing equality between fractions to mean that they have the same value, we will write equality between a fraction and a mixed fraction to mean that they have the same value. The preceding

theorem says that if $n = d \cdot q + r$, then
$$\frac{n}{d} = q\frac{r}{d}$$
How many ones in $\frac{14}{4}$? The number of ones in $\frac{14}{4}$ is the quotient of $14 \div 4$, which is 3. The remainder in this long division is 2. What is the meaning of this remainder? It is the number of fourths remaining. It is not enough to make another one, because the remainder in a long division is always smaller than the divisor. We conclude that in $\frac{14}{4}$ there are 3 ones and 2 fourths, which agrees with the statement that $\frac{14}{4}$ has the same value as the mixed fraction $3\frac{2}{4}$, that is, $\frac{14}{4} = 3\frac{2}{4}$, as pictured:

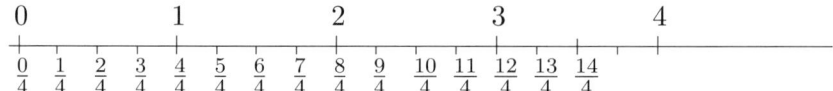

Oral exercise 5.10. (1) How many ones in 9/3? In 48/6? In 81/9?

(2) How many ones in 8/3? Express 8/3 as a mixed fraction.

(3) Express 104/7 as a mixed fraction.

(4) Express $3\frac{5}{7}$ as an improper fraction. Same for $7\frac{1}{7}$.

5.1.2. Applying Fractions. We have defined the value of fractions in terms of dividing the unit interval into a given number of equal subintervals. This serves to simplify the discussion and to relate fractions to an extension of the set of whole numbers called the positive rational numbers. To understand how different fractions can have the same value, we divided these subintervals into a given number of equal subintervals. In order to apply fractions to a wider range of problems, we need to see that any object can be substituted for line segments in this discussion. We can divide into a given number of equal parts any object, such as a pizza, the miles in a journey, the length of a day, the duration of a job, a plot of land, a glass of juice, or a set of elements.

The quantity contained in a given fraction of an object is the value of the fraction with the object's name as a label on the quantity. For example, $\frac{3}{5}$ of a pizza, $\frac{6}{10}$ of a pizza, or $\frac{9}{15}$ of a pizza are all the same amount of pizza, since these fractions all have the same value.

If the object under consideration is a set of n elements, then the set is the whole and each element is the result of dividing the whole into n equal parts. Thus, each element is $\frac{1}{n}$ of the set. If in a classroom of 20 children, 4 of them have red hair, then the fraction of children in this classroom with red hair is $\frac{4}{20}$, because each child is $\frac{1}{20}$ of the classroom, so 4 of these parts is $\frac{4}{20}$. If half the children in this classroom are boys, how many are boys? Half of the whole is the result of dividing it into 2 equal parts. Dividing 20

5.1. Representation by Fractions

children into 2 equal parts then means that each part is half of the children in the classroom, and this is $20 \div 2 = 10$ children in each part. This shows that $\frac{1}{2}$ the children in the classroom is the same number as $\frac{10}{20}$ the children in the classroom. In general, if fractions of the same value are taken of a whole, the resulting amounts are the same.

Example 5.11. When Gavin died he left $\frac{1}{2}$ of his money to his wife and divided the other $\frac{1}{2}$ equally among his 3 children. What fraction of his money did each child receive? This is a question about fractions. If half of a unit interval is divided into 3 equal subintervals, then 2 halves will contain $2 \cdot 3 = 6$ of these subintervals, and this is the whole unit interval. Therefore, these subintervals are sixths. We conclude that each child received $\frac{1}{6}$ of Gavin's money.

Gavin's money
$$\underbrace{\hspace{3cm}}_{\frac{1}{2}} \; \underbrace{\hspace{1cm}}_{\frac{1}{6}} \; \underbrace{\hspace{1cm}}_{\frac{1}{6}} \; \underbrace{\hspace{1cm}}_{\frac{1}{6}}$$
wife — children

Suppose that Gavin had $30,000 when he died. How much money did each child receive?

Method 1. We determined that each child received $\frac{1}{6}$ of his money, which is the amount in each part if $30,000 were divided into 6 equal parts. This is $30000 \div 6 = 5000$ in each part. Each child received $5,000.

Method 2. The wife received $\frac{1}{2}$ of $30,000, which is $30000 \div 2 = 15000$ dollars. This leaves $30000 - 15000 = 15000$ for the children. Each child receives $\frac{1}{3}$ of this, which is $15000 \div 3 = 5000$ dollars.

Example 5.12. Joshua has 60 cents and Caitlin has $\frac{4}{5}$ as many. How many cents has Caitlin?

Joshua's 60 cents $\underbrace{\overbrace{\hspace{2cm}}^{?}}_{12}$

5 equal parts is 60 cents.

1 part is $60 \div 5 = 12$ cents.

4 parts is $4 \cdot 12 = 48$ cents. $\frac{4}{5}$ of 60 cents is 48 cents.

Oral exercise 5.13. Margaret has 60 cents and Martha has $\frac{5}{4}$ as many. How many cents has Martha?

Oral exercise 5.14. Morris had 600 bushels of wheat and sold $\frac{3}{10}$ of it. How many bushels did he sell?

Oral exercise 5.15. In a classroom of 24 children, 9 of them are girls. What fraction are girls?

Oral exercise 5.16. A quarter of a cake is divided into 3 equal parts. What fraction of a cake is in each part?

§5.1 Exercises

(1) Describe the value of $\frac{4}{7}$. Of $\frac{3}{2}$. Of $\frac{5}{8}$. Of $\frac{12}{15}$. For each fraction draw a number line to illustrate its value.

(2) State and prove Theorem 5.9 for the case of $1\frac{3}{5}$ and $\frac{8}{5}$. Draw a number line to illustrate the argument.

(3) State and prove Theorem 5.9 for the fraction $\frac{59}{7}$ and write the proof for this case. Draw a number line to illustrate the argument.

(4) State and prove Theorem 5.9 for the mixed fraction $4\frac{3}{10}$ and write the proof for this case. Draw a number line to illustrate the argument.

(5) What fraction of a pound is 1 ounce? What fraction of a foot is 3 inches? What fraction of a mile is 100 yards?

(6) How many ounces are in $\frac{3}{8}$ of a pound? How many inches in $\frac{2}{3}$ of a foot? How many feet in $\frac{1}{3}$ of a yard? How many feet in $\frac{1}{10}$ of a mile?

(7) Niels gave a third of his pizza to Leah, who gave half of that to Anthony. What fraction of Niels's pizza did Anthony receive? Illustrate with a disk and with a line segment.

(8) It is 10 miles to Ragna's house and we have gone $\frac{4}{5}$ of the way. How many miles have we gone?

(9) David spent $\frac{3}{5}$ of his money on a book of cattle brands registered in Montana. The book cost $24. How much money did he have at first?
Hint: Use the method of Example 5.12.

(10) George has driven 32 miles, which is $\frac{4}{7}$ of the way. How many miles in the whole way?

5.2. Comparing Fractions

Knowing that different fractions can have the same value, we need a simple way to determine when two fractions have the same value, or to determine which one has the larger value. We also want a simple description of all the fractions with the same value as some given fraction.

5.2.1. Fractions of the Same Value. We saw in examples 5.3 and 5.4 that different fractions can have the same value. In example 5.3 we saw that $1/2 = 2/4$. The second fraction is obtained from the first by multiplying the numerator and denominator each by 2, as in $\frac{1 \cdot 2}{2 \cdot 2} = \frac{2}{4}$. The next theorem explains why multiplying numerator and denominator of a fraction by any natural number always produces a fraction of the same value.

5.2. Comparing Fractions

Theorem 5.17. *If n is a natural number, then the fractions $\frac{p}{q}$ and $\frac{p \cdot n}{q \cdot n}$ have the same value, that is, $\frac{p}{q} = \frac{p \cdot n}{q \cdot n}$.*

Some are tempted to say that the proof is just $\frac{p \cdot n}{q \cdot n} = \frac{p}{q} \cdot \frac{n}{n} = \frac{p}{q} \cdot 1 = \frac{p}{q}$. Such an argument is putting the cart before the horse, because it uses multiplication of fractions, something not defined yet. The result of this theorem is very basic, and can be understood directly from the definition of the value of a fraction together with the definition of multiplication of whole numbers.

Proof. The proof can be illustrated by the case of $\frac{4}{7}$ and $\frac{4 \cdot 3}{7 \cdot 3}$. The value of $\frac{4}{7}$ is the point obtained by dividing the unit interval into 7 equal parts, each called a seventh, and going 4 of these sevenths from 0.

Now divide a seventh into 3 equal parts. As it takes 7 sevenths to go from 0 to 1, and each seventh contains 3 of the new parts, it will take $7 \cdot 3$ of the new parts to go from 0 to 1. Therefore, the new parts are what you get when you divide the unit interval into $7 \cdot 3 = 21$ equal parts, called twenty-firsts. Since 3 of these twenty-firsts make 1 seventh, the value of $\frac{4}{7}$ is the point reached by going 4 sevenths from 0, which is $4 \cdot 3$ twenty-firsts from 0, which is the value of $\frac{4 \cdot 3}{7 \cdot 3}$. Therefore, $\frac{4}{7} = \frac{4 \cdot 3}{7 \cdot 3}$.

Here is the general explanation. The value of p/q is the point obtained by dividing the unit interval into q equal parts, called $q$ths, and going p of these $q$ths from 0. Now divide a qth into n equal parts. As it takes q $q$ths to go from 0 to 1, and each qth contains n of the new parts, it will take $q \cdot n$ of the new parts to go from 0 to 1. Therefore, the new parts are what you get when you divide the unit interval into $q \cdot n$ equal parts, and n of these parts makes one qth. Now, the value of p/q is the point reached by going p $q$ths from 0, which is $p \cdot n$ of these new parts from 0, which is the value of $\frac{p \cdot n}{q \cdot n}$. Therefore, $\frac{p}{q} = \frac{p \cdot n}{q \cdot n}$. □

Example 5.18. $3/5 = 6/10 = 126/210$, since $6/10 = (3 \cdot 2)/(5 \cdot 2)$ and $126/210 = (6 \cdot 21)/(10 \cdot 21)$.

Corollary 5.19. *If p and q are both divisible by the natural number n, then $\frac{p \div n}{q \div n} = \frac{p}{q}$.*

Proof. By the Theorem, $\frac{p \div n}{q \div n} = \frac{(p \div n) \cdot n}{(q \div n) \cdot n}$, and this last fraction is $\frac{p}{q}$, because $(p \div n) \cdot n = p$ and $(q \div n) \cdot n = q$, by definition of long division. □

Definition 5.20 (Lowest Terms). The process in Corollary 5.19 of dividing out a common factor from the numerator and denominator is called *reducing the terms of the fraction by a common factor*. A fraction is *expressed in lowest terms* if its numerator and denominator are relatively prime.

Recall that Theorem 4.40 says that if $d = \gcd(p, q)$, then
$$\gcd(p \div d, q \div d) = 1$$
This means that in order to express a fraction in lowest terms, divide numerator and denominator by their greatest common factor, that is, reduce it by the greatest common factor of its numerator and denominator.

Oral exercise 5.21. (1) How many sixths in 2/3? SOLUTION. Since $6 = 2 \cdot 3$, we have $\frac{2}{3} = \frac{2 \cdot 2}{2 \cdot 3} = \frac{4}{6}$, which means there are 4 sixths in 2 thirds.

(2) How many eighths in 1/2? In 3/4?

(3) How many twelfths in 1/4? In 2/3?

(4) Express 3/5 in twentieths. In thirtieths.

(5) Express $\frac{8}{12}$ in lowest terms.

(6) Express $\frac{10}{16}$ in lowest terms.

(7) Express $\frac{27}{15}$ in lowest terms.

5.2.2. Common Denominator. It is easy to compare the values of two fractions which have the same denominators. For example, 2/5 and 3/5 have different values, because going 2 fifths from 0 is not as far as going 3 fifths from 0.

Theorem 5.22. $\frac{a}{d} = \frac{b}{d}$ *if and only if* $a = b$

Proof. a/d represents the point obtained by dividing the unit interval into d equal parts and going a of them from 0, while b/d represents the point obtained by going b of these parts from 0. The same point is reached if and only if the number of parts is the same, namely, if and only if $a = b$. □

Example 5.23 (Cross multiplication). In the light of this theorem, a strategy for comparing the values of two fractions, such as 5/9 and 11/16, whose denominators are different, is to use Theorem 5.17 to find fractions with the same value as these and with a common denominator. This is accomplished by *cross multiplication*, which means to multiply numerator and denominator of one fraction by the denominator of the other, and vice-versa. After doing this, both fractions will have the same denominator, which will be the product of the original denominators.

$$\frac{5}{9} = \frac{5 \cdot 16}{9 \cdot 16} = \frac{80}{144} \quad \text{and} \quad \frac{11}{16} = \frac{9 \cdot 11}{9 \cdot 16} = \frac{99}{144}$$

which shows that these fractions have different values, with the value of 11/16 lying to the right of the value of 5/9 on the number line, since 99 > 80 means that 99 of the one-hundred-forty-fourths goes farther than 80 of them. Notice that the new denominators are both the product of the original

5.2. Comparing Fractions

two denominators, and therefore are the same. To express more than two fractions with a common denominator, multiply both terms of each fraction by all the denominators except its own. For example, to express 4/5, 2/7, and 1/3 in a common denominator, calculate

$$\frac{4}{5} = \frac{4 \cdot 7 \cdot 3}{5 \cdot 7 \cdot 3} = \frac{84}{105}$$
$$\frac{2}{7} = \frac{5 \cdot 2 \cdot 3}{5 \cdot 7 \cdot 3} = \frac{30}{105}$$
$$\frac{1}{3} = \frac{5 \cdot 7 \cdot 1}{5 \cdot 7 \cdot 3} = \frac{35}{105}$$

The common denominator is the product of the given denominators, $5 \cdot 7 \cdot 3 = 105$.

Oral exercise 5.24. Express in a common denominator, by which we mean find fractions of the same values as those given so that the new fractions have a common denominator.

(1) 1/2 and 2/3.

(2) 2/5 and 3/4.

(3) 5/6, 4/5, and 1/10.

(4) 1/2, 7/16, and 3/8.

The idea of cross multiplication leads to the following process for determining if two fractions have the same value.

Theorem 5.25. $\frac{a}{b} = \frac{p}{q}$ if and only if $a \cdot q = b \cdot p$.

Proof. $\frac{a \cdot q}{b \cdot q} = \frac{a}{b}$ and $\frac{b \cdot p}{b \cdot q} = \frac{p}{q}$, by Theorem 5.17. By Theorem 5.22, the fractions with the common denominator $b \cdot q$ have the same value if and only if their numerators are equal, that is, $a \cdot q = b \cdot p$. $\square$

Example 5.26. Determine whether 15/21 and 40/56 have the same value. By the preceding theorem, we calculate $15 \cdot 56 = 840$ and $21 \cdot 40 = 840$. These products being equal implies that the two given fractions have the same value.

Oral exercise 5.27. Do the fractions 8/12 and 12/18 have the same value? Explain how to determine the answer. How would you express these fractions in a common denominator?

The preceding exercise indicates that it is not always most efficient to use cross multiplication to express two fractions in common denominator or to determine whether or not they are equal. At times it is convenient to express two or more fractions in terms of their *least common denominator*, which is the lcm of the given denominators. In the preceding exercise, the lcm(12, 18) = 36, and we can express each fraction with the denominator 36 by multiplying numerator and denominator of 8/12 by 36 ÷ 12 = 3, arriving at 8/12 = 24/36, and by multiplying numerator and denominator of 12/18 by 36 ÷ 18 = 2, arriving at 12/18 = 24/36.

Oral exercise 5.28. Express 13/30 and 43/105 in terms of their least common denominator.

We observed above that 8/12 and 12/18, expressed in terms of their least common denominator, are both 24/36. But gcf(24, 36) = 12 allows us to use Corollary 5.19 to conclude that

$$\frac{24}{36} = \frac{24 \div 12}{36 \div 12} = \frac{2}{3}$$

Oral exercise 5.29. Express each of 9/12 and 34/16 in lowest terms, then with a common denominator.

Theorem 5.30. *If $\frac{a}{b}$ is expressed in lowest terms, and if $a/b = p/q$, then there is a natural number n such that $p = a \cdot n$ and $q = b \cdot n$.*

Proof. The case $\frac{a}{b} = \frac{6}{35}$ illustrates the essential features of the proof. Notice that gcf(6, 35) = 1, so $\frac{6}{35}$ is in lowest terms. We need to prove that if $\frac{6}{35} = \frac{p}{q}$, then there is a natural number n such that $p = n \cdot 6$ and $q = n \cdot 35$. Begin by observing that $q \cdot 6 = 35 \cdot p$, by Theorem 5.25. Since gcf(6, 35) = 1, the prime factorizations $6 = 2 \cdot 3$ and $35 = 5 \cdot 7$ have no primes in common. Therefore, the equation

(5.1) $$q \cdot 2 \cdot 3 = 5 \cdot 7 \cdot p$$

shows that the prime factorization of q must contain 5 and 7, which means that $q = 5 \cdot 7 \cdot m = 35 \cdot m$, for some natural number m, and that the prime factorization of p must contain 2 and 3, which means that $p = 2 \cdot 3 \cdot n = 6 \cdot n$, for some natural number n. Substituting these factorizations of p and q into equation 5.1, we have

(5.2) $$35 \cdot m \cdot 6 = 35 \cdot 6 \cdot n$$

from which we conclude that $m = n$. Therefore, $p = 6 \cdot n$ and $q = 35 \cdot n$. □

In this theorem it is essential that a/b be expressed in lowest terms. For example, 4/6 = 6/9 but there is no natural number n for which $6 = 4 \cdot n$ and $9 = 6 \cdot n$.

We now have the following simple description of all the fractions with the same value as a given fraction a/b. If p/q is the expression of a/b in lowest terms, then the set of all fractions with the same value as a/b is

(5.3) $$\{\frac{p \cdot n}{q \cdot n} : n = 1, 2, 3, \ldots\}$$

Notice that if p/q is not in lowest terms, then this list will omit some fractions with the same value as p/q. For example, $\frac{8}{12} = \frac{4}{6}$, but the list

$$\{\frac{4 \cdot n}{6 \cdot n} : n = 1, 2, 3, \ldots\}$$

does not contain all fractions of the same value as 8/12. For example, it omits 10/15, because there is no natural number n such that $10 = 4 \cdot n$ and $15 = 6 \cdot n$.

Theorem 5.31. *If x is a rational number, then it is the value of exactly one fraction expressed in lowest terms.*

To say that x is the value of exactly one fraction expressed in lowest terms means that two different fractions, each expressed in lowest terms, must have different values.

If two fractions have the same value, then they must reduce to the same fraction in lowest terms.

Proof. A positive rational number x is the value of some fraction, say a/b, where a and b are natural numbers, because that is the definition of positive rational number. For example, suppose x is the value of $\frac{273}{357}$. From prime factorization or by Euclid's Algorithm, find $\gcd(273, 357) = 21$. Then

$$273 \div 21 = 13 \quad \text{and} \quad 357 \div 21 = 17$$

must be relatively prime, by Theorem 4.40. Then the Corollary of Theorem 5.22 implies that $\frac{13}{17} = \frac{273}{357}$. Therefore, x is the value of $\frac{13}{17}$, which is expressed in lowest terms. We have proved that x is the value of at least one fraction expressed in lowest terms.

Can there be another fraction expressed in lowest terms whose value is x? No, because Theorem 5.30 says that all fractions of the same value as $\frac{13}{17}$ must be of the form $\frac{13 \cdot n}{17 \cdot n}$ for some natural number n. Only the case $n = 1$ is a fraction in lowest terms. □

Oral exercise 5.32. (1) How many eighths in $\frac{1}{2}$? In $\frac{1}{4}$? In $\frac{3}{4}$?

(2) How many twelfths in $\frac{1}{3}$? In $\frac{1}{4}$? In $\frac{2}{3}$? In $\frac{3}{4}$?

(3) How many thirds in $\frac{4}{6}$? In $\frac{8}{12}$? In $\frac{12}{6}$? In $\frac{24}{18}$?

(4) List all fractions of the same value as 6/8. As 13/39.

5.2.3. Inequalities. Recall that we defined $a < b$ for whole numbers a and b to mean that a set with a elements can be put in one-to-one correspondence with a proper subset of a set with b elements. We then proved that if $a < b$, then a lies to the left of b on the number line. We shall use this latter characterization as our definition of $x < y$ for rational numbers.

Definition 5.33. The rational number x is *less than* the rational number y if the point x lies to the left of the point y on the number line.

Write $x < y$ to indicate that x is less than y. Write $x \leq y$ to denote that either $x < y$ or $x = y$. We also write $y > x$ to mean the same thing as $x < y$. Read it as y is greater than x. For fractions $\frac{p}{q}$ and $\frac{a}{b}$ write $\frac{p}{q} < \frac{a}{b}$ to mean that the value of $\frac{p}{q}$ is less than the value of $\frac{a}{b}$.

Example 5.34. $\frac{3}{7} < \frac{5}{7}$, because dividing the unit into 7 equal pieces and going 3 of them from 0 lies to the left of going 5 of them from 0. In other words, $3 < 5$ implies that 3 sevenths is fewer than 5 sevenths. A set of 3 sevenths can be put into one-to-one correspondence with a proper subset of a set of 5 sevenths.

Example 5.35. Which is smaller: 4/7 or 3/5? To make the comparison, represent both rational numbers by fractions with a common denominator. In this case, $\frac{4}{7} = \frac{4 \cdot 5}{7 \cdot 5} = \frac{20}{35}$ and $\frac{3}{5} = \frac{7 \cdot 3}{7 \cdot 5} = \frac{21}{35}$. Dividing the unit interval into 35 equal pieces and going 20 of them from 0 lies to the left of going 21 of them from 0, because $20 < 21$. Therefore, $4/7 < 3/5$. An explanation such as this should be given together with a number line picture like the following, because the picture succeeds in demonstrating the correct order only after the thirty-fifths are marked off.

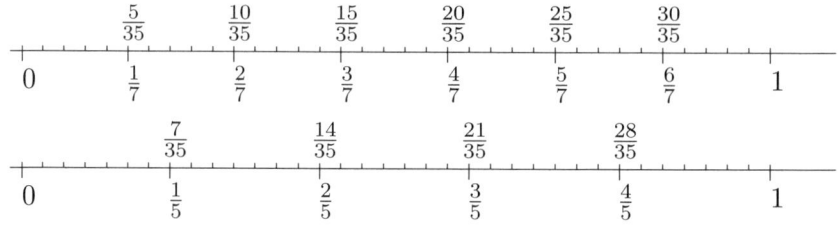

Example 5.36. Niels has finished $\frac{1}{3}$ of his job and Leah has finished $\frac{2}{5}$ of her job. Which one has finished the larger fraction of his or her job? The question asks which of $\frac{1}{3}$ and $\frac{2}{5}$ has the larger value. Reducing these fractions to a common denominator of 15, we have $\frac{1}{3} = \frac{5}{15}$ and $\frac{2}{5} = \frac{6}{15}$, from which we conclude that $\frac{1}{3} = \frac{5}{15} < \frac{6}{15} = \frac{2}{5}$, because $5 < 6$. Niels has finished a smaller fraction of his job than Leah has finished of her job.

Example 5.37. Which is more, $\frac{2}{3}$ of the people in the room or $\frac{7}{12}$ of the people in the room? Substituting the unit interval for the number of people in the room, we see that the question asks which of these two fractions has the larger value. Reducing them to a common denominator, we have

$\frac{2}{3} = \frac{8}{12} > \frac{7}{12}$. Therefore, $\frac{8}{12}$ of the people in the room is more than $\frac{7}{12}$ of the people in the room.

Theorem 5.38. *(1) The order of fractions with a common denominator is the same as the order of their numerators. In symbols, $\frac{a}{d} < \frac{b}{d}$ if and only if $a < b$.*

(2) In general, $\frac{a}{b} < \frac{p}{q}$ if and only if $a \cdot q < b \cdot p$

(3) The order of fractions with a common numerator is the opposite of the order of their denominators. In symbols, $\frac{n}{a} < \frac{n}{b}$ if and only if $a > b$.

Proof. (1) For fractions with a common denominator, such as $\frac{5}{17}$ and $\frac{11}{17}$, the order is the same as the order of the numerators, because $5 < 11$ means that 5 seventeenths is fewer than 11 seventeenths, and therefore the value of $\frac{5}{17}$ lies to the left of the value of $\frac{11}{17}$.

(2) We reduce the fractions to common denominator by cross multiplication to arrive at $\frac{a}{b} = \frac{a \cdot q}{b \cdot q}$ and $\frac{p}{q} = \frac{b \cdot p}{b \cdot q}$. By part (1), the order of these fractions is the same as the order of the numerators $a \cdot q$ and $b \cdot p$, as claimed.

(3) Apply part (2) to the given fractions to conclude that $\frac{n}{a} < \frac{n}{b}$ if and only if $n \cdot b < a \cdot n$, and this last inequality is true if and only if $b < a$ by the order property of multiplication. $\square$

Oral exercise 5.39. Which is larger?

(1) $\frac{5}{17}$ or $\frac{7}{17}$? $\frac{8}{5}$ or $\frac{8}{7}$?

(2) $\frac{17}{5}$ or $\frac{17}{11}$? $\frac{3}{5}$ or $\frac{3}{7}$?

(3) $\frac{7}{8}$ or $\frac{13}{16}$? $\frac{13}{8}$ or $\frac{13}{16}$?

(4) $\frac{5}{11}$ or $\frac{6}{13}$? $\frac{6}{11}$ or $\frac{5}{13}$?

§5.2 Exercises

(1) Prove Theorem 5.17 for $\frac{2}{5}$ and $\frac{4}{10}$. Show appropriate intervals on the number line to illustrate your argument.

(2) Prove Theorem 5.17 for $\frac{3}{5}$ and $\frac{12}{20}$. Show appropriate intervals on the number line to illustrate your argument.

(3) Explain, as in the proof of Theorem 5.17, why $\frac{17}{37}$ has the same value as $\frac{187}{407}$. In this case a diagram is impractical, so your explanation will

have to be entirely verbal. Make sure that *you* find your explanation convincing.

(4) Explain why $\frac{4}{6}$ and $\frac{6}{9}$ have the same value. Illustrate your explanation with a number line drawing. Does Theorem 5.17 apply to this case?

(5) Express $\frac{9}{12}$ in lowest terms and explain, using number line diagrams and an argument like the proof of Theorem 5.17, why these two fractions have the same value.

(6) Use a number line diagram to explain which of $\frac{3}{5}$ and $\frac{3}{7}$ is larger. Explain verbally why a seventh is smaller than a fifth. (More of the smaller one is needed to make the unit interval).

(7) Use a number line diagram to explain which of $\frac{5}{9}$ and $\frac{7}{9}$ is larger.

(8) Use a number line diagram to illustrate the proof of part 3 of Theorem 5.38 that $\frac{1}{2} < \frac{2}{3}$.

(9) Express in a common denominator. (a) 5/11 and 6/13. (b) 4/5, 8/9, and 3/15. (c) 2/3, 3/4, and 4/5. (d) 2/7, 5/14, and 6/5. (e) $1\frac{1}{2}$ and $2\frac{3}{4}$.

(10) Express in the least common denominator. (a) 5/6 and 5/9. (b) 7/12 and 6/8. (c) 7/12, 6/14, and 8/15.

(11) Express in lowest terms, then with a common denominator. (a) 18/20 and 15/45. (b) 12/18 and 25/30.

(12) Write the proof of part (1) of Theorem 5.38 for the fractions $\frac{5}{9}$ and $\frac{6}{9}$. Illustrate the argument on the number line.

(13) Write the proof of part (2) of Theorem 5.38 for the fractions $\frac{11}{16}$ and $\frac{7}{15}$.

(14) Write the proof of part (3) of Theorem 5.38 for the fractions $\frac{17}{5}$ and $\frac{17}{11}$. Illustrate the proof for this case on the number line.

(15) Prove the following Corollary to Theorem 5.25: $\frac{a}{b} = \frac{p}{q}$ if and only if $\frac{a}{a+b} = \frac{p}{p+q}$. For example, $\frac{2}{6} = \frac{4}{12}$ so $\frac{2}{2+6} = \frac{4}{4+12}$. Give another example.

(16) Express in lowest terms, then with a common denominator. (a) 120/170, 792/864. (b) 81/189, 360/576. (c) 576/432, 567/783. (d) Express $\frac{2 \cdot 3 \cdot 5 \cdot 7 \cdot 11 \cdot 19}{3 \cdot 5 \cdot 13 \cdot 19}$ in lowest terms.

(17) David spent $\frac{2}{5}$ of his money on a cattle-brand book. The book cost \$20. How much money did he have at first? *Hint:*

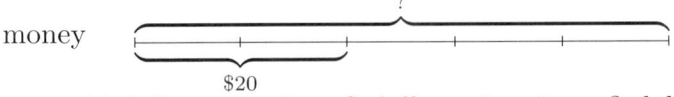

2 units = 20 dollars. 1 unit = ? dollars. 5 units = ? dollars.

(18) Nels and Gavin each received the same amount of money. Nels had to pay a tax of $\frac{11}{100}$ of his money while Gavin had to pay a tax of $\frac{13}{125}$ of his money. Who paid the larger tax?

(19) In the preceding problem, if each received $3,000, how much tax did each pay? Answer by explaining how to find $\frac{11}{100}$ of 3000 and $\frac{13}{125}$ of 3000, from the definition of fraction of something. Do not use multiplication of fractions, as we have not yet defined that.

(20) Nicholas, Joshua, and Anthony put their marbles together in a bag. Three-fifths of these marbles belong to Nicholas and he has twice as many as Joshua, who has 10 more than Anthony. How many marbles does each have? *Hint:*

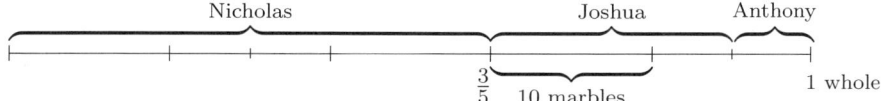

5.3. Addition of Rational Numbers

The number line interpretation of addition of whole numbers can be used as the definition of addition of rational numbers.

Definition 5.40. For rational numbers x and y, their sum $x+y$ is the point on the number line reached by going a distance y to the right of x.

We shall verify below that the point $x+y$ is a rational number. If x and y are whole numbers, this definition of their sum certainly agrees with the whole number sum, by the number line interpretation of such sums given in Theorem 2.7. We shall say and write $\frac{a}{b} + \frac{p}{q}$ to mean the value of $\frac{a}{b}$ plus the value of $\frac{p}{q}$. The definition is stated in terms of values to emphasize the independence of this sum from the choice of fractions used to represent these values.

Example 5.41. $5 + \frac{3}{8} = 5\frac{3}{8}$, because to start at 5 and go a distance 3/8 to the right is to end at $5\frac{3}{8}$, by definition of mixed fractions.

Example 5.42. $1/2 + 1/2 = 2/2 = 1$, because to start at 1/2 and go a distance 1/2 to the right is to end at 1, there being 2 halves in 1.

Example 5.43. $2/5 + 4/5 = 6/5$, because to start at 2/5 and go a distance of 4/5 to the right is to go a total distance of $2+4 = 6$ fifths to the right of 0, which is to the point represented by 6/5.

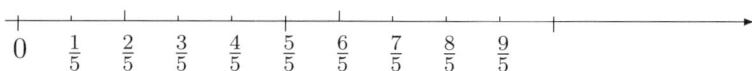

Example 5.44. $\frac{1}{2} + \frac{3}{4} = \frac{2}{4} + \frac{3}{4} = \frac{5}{4}$, because $\frac{1}{2} = \frac{2}{4}$, and to start at $\frac{2}{4}$ and go a distance of $\frac{3}{4}$ to the right is to go $2 + 3 = 5$ fourths to the right of 0.

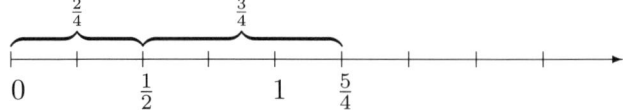

Example 5.45. $1/2 + 2/3 = 7/6$, because $1/2 = 3/6$ and $2/3 = 4/6$, so to go four sixths to the right of $3/6$ is to go $3 + 4 = 7$ sixths to the right of 0.

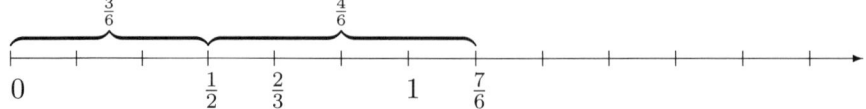

The preceding examples combined with Theorem 5.9 show the connection between long division and fractions, which we state as follows.

Theorem 5.46. *(Long Division and Fractions) (1). If q, r, and $d \neq 0$ are whole numbers, then $q\frac{r}{d} = q + \frac{r}{d}$.*

(2). If a is a whole number such that $a \div d$ has quotient q and remainder r, that is, if $a = q \cdot d + r$, then $\frac{a}{d} = q\frac{r}{d} = q + \frac{r}{d}$.

Proof. (1). For example, the mixed fraction $5\frac{3}{8}$ represents the point on the number line obtained by dividing the unit into 8 equal parts and going 3 of these parts to the right from 5. That is precisely the definition of $5 + \frac{3}{8}$. Therefore, $5\frac{3}{8} = 5 + \frac{3}{8}$.

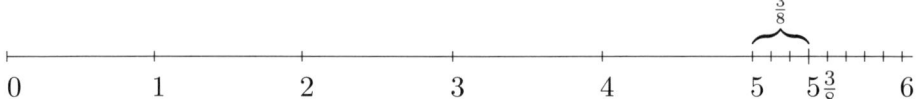

The second statement of the theorem just combines the first statement with Theorem 5.9. □

Theorem 5.47. *(1) If fractions $\frac{a}{d}$ and $\frac{c}{d}$ have a common denominator, then*

(5.4) $$\frac{a}{d} + \frac{c}{d} = \frac{a+c}{d}$$

Just add the numerators.

(2) If $\frac{a}{b}$ and $\frac{c}{d}$ are fractions, then their sum is calculated by first reducing these fractions to a common denominator and then adding the resulting numerators. For example, if they are reduced to common denominator by cross multiplying, then

(5.5) $$\frac{a}{b} + \frac{c}{d} = \frac{a \cdot d + b \cdot c}{b \cdot d}$$

(3) If x and y are rational numbers, then $x + y$ is a rational number.

5.3. Addition of Rational Numbers

Proof. (1) To find $\frac{a}{d} + \frac{c}{d}$, start at $\frac{a}{d}$, which is a of $\frac{1}{d}$ to the right from 0, and from there go c of $\frac{1}{d}$ to the right, to end at $a + c$ of $\frac{1}{d}$ to the right of 0. This is the point $\frac{a+c}{d}$.

As an example, to find $\frac{4}{11} + \frac{3}{11}$, start at $\frac{4}{11}$, the point arrived at by dividing the unit into 11 equal parts, called elevenths, and go 4 of $\frac{1}{11}$ to the right from 0. From there go 3 of $\frac{1}{11}$ to the right, for a total of $4 + 3$ of $\frac{1}{11}$ to the right from 0. This is the point $\frac{4+3}{11}$.

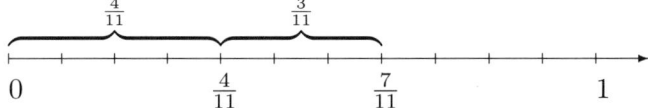

(2) Reduce the fractions to a common denominator by cross multiplication to obtain $\frac{a}{b} = \frac{a \cdot d}{b \cdot d}$ and $\frac{c}{d} = \frac{b \cdot c}{b \cdot d}$. By the first part, the sum of the values of these fractions is obtained by adding the numerators, as claimed.

(3) Notice that the first two parts show that the sum of two fractions is the value of some fraction. If x and y are rational numbers, then $x = \frac{a}{b}$ and $y = \frac{c}{d}$, for some fractions. The sum $x + y$ is then the sum of these two fractions, which we have just seen will be the value of a fraction, and therefore is a rational number. □

The sum of fractions with a common denominator is the whole number sum of their numerators. For example, $\frac{3}{8} + \frac{7}{8}$ means the sum of 3 eighths and 7 eighths, which would be $3 + 7$ eighths, which is the number of eighths in the union of a set of 3 eighths with a set of 7 eighths.

Oral exercise 5.48. Find the sums. Explain the reasoning in each case.

(1) $3/8 + 5/8$.

(2) $7/8 + 5/16$.

(3) $4/5 + 7/10$ and $7/10 + 4/5$.

(4) $(2/7 + 3/7) + 1/7$.

(5) $1/2 + (1/2 + 1/4)$.

Theorem 5.49. *Addition of rational numbers x, y, and z has the following properties.*

(1) Whole numbers added as rational numbers have the same sum as when added as whole numbers.

(2) Commutative property: $x + y = y + x$.

(3) Associative property: $(x + y) + z = x + (y + z)$.

(4) Identity of 0: $0 + x = x + 0 = x$.

(5) Order property: If $x < y$, then $z + x < z + y$, for any rational number z.

Proof. (1) The whole numbers 3 and 7 are represented as fractions by $\frac{3}{1}$ and $\frac{7}{1}$, respectively. Their sum as rational numbers is represented by $\frac{3}{1}+\frac{7}{1}=\frac{3+7}{1}$, which represents the whole number $3+7$. Therefore, the sum of 3 and 7 as rational numbers is the same as their sum as whole numbers.

(2) The commutative property follows from the commutative property of addition, as follows. Represent x and y by fractions with a common denominator, say $x=\frac{a}{d}$ and $y=\frac{b}{d}$. Then

$$x+y = \frac{a}{d} + \frac{b}{d},$$
$$= \frac{a+b}{d}, \text{ by Theorem 5.47}$$
$$= \frac{b+a}{d}, \text{ by comm. property of add. of whole numbers}$$
$$= \frac{b}{d} + \frac{a}{d} = y+x$$

(3) The associative property follows from the associative property of addition of whole numbers. Represent x, y, and z by fractions with a common denominator, say $x=\frac{a}{d}$, $y=\frac{b}{d}$, and $z=\frac{c}{d}$. Then

$$(x+y)+z = (\frac{a}{d} + \frac{b}{d}) + \frac{c}{d},$$
$$= \frac{a+b}{d} + \frac{c}{d} = \frac{(a+b)+c}{d}, \text{ by Theorem 5.47}$$
$$= \frac{a+(b+c)}{d}, \text{ by assoc. property of add. of whole numbers}$$
$$= \frac{a}{d} + \frac{b+c}{d} = \frac{a}{d} + (\frac{b}{d} + \frac{c}{d}), \text{ by Theorem 5.47}$$
$$= x+(y+z)$$

(4) The identity property of 0 is clear from the definition of addition.

(5) The proof of the order property of addition of rational numbers is the same as it was for whole numbers (see Theorem 2.4). If $x < y$, then x lies to the left of y on the number line. After moving a fixed distance to the right from each number, the resulting numbers will still lie in the same order from left to right. □

The commutative and associative properties of addition imply that the sum of three numbers can be done in any order. For example, the sum $\frac{1}{2} + \frac{3}{7} + \frac{2}{5}$ can be done by choosing any of the numbers to be first, add to it one of the remaining two numbers and to this sum add the remaining

5.3. Addition of Rational Numbers

number. There are six ways to do this:

(1) $(\frac{1}{2} + \frac{3}{7}) + \frac{2}{5}$, (2) $(\frac{3}{7} + \frac{1}{2}) + \frac{2}{5}$, (3) $(\frac{1}{2} + \frac{2}{5}) + \frac{3}{7}$
(4) $(\frac{2}{5} + \frac{1}{2}) + \frac{3}{7}$, (5) $(\frac{3}{7} + \frac{2}{5}) + \frac{1}{2}$, (6) $(\frac{2}{5} + \frac{3}{7}) + \frac{1}{2}$

but the sum is the same in all six cases. For example, the first and sixth cases are the same because

$$\begin{aligned}(\frac{1}{2} + \frac{3}{7}) + \frac{2}{5} &= \frac{1}{2} + (\frac{3}{7} + \frac{2}{5}), \text{ by associative property}\\ &= (\frac{3}{7} + \frac{2}{5}) + \frac{1}{2}, \text{ by commutative property}\\ &= (\frac{2}{5} + \frac{3}{7}) + \frac{1}{2}, \text{ by commutative property}\end{aligned}$$

In practice, we use this freedom of order and grouping to do mental calculations, to derive algorithms and to do algebra.

Example 5.50. To calculate $\frac{3}{5} + \frac{4}{7} + \frac{2}{5}$, we can add the first and third and to this sum add the second fraction:

$$(\frac{3}{5} + \frac{2}{5}) + \frac{4}{7} = \frac{5}{5} + \frac{4}{7} = 1\frac{4}{7}$$

Example 5.51. $5\frac{2}{3} + 3\frac{7}{8}$ can be computed in two ways. One way is to express each mixed fraction as a fraction, and then compute using Theorem 5.47, as follows.

$$\begin{aligned}5\frac{2}{3} + 3\frac{7}{8} &= \frac{5 \cdot 3 + 2}{3} + \frac{3 \cdot 8 + 7}{8}, \text{ Theorem 5.9}\\ &= \frac{17 \cdot 8 + 3 \cdot 31}{3 \cdot 8}, \text{ Theorem 5.47}\\ &= \frac{229}{24} = 9\frac{13}{24}, \text{ Theorem 5.9}\end{aligned}$$

Another method is to use the commutative and associative properties of Theorem 5.46 in order to choose the order and grouping of the sums in the most advantageous way.

$$5\frac{2}{3} + 3\frac{7}{8} = (5 + \frac{2}{3}) + (3 + \frac{7}{8}), \text{ Theorem 5.46}$$

$$= 5 + \frac{2}{3} + 3 + \frac{7}{8}, \text{ free to choose order and grouping}$$

$$= (5 + 3) + (\frac{2}{3} + \frac{7}{8}), \text{ our choice of order and grouping}$$

$$= 8 + (\frac{2 \cdot 8 + 3 \cdot 7}{3 \cdot 8}), \text{ Theorem 5.47}$$

$$= 8 + \frac{37}{24}, \text{ Theorem 5.47}$$

$$= 8 + (1 + \frac{13}{24}), \text{ Theorem 5.46}$$

$$= 9 + \frac{13}{24}, \text{ Assoc. Property}$$

Oral exercise 5.52. Compute and explain your method.
 (1) $\frac{5}{13} + \frac{8}{13} + \frac{2}{3}$
 (2) $\frac{13}{5} + \frac{4}{9} + \frac{7}{5}$
 (3) $6\frac{2}{7} + 4\frac{3}{7}$
 (4) $8\frac{3}{11} + 3\frac{2}{5} + 2\frac{8}{11}$

Example 5.53. If Jen walks $12\frac{3}{5}$ miles one day and $9\frac{2}{3}$ miles the next, how far does she walk in both days? *Solution:*

$11\frac{3}{5}$ miles	$9\frac{2}{3}$ miles
Day 1	Day 2

In both days together she walks $12\frac{3}{5} + 9\frac{2}{3} = 22\frac{4}{15}$ miles.

§5.3 Exercises

(1) On a number line mark the values of $\frac{13}{19}$ and $\frac{28}{8}$, respectively.
 (a) Use your drawing to explain why $\frac{13}{19} < \frac{28}{8}$.
 (b) Use Theorem 5.38 to show that $\frac{13}{19} < \frac{28}{8}$.
 (c) On your drawing mark the points represented by $\frac{7}{5} + \frac{13}{19}$ and $\frac{7}{5} + \frac{28}{8}$. From this picture explain which is larger and why that would be the case for any fraction used in place of $\frac{7}{5}$.
 (d) Carry out the additions to find $\frac{a}{b} = \frac{7}{5} + \frac{13}{19}$ and $\frac{c}{d} = \frac{7}{5} + \frac{28}{8}$.
 (e) Verify that $a \cdot d < b \cdot c$. From this, what does Theorem 5.38 allow you to conclude?

5.4. Subtraction of Rational Numbers

(2) Follow the proof of Theorem 5.49 to prove the following.
   (a) The commutative property for $\frac{2}{3} + \frac{7}{8}$.
   (b) The associate property for $(\frac{5}{8} + \frac{9}{4}) + \frac{11}{3}$.

(3) Describe a demonstration which shows that if point x lies to the left of point y on the number line, then going a distance z to the right of each point results in points lying in the same order.

(4) Find the sum of $\frac{1}{2}$, $\frac{1}{3}$, and $\frac{1}{4}$. Explain how you use the associative property to do this.

(5) Find $3\frac{1}{2} + 2\frac{4}{5}$ by the two methods given in example 5.51. Give detailed explanations for each step, in the way it was done in the example.

(6) Explain how you might use wooden rods for line segments to demonstrate:
   (a) The commutative property $\frac{2}{3} + \frac{7}{8} = \frac{7}{8} + \frac{2}{3}$.
   (b) The associative property $(\frac{5}{8} + \frac{9}{4}) + \frac{11}{3} = \frac{5}{8} + (\frac{9}{4} + \frac{11}{3})$.

(7) Write the order and grouping you would use to calculate in your head the sum $2\frac{4}{13} + 5\frac{4}{5} + 3\frac{9}{13}$.

(8) Joshua has $\frac{3}{5}$ of a quart of chokecherries; Caitlin, $\frac{2}{3}$ quart; and Nicholas, $\frac{5}{6}$ of a quart: how many quarts have they in all? Use a line segment diagram to illustrate what to do.

(9) Niels sold $\frac{1}{5}$ of his grain to Morris, $\frac{1}{4}$ to Tony, and $\frac{1}{3}$ to Gavin: what fraction of his grain did he sell? Use a line segment diagram to illustrate what to do.

(10) David spent $\frac{1}{2}$ of his money for a coat, $\frac{1}{4}$ for a hat, and $\frac{1}{5}$ for shoes: what fraction of his money did he spend? Use a line segment diagram to illustrate what to do.

(11) Morris can unload a truck load of sugar at a constant rate in 3 hours. Tony can unload the same load at a constant rate in 4 hours. What fraction of the truck can they unload together in 1 hour, working independently, at their usual rates and without getting in each other's way? *Hint:* What fraction of the truck can each unload in 1 hour?

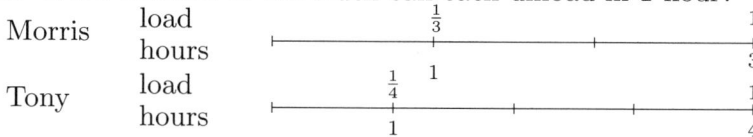

5.4. Subtraction of Rational Numbers

Whole number subtraction was defined as the inverse to addition. This was done in two ways, as comparative subtraction and as take-away subtraction, both of which gave the same difference because addition is commutative. The exact same definitions carry over to the set of rational numbers.

Definition 5.54 (Comparative subtraction). If x and y are rational numbers and if $y \leq x$, then $x - y$ is that rational number z for which $y + z = x$. In symbols, $x - y = z$ means that $x = y + z$.

Definition 5.55 (Take-away subtraction). If x and y are rational numbers and if $y \leq x$, then $x - y$ is that rational number z for which $z + y = x$. In symbols, $x - y = z$ means that $x = z + y$.

In comparative subtraction, $x - y$ is that number which must be added to y to make x. In take-away subtraction, $x - y$ is that number which y must be added to in order to make x.

We must verify below in Theorem 5.60 that $x - y$ is a rational number. If x and y are whole numbers, these definitions of subtraction certainly agree with the whole number definitions. The statement $x - y$ is read x *minus* y and the result is called the *difference*. The number x is called the *minuend* and the number y is called the *subtrahend*. We use the same notation for both kinds of subtraction, because both operations produce the same difference.

Theorem 5.56. *If x and y are rational numbers and $y \leq x$, then $x - y$ in comparative subtraction has the same value as $x - y$ in take-away subtraction.*

Proof. If $z = x - y$ for comparative subtraction, then $y + z = x$. But $y + z = z + y$, by the commutative property of addition, and therefore $z + y = x$ and $z = x - y$ for take-away subtraction as well. $\square$

Comparative subtraction is called that because it answers the question: How much larger than y is x? The answer is $z = x - y$, because z must be added to y to make x.

Example 5.57 (Comparative subtraction). How much larger than $\frac{1}{2}$ is $\frac{2}{3}$? The answer is $\frac{2}{3} - \frac{1}{2} = \frac{1}{6}$, because $\frac{1}{2} + \frac{1}{6} = \frac{3+1}{6} = \frac{2}{3}$ shows that $\frac{1}{6}$ added to $\frac{1}{2}$ makes $\frac{2}{3}$.

Example 5.58 (Comparative subtraction). If Ragna has already gone $3\frac{1}{2}$ miles of the $9\frac{3}{4}$ miles to her grandmother's house, how much farther has she to go? The answer is $9\frac{3}{4} - 3\frac{1}{2} = 6\frac{1}{4}$ miles, because this is the number of miles which must be added to $3\frac{1}{2}$ miles to make $9\frac{3}{4}$ miles.

Example 5.59 (Take-away subtraction). If Denise has a board 9 ft. long and she cuts $2\frac{3}{8}$ ft. from the end of it, how long is the remaining board? The answer is $9 - 2\frac{3}{8} = 6\frac{5}{8}$ ft. because putting back the sawed off piece makes the whole board, $6\frac{5}{8} + 2\frac{3}{8} = 9$.

5.4. Subtraction of Rational Numbers

Theorem 5.60. *If x and y are rational numbers such that $y \leq x$, and if $x = \frac{a}{d}$ and $y = \frac{c}{d}$ are represented by fractions with a common denominator, then $c \leq a$, and their difference is the value of the fraction*

$$(5.6) \qquad \frac{a}{d} - \frac{c}{d} = \frac{a-c}{d}$$

obtained by subtracting the numerators.

If $x = \frac{a}{b}$ and $y = \frac{c}{d}$, then their difference can be calculated from (5.6) after first reducing them to a common denominator. For example, if they are reduced to common denominator by cross multiplication, then their difference is the value of the fraction

$$(5.7) \qquad \frac{a}{b} - \frac{c}{d} = \frac{a \cdot d - b \cdot c}{b \cdot d}$$

In particular, the difference of rational numbers, $x - y$ for which $y \leq x$, is always a rational number.

Proof. By Theorem 5.56 we need only prove that the fractions given in (5.6) and (5.7) satisfy the definition of comparative subtraction. To that end, for the common denominator case we compute

$$\begin{aligned}
\frac{c}{d} + \frac{a-c}{d} &= \frac{c+(a-c)}{d}, \text{ by Theorem 5.47} \\
&= \frac{a}{d}, \text{ by whole number comparative subtraction}
\end{aligned}$$

In the general case, if $\frac{c}{d} \leq \frac{a}{b}$, then Theorem 5.38 tells us that $b \cdot c \leq a \cdot d$, so that the whole number difference $a \cdot d - b \cdot c$ in (5.7) is defined. Then

$$\begin{aligned}
\frac{c}{d} + \frac{a \cdot d - b \cdot c}{b \cdot d} &= \frac{b \cdot c}{b \cdot d} + \frac{a \cdot d - b \cdot c}{b \cdot d}, \text{ reduce to common denominator} \\
&= \frac{b \cdot c + a \cdot d - b \cdot c}{b \cdot d}, \text{ by Theorem 5.47} \\
&= \frac{a \cdot d}{b \cdot d}, \text{ by take-away subtraction of whole numbers} \\
&= \frac{a}{b}, \text{ by Theorem 5.19}
\end{aligned}$$

This verifies that the right side of (5.7) is the difference $\frac{a}{b} - \frac{c}{d}$ for comparative subtraction. $\square$

Example 5.61. Calculate the differences and express the answer in lowest terms. Check the answer by using the definition of comparative or take-away subtraction.

(1) How much remains if $\frac{3}{8}$ is taken from $\frac{7}{8}$? It is the take-away difference $\frac{7}{8} - \frac{3}{8} = \frac{7-3}{8} = \frac{4}{8} = \frac{1}{2}$. Check: $\frac{4}{8} + \frac{3}{8} = \frac{7}{8}$.

(2) How much larger than $\frac{2}{7}$ is $\frac{12}{13}$? It is the comparative difference $\frac{12}{13} - \frac{2}{7} = \frac{12 \cdot 7 - 13 \cdot 2}{13 \cdot 7} = \frac{58}{91}$. Check: $\frac{2}{7} + \frac{58}{91} = \frac{13 \cdot 2}{13 \cdot 7} + \frac{58}{91} = \frac{26+58}{91} = \frac{84}{91} = \frac{12 \cdot 7}{13 \cdot 7} = \frac{12}{13}$.

Oral exercise 5.62. Do the indicated calculations. Express answers in lowest terms. Check your answer by the definition of subtraction.

(1) How much larger than $\frac{3}{12}$ is $\frac{9}{12}$? What is 1 take away $\frac{3}{12}$?

(2) What is $\frac{9}{19}$ take away $\frac{4}{19}$? How much larger than $\frac{3}{9}$ is $\frac{8}{9}$?

(3) How much larger than $\frac{3}{10}$ is $\frac{2}{5}$? What is $\frac{11}{16}$ take away $\frac{3}{8}$?

Theorem 5.63 (Comparative subtraction on the number line). *If x and y are rational numbers and if $x \geq y$, then $x - y$ is the length of the line segment from y to x, by which we mean that if we were to start at y and go $x - y$ to the right we would end at x.*

Proof. Since $x \geq y$, we know that x lies to the right of y on the number line. If $z = x - y$, then $x = y + z$ shows that if we start at y and go z to the right we end at x. □

Here is an illustration of the comparative subtraction $\frac{6}{7} - \frac{5}{9} = \frac{19}{63}$ on the number line. Start at $\frac{5}{9}$ and go 19 sixty-thirds to the right to reach $\frac{6}{7}$.

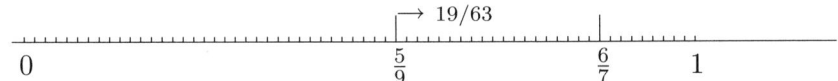

Theorem 5.64 (Take-away subtraction on the number line). *If x and y are rational numbers with $x \geq y$, then the take-away difference $x - y$ is the point on the number line arrived at by starting at x and going y to the left.*

Proof. If we start at x and go y to the left, we end at a point z. If we start at z and go y to the right, we certainly end up at x, our starting point. But starting at z and going y to the right is $z + y$, so we have shown that $z + y = x$, which shows that z is the take-away difference $x - y$. □

Here is an illustration of the take-away subtraction $\frac{13}{4} - \frac{5}{3} = \frac{19}{12}$ on the number line. Start at $\frac{13}{4}$ and go $\frac{5}{3} = 20$ twelfths to the left to reach $\frac{19}{12}$.

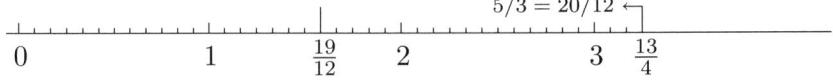

Theorem 5.65 (Subtraction Properties). *Subtraction of rational numbers has the following properties. Here x, y, and z denote rational numbers. In each identity it is assumed that the indicated subtractions are defined.*

(1) Identity property of 0: $x - 0 = x$, for any rational number x.

(2) Order property: If $x < y$ and if $z < x$, then $x - z < y - z$.

(3) $x - y = (x + z) - (y + z)$

5.4. Subtraction of Rational Numbers

(4) $(x+y) - z = x + (y-z)$
(5) $x - (y-z) = (x-y) + z$
(6) $x - (y+z) = (x-y) - z$
(7) $x - (y-z) = (x+z) - y$
(8) $x - y = (x-z) - (y-z)$

Proof. The statement of this theorem is identical to the statement of Theorem 2.33, except that now any rational numbers can be used instead of just whole numbers. The exact same proof works here, because the proof of Theorem 2.33 used only the properties of addition of whole numbers, and Theorem 5.49 shows that all of these properties also hold for addition of rational numbers. $\square$

5.4.1. Applications and Examples of the Subtraction Properties.

Example 5.66 (Order Property). $\frac{7}{8} < \frac{13}{5}$ and $\frac{2}{3} < \frac{7}{8}$, so $\frac{7}{8} - \frac{2}{3} < \frac{13}{5} - \frac{2}{3}$.

Example 5.67 (Subtraction Associative Property (4)). The following are all equal.

$$(\frac{23}{9} + \frac{8}{3}) - \frac{2}{5}, \qquad \frac{23}{9} + (\frac{8}{3} - \frac{2}{5}), \qquad (\frac{8}{3} - \frac{2}{5}) + \frac{23}{9}$$
$$(\frac{8}{3} + \frac{23}{9}) - \frac{2}{5}, \qquad \frac{8}{3} + (\frac{23}{9} - \frac{2}{5}), \qquad (\frac{23}{9} - \frac{2}{5}) + \frac{8}{3}$$

In fact, the first two are equal by (4), the second and third are equal by the commutative property, the first and fourth are equal by the commutative property, the fourth and fifth are equal by (4), and finally the fifth and sixth are equal by the commutative property. There is, therefore, no ambiguity in writing these as

(5.8) $\qquad \frac{23}{9} + \frac{8}{3} - \frac{2}{5}$ or as $\frac{23}{9} - \frac{2}{5} + \frac{8}{3}$, etc.

In this context, care must be taken only when one of the subtractions is not defined in rational numbers. For example, $\frac{1}{3} + \frac{5}{2} - \frac{5}{6}$ is equal to $\frac{1}{3} + (\frac{5}{2} - \frac{5}{6})$, but not to $\frac{5}{2} + (\frac{1}{3} - \frac{5}{6})$, because $\frac{1}{3} < \frac{5}{6}$ means that $\frac{1}{3} - \frac{5}{6}$ is not defined in rational numbers.

In summary, Property (4) allows two sums and a subtraction to be done in any order in which each operation is defined. As an application, consider

$$\frac{22}{9} + \frac{8}{3} - \frac{4}{9} = (\frac{22}{9} - \frac{4}{9}) + \frac{8}{3} = \frac{18}{9} + \frac{8}{3} = 2\frac{8}{3} = 2 + 2\frac{2}{3} = 4\frac{2}{3}$$

Example 5.68 (Subtraction Associative Property (5)). $\frac{2}{3} < \frac{4}{5} < \frac{7}{6}$, so $\frac{4}{5} - \frac{2}{3} < \frac{7}{6}$ and $\frac{7}{6} - (\frac{4}{5} - \frac{2}{3}) = (\frac{7}{6} - \frac{4}{5}) + \frac{2}{3}$. As an application, consider

$$\frac{18}{13} - (\frac{5}{13} - \frac{1}{4}) = (\frac{18}{13} - \frac{5}{13}) + \frac{1}{4} = 1\frac{1}{4}$$

Example 5.69 (Subtraction Associative Property (6)). The following are all equal by Property (6), as you should verify as was done above.

$$\frac{13}{8} - (\frac{1}{5} + \frac{2}{3}), \quad (\frac{13}{8} - \frac{1}{5}) - \frac{2}{3}, \quad (\frac{13}{8} - \frac{2}{3}) - \frac{1}{5}$$

Without ambiguity, we can write these as $\frac{13}{8} - \frac{1}{5} - \frac{2}{3}$. The resulting freedom of choice in which the order of operations takes place aids in the mental calculation of problems like

$$\frac{13}{8} - \frac{1}{5} - \frac{7}{8} = (\frac{13}{8} - \frac{7}{8}) - \frac{1}{5} = \frac{6}{8} - \frac{1}{5} = \frac{3}{4} - \frac{1}{5} = \frac{11}{20}$$

and

$$\frac{9}{2} - \frac{4}{7} - \frac{17}{7} = \frac{9}{2} - (\frac{4}{7} + \frac{17}{7}) = \frac{9}{2} - 3 = \frac{9}{2} - \frac{6}{2} = \frac{3}{2}$$

Example 5.70 (Subtraction Associative Property (7)). The following are all equal by Property (7), as you should verify as was done above.

$$\frac{13}{8} - (\frac{2}{3} - \frac{1}{8}), \quad (\frac{13}{8} + \frac{1}{8}) - \frac{2}{3}, \quad (\frac{13}{8} - \frac{2}{3}) + \frac{1}{8}$$

Example 5.71 (Mixed fractions). Subtraction problems with mixed fractions can be done either by converting them to fractions or by using subtraction associative properties. The latter method requires care in making sure that one is subtracting a smaller number from a larger number.

Find $4\frac{1}{3} - 2\frac{1}{5}$ by first converting the mixed fractions to fractions. We have $4\frac{1}{3} = \frac{4 \cdot 3 + 1}{3} = \frac{13}{3}$ and $2\frac{1}{5} = \frac{2 \cdot 5 + 1}{5} = \frac{11}{5}$. Therefore,

$$\begin{aligned} 4\frac{1}{3} - 2\frac{1}{5} &= \frac{13}{3} - \frac{11}{5} \\ &= \frac{13 \cdot 5 - 3 \cdot 11}{3 \cdot 5}, \text{ by equation 5.7} \\ &= \frac{32}{15} = 2\frac{2}{15} \end{aligned}$$

5.4. Subtraction of Rational Numbers

To do the same problem using subtraction associative properties, we have

$$4\frac{1}{3} - 2\frac{1}{5} = (4 + \frac{1}{3}) - (2 + \frac{1}{5}), \text{ by Theorem 5.46}$$
$$= ((4 + \frac{1}{3}) - 2) - \frac{1}{5}, \text{ by Sub. Assoc. Prop. 3(c)}$$
$$= ((\frac{1}{3} + 4) - 2) - \frac{1}{5}, \text{ by Comm. Prop.}$$
$$= ((\frac{1}{3} + (4 - 2)) - \frac{1}{5}, \text{ by Sub. Assoc. Prop. 3(a)}$$
$$= 2 + (\frac{1}{3} - \frac{1}{5}), \text{ Comm. and Assoc. Prop. 3(a)}$$
$$= 2 + \frac{5-3}{3 \cdot 5}, \text{ by equation 5.7}$$
$$= 2 + \frac{2}{15} = 2\frac{2}{15}$$

In effect, do the subtraction of the whole number parts and the subtraction of the fractional parts and add the results. Our preceding discussion reduces the steps of this computation to

$$4\frac{1}{3} - 2\frac{1}{5} = (4 + \frac{1}{3}) - (2 + \frac{1}{5})$$
$$= 4 + \frac{1}{3} - 2 - \frac{1}{5} = 4 - 2 + \frac{1}{3} - \frac{1}{5}$$
$$= 2 + \frac{2}{15} = 2\frac{2}{15}$$

If the problem is changed to $4\frac{1}{5} - 2\frac{1}{3}$, a difficulty arises in trying to use the subtraction associative properties, because $\frac{1}{5} < \frac{1}{3}$. In this case, rewrite $4\frac{1}{5}$ as $(3+1) + \frac{1}{5} = 3 + (\frac{5}{5} + \frac{1}{5}) = 3\frac{6}{5}$ and then do the subtraction as before. Think of this as taking a unit from 4 and decomposing it into 5 fifths.

$$4\frac{1}{5} - 2\frac{1}{3} = 3\frac{6}{5} - 2\frac{1}{3} = (3 - 2) + (\frac{6}{5} - \frac{1}{3}) = 1\frac{13}{15}$$

Oral exercise 5.72. Calculate mentally and explain your procedure.

(1) Which is smaller: $\frac{11}{8} - \frac{2}{3}$ or $\frac{15}{16} - \frac{2}{3}$?

(2) Which is smaller: $\frac{101}{102}$ or $\frac{1001}{1002}$? Hint: Both are smaller than 1. Why? Thus, the larger number is closer to 1. Which is closer to 1? The larger of $1 - \frac{101}{102}$ and $1 - \frac{1001}{1002}$.

(3) $\frac{3}{2} + \frac{1}{4} - \frac{1}{2}$

(4) $\frac{15}{16} + \frac{3}{8} - \frac{7}{8}$

(5) $\frac{7}{10} - (\frac{9}{13} - \frac{3}{10})$

(6) $\frac{11}{8} + \frac{5}{7} - \frac{3}{8}$

(7) $\frac{11}{3} - \frac{2}{5} - \frac{8}{5}$

(8) $5\frac{3}{8} - 2\frac{1}{8}$

(9) $5\frac{1}{6} - 3\frac{3}{12}$

(10) $\frac{23}{5} - \frac{2}{3} - \frac{8}{5}$

(11) $\frac{1}{2} - \frac{1}{5} + \frac{3}{5}$

(12) $7\frac{3}{4} - 2\frac{1}{5} + 1\frac{1}{5}$

(13) $6\frac{3}{4} - (3\frac{1}{4} - 1\frac{1}{2})$

§5.4 Exercises

Explain your solution in each problem.

(1) Find $\frac{5}{6} - \frac{4}{5}$. Find $\frac{6}{7} - \frac{5}{9}$. Use the definition of subtraction to check that your answer is correct.

(2) How much larger than $\frac{4}{15}$ is $\frac{7}{10}$? How much smaller than $\frac{5}{8}$ is $\frac{7}{12}$?

(3) Use the number line interpretation of subtraction to show that $\frac{7}{8} - \frac{3}{8} = \frac{1}{2}$.

(4) Compute $36\frac{2}{7} - 18\frac{2}{5}$ and explain your method. Use the definition of subtraction to verify that your answer is correct.

(5) Compute $\frac{19}{21} - \frac{11}{14}$ and explain your method. Use the definition of subtraction to verify that your answer is correct.

(6) Do the following computations and explain your method in each case.
 (a) $7\frac{1}{2} - 2\frac{4}{9}$ and $9\frac{1}{6} - 8\frac{8}{9}$
 (b) $\frac{1}{2} + \frac{1}{4} - \frac{1}{3}$ and $\frac{2}{3} + \frac{3}{4} - \frac{5}{6}$
 (c) $\frac{1}{2} + \frac{7}{8} - \frac{1}{16}$ and $\frac{3}{6} - \frac{3}{15} + \frac{1}{2}$
 (d) $5\frac{3}{8} + 2\frac{1}{8} - 3\frac{1}{4}$ and $4\frac{3}{4} - 2\frac{1}{5} + 1\frac{1}{3}$
 (e) $14 + 6\frac{1}{2} - 9\frac{3}{4}$ and $2\frac{3}{4} + 3\frac{7}{8} - 5\frac{1}{2}$

(7) David spent $\frac{1}{2}$ of his money for a coat, $\frac{1}{4}$ for a hat, and $\frac{1}{5}$ for shoes: what fraction of his money remained? *David's money:*

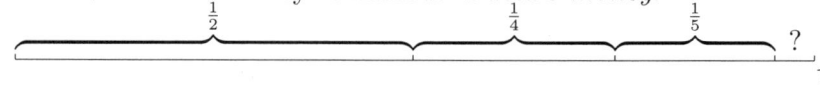

(8) Samantha is $9\frac{1}{2}$ years old and Andrea $6\frac{1}{5}$ years: how much older is Samantha than Andrea? How much older was she $2\frac{1}{2}$ years ago? Explain your answer.

5.5. Multiplication of Rational Numbers

(9) A piece of carpet contains $65\frac{1}{2}$ yards: if Janet sells to Sue $22\frac{1}{4}$ yards, and to Sigrid $23\frac{1}{8}$ yard, how much remains?

5.5. Multiplication of Rational Numbers

The process of multiplying fractions is simpler than the process of adding or subtracting fractions, because the concepts have already been encountered. To multiply a number by a fraction is to take that fraction of the number, where taking a fraction of something was defined in Definition 2.104. We must check then that the product depends only on the value of the fraction.

Definition 5.73 (Multiplication by a fraction). For a fraction $\frac{a}{b}$ and a rational number y, the product $\frac{a}{b} \cdot y$ is $\frac{a}{b}$ of y, which is the number obtained from dividing y into b equal parts and going a of these parts to the right from 0.

Recall that a fraction such as $\frac{4}{7}$ is 4 sevenths. The numerator is the number of objects, and the denominator denotes which objects, sevenths in this case. The same value is maintained if we divide each object into n equal parts, provided that we take n times as many of these new parts. For the example of $\frac{4}{7}$, divide each seventh into 3 equal parts making each part a twenty-first. Then we have 4 groups of 3 of these new parts, as in $\frac{4}{7} = \frac{4 \cdot 3}{7 \cdot 3}$, which is $4 \cdot 3 = 12$ twenty-firsts.

Example 5.74 (Multiplication by a whole number). The product $3 \cdot \frac{7}{8}$ is 3 of 7 eighths, which is $3 \cdot 7$ eighths, which is $\frac{3 \cdot 7}{8}$, by definition of the value of fractions. Therefore, $3 \cdot \frac{7}{8} = \frac{3 \cdot 7}{8}$. Notice that multiplication by 3 is to add the multiplicand 3 times: $\frac{7}{8} + \frac{7}{8} + \frac{7}{8} = \frac{7+7+7}{8} = \frac{3 \cdot 7}{8} = 3 \cdot \frac{7}{8}$.

Example 5.75 (Multiplication of a whole number). To multiply $\frac{3}{4} \cdot 8$, we divide 8 into 4 equal parts, giving 2 in each part, and take 3 of these parts, which makes $3 \cdot 2 = 6$. Therefore, $\frac{3}{4} \cdot 8 = 6$. The unitary method views this as follows.

4 units = 8

1 unit = 8/4 = 2

3 units = $3 \cdot 2 = 6$

To multiply $\frac{2}{3} \cdot 4$ we need to divide 4 units into 3 equal parts and go 2 of them from 0. Now 4 things cannot be divided into 3 equal parts without dividing each thing, as expressed by $4 = \frac{4 \cdot 3}{3}$, which is $4 \cdot 3$ thirds. A set of $4 \cdot 3$ thirds can be divided into 3 equal parts, with $4 \cdot 3 \div 3 = 4$ thirds in each part. To go 2 of these from 0 is to go $2 \cdot 4$ thirds from 0, which ends at the value of $\frac{2\cdot 4}{3}$. Therefore, $\frac{2}{3} \cdot 4 = \frac{2\cdot 4}{3}$.

We divided the interval from 0 to 4 into 3 equal subintervals by dividing each unit interval into 3 equal subintervals and then applied whole number partitive division to divide the resulting set of $4 \cdot 3 = 12$ thirds into 3 equal parts. This can be pictured as follows. The subintervals of length $\frac{4}{3}$ divide the interval from 0 to 4 into 3 equal parts.

By the unitary method:

    3 units $= 4$

    1 unit $= \frac{4}{3}$

    2 units $= 2 \cdot \frac{4}{3} = \frac{8}{3}$

Example 5.76 (Multiplication of a fraction by a fraction). To multiply $\frac{2}{3} \cdot \frac{4}{5}$ we need to divide the interval from 0 to $\frac{4}{5}$ into 3 equal parts and go 2 of them from 0. A set of 4 fifths cannot be divided into 3 equal parts without dividing the fifths. To divide each fifth into 3 equal parts is to express $\frac{4}{5} = \frac{4\cdot 3}{5\cdot 3}$, which is $4 \cdot 3$ fifteenths. We can divide a set of $4 \cdot 3$ fifteenths into 3 equal parts, by the partitive division $4 \cdot 3 \div 3 = 4$ fifteenths in each part. Taking 2 of these from 0 is to go $2 \cdot 4$ fifteenths from 0, which ends at the value of $\frac{2\cdot 4}{3\cdot 5}$. Therefore, $\frac{2}{3} \cdot \frac{4}{5} = \frac{2\cdot 4}{3\cdot 5}$.

We divided 4 fifths into 3 equal parts by dividing each fifth into 3 equal parts, each of which is a fifteenth, since $5 \cdot 3 = 15$ of them make the unit interval. Then $\frac{4}{5} = \frac{4\cdot 3}{5\cdot 3} = 4 \cdot 3$ fifteenths. This can be divided into 3 equal parts by taking $(4 \cdot 3) \div 3 = 4$ fifteenths in each part, illustrated as follows.

By the unitary method:

    3 units $= \frac{4}{5}$

    1 unit $= \frac{4}{5} \div 3 = \frac{4}{15}$

    2 units $= 2 \cdot \frac{4}{15} = \frac{8}{15}$

Oral exercise 5.77. Answer in the style of the preceding examples.

    (1) What is $\frac{1}{2}$ of 4? What is $\frac{1}{2} \cdot 6$?

    (2) What is $\frac{1}{2}$ of 3? What is $\frac{1}{2} \cdot 5$?

    (3) What is $\frac{1}{2}$ of $\frac{6}{5}$? of $\frac{10}{7}$? of $\frac{11}{12}$?

5.5. Multiplication of Rational Numbers

(4) What is $\frac{1}{3} \cdot \frac{3}{5}$? $\frac{1}{3} \cdot \frac{6}{7}$? $\frac{1}{3} \cdot \frac{7}{3}$?

(5) What is $\frac{4}{7}$ of $\frac{14}{15}$? Of $\frac{1}{2}$?

(6) What is $\frac{6}{5} \cdot \frac{10}{7}$? $\frac{6}{5} \cdot \frac{3}{4}$?

Theorem 5.78. *For a fraction $\frac{a}{b}$ and a rational number y represented by the fraction $\frac{p}{q}$, their product $\frac{a}{b} \cdot y$ is the rational number represented by the fraction $\frac{a \cdot p}{b \cdot q}$. Namely,*

$$\frac{a}{b} \cdot \frac{p}{q} = \frac{a \cdot p}{b \cdot q}$$

Proof. $\frac{a}{b} \cdot \frac{p}{q}$ is a of $\frac{1}{b}$ of $\frac{p}{q}$. Since $\frac{p}{q} = \frac{b \cdot p}{b \cdot q} = b \cdot p \, \frac{1}{b \cdot q}$'s, this divided into b equal parts has $b \cdot p \div b = p \, \frac{1}{b \cdot q}$'s in each part. To go a of these from 0 is to go $a \cdot p \, \frac{1}{b \cdot q}$'s from 0, which is $\frac{a \cdot p}{b \cdot q}$, by definition of the value of fractions. □

Next we verify that the value of $\frac{a}{b} \cdot y$ is the same if $\frac{a}{b}$ is replaced by any fraction of the same value.

Theorem 5.79. *If $\frac{a}{b} = \frac{c}{d}$ and if y is any rational number, then $\frac{a}{b} \cdot y = \frac{c}{d} \cdot y$.*

Proof. Consider the case of $\frac{4}{6} = \frac{6}{9}$ and $y = \frac{7}{5}$. Expressed in lowest terms, $\frac{4}{6} = \frac{2 \cdot 2}{2 \cdot 3} = \frac{2}{3}$ and $\frac{6}{9} = \frac{3 \cdot 2}{3 \cdot 3} = \frac{2}{3}$. Then

(5.9)
$$\begin{aligned}
\frac{4}{6} \cdot y &= \frac{4 \cdot 7}{6 \cdot 5}, \text{ by Theorem 5.78} \\
&= \frac{(2 \cdot 2) \cdot 7}{(2 \cdot 3) \cdot 5}, \text{ factor 2 from 4 and 6} \\
&= \frac{2 \cdot (2 \cdot 7)}{2 \cdot (3 \cdot 5)}, \text{ associative property for whole numbers} \\
&= \frac{2 \cdot 7}{3 \cdot 5}, \text{ by Theorem 5.17} \\
&= \frac{2}{3} \cdot y, \text{ by Theorem 5.78}
\end{aligned}$$

and

(5.10)
$$\begin{aligned}
\frac{6}{9} \cdot y &= \frac{6 \cdot 7}{9 \cdot 5}, \text{ by Theorem 5.78} \\
&= \frac{(3 \cdot 2) \cdot 7}{(3 \cdot 3) \cdot 5}, \text{ factor 3 from 6 and 9} \\
&= \frac{3 \cdot (2 \cdot 7)}{3 \cdot (3 \cdot 5)}, \text{ associative property} \\
&= \frac{2 \cdot 7}{3 \cdot 5}, \text{ by Theorem 5.17} \\
&= \frac{2}{3} \cdot y, \text{ by Theorem 5.78}
\end{aligned}$$

Both products being equal to $\frac{2}{3} \cdot y$, we conclude that they are equal to each other, as was to be proved.

The general argument goes in the same way. Multiplication by a fraction gives the same value as multiplication by this fraction's expression in lowest terms. Since any two fractions with the same value have the same fraction expressing them in lowest terms, it follows that multiplication by these two fractions will give the same value. □

Corollary 5.80 (Cancellation). *If b is a factor of p, then*

$$(5.11) \qquad \frac{a}{b} \cdot \frac{p}{q} = \frac{a \cdot (p \div b)}{q} = \frac{a \cdot \overset{p/b}{\cancel{p}}}{\cancel{b} \cdot q}$$

If q is a factor of a, then

$$(5.12) \qquad \frac{a}{b} \cdot \frac{p}{q} = \frac{(a \div q) \cdot p}{b} = \frac{\overset{a/q}{\cancel{a}} \cdot p}{b \cdot \cancel{q}}$$

Proof. The first statement follows directly from the definition of multiplication by a fraction. We can divide p $q$ths into b equal parts, with $p \div b$ $q$ths in each part, because we have assumed that b is a factor of p. Each part is $\frac{p/b}{q}$. Then a of these equals $a \cdot (p \div b)$ $q$ths, as stated in (5.11).

For the second statement, use the preceding theorem to write $\frac{a}{b} \cdot \frac{p}{q} = \frac{a \cdot p}{b \cdot q}$. This last fraction is equal to $\frac{p \cdot a}{q \cdot b}$, by the commutative property of multiplication of whole numbers. But then the preceding theorem says that this last fraction equals $\frac{p}{q} \cdot \frac{a}{b}$, and then the first statement applied to this gives (5.12). □

Example 5.81.

$$\frac{3}{5} \cdot \frac{15}{8} = \frac{3 \cdot (15 \div 5)}{8} = \frac{3 \cdot 3}{8} = \frac{9}{8}$$

$$\frac{6}{7} \cdot \frac{4}{3} = \frac{\overset{2}{\cancel{6}} \cdot 4}{7 \cdot \cancel{3}} = \frac{8}{7}$$

Definition 5.82 (Multiplication of rational numbers). For rational numbers x and y, their product, $x \cdot y = \frac{a}{b} \cdot y$, where $\frac{a}{b}$ is any fraction whose value is x.

Notice that Theorem 5.79 shows that any choice of fraction $\frac{a}{b}$ representing x gives the same value for $\frac{a}{b} \cdot y$. For example, to find the product of the rational numbers represented by $\frac{2}{3}$ and $\frac{3}{5}$, we get the same value from $\frac{2}{3} \cdot \frac{3}{5}$ as from $\frac{4}{6} \cdot \frac{12}{20}$. In fact, we get the same value from the product of any two fractions whose values are the same as $\frac{2}{3}$ and $\frac{3}{5}$, respectively.

5.5. Multiplication of Rational Numbers

Theorem 5.83. *Multiplication of rational numbers has the following properties. If x, y, and z are rational numbers, then:*

(1) Multiplication of rational numbers, when applied to whole numbers, agrees with multiplication of whole numbers;

(2) $x \cdot y = y \cdot x$, commutative property;

(3) $x \cdot (y \cdot z) = (x \cdot y) \cdot z$, associative property;

(4) $x \cdot 1 = 1 \cdot x = x$, identity property of 1;

(5) $x \cdot (y + z) = x \cdot y + x \cdot z$, distributive property of multiplication over addition;

(6) If $z < y$, and $x \neq 0$, then $x \cdot z < x \cdot y$, order property;

(7) If $z < y$, then $x \cdot (y-z) = x \cdot y - x \cdot z$, distributive property of multiplication over subtraction.

Proof. We illustrate the proof with specific numbers in such a way as to illustrate the general arguments. Observe how each of these properties is true ultimately because it is true for whole numbers.

(1) If $x = 5$ and $y = 7$ are whole numbers, then $5 = \frac{5}{1}$ and $7 = \frac{7}{1}$, so Theorem 5.78 says that $x \cdot y = \frac{5}{1} \cdot \frac{7}{1} = \frac{5 \cdot 7}{1 \cdot 1}$, which has the same value as the whole number $5 \cdot 7$.

(2) The commutative property of multiplication of fractions was illustrated in Example 2.107. If $x = \frac{a}{b}$ and $y = \frac{c}{d}$, then

$$x \cdot y = \frac{a \cdot c}{b \cdot d} \quad \text{and} \quad y \cdot x = \frac{c \cdot a}{d \cdot b}$$

which are equal because the multiplications in numerator and denominator are between whole numbers, and multiplication of whole numbers is commutative.

(3) The associative property of multiplication of fractions was also illustrated in Example 2.107, which shows that $\frac{2}{3}$ of $\frac{4}{5}$ of the rectangle, so $\frac{2}{3} \cdot (\frac{4}{5} \cdot \text{whole})$, is equal to $\frac{2}{3} \cdot \frac{4}{5} = \frac{8}{15}$ of the rectangle, so $(\frac{2}{3} \cdot \frac{4}{5}) \cdot \text{whole}$.

If $x = \frac{3}{5}$, $y = \frac{9}{8}$, and $z = \frac{2}{7}$, then

$$x \cdot (y \cdot z) = \frac{3}{5} \cdot (\frac{9}{8} \cdot \frac{2}{7}), \quad \text{by Definition 5.82}$$

$$= \frac{3}{5} \cdot \frac{9 \cdot 2}{8 \cdot 7}, \quad \text{by Theorem 5.78}$$

$$= \frac{3 \cdot (9 \cdot 2)}{5 \cdot (8 \cdot 7)}, \quad \text{by Theorem 5.78}$$

and, for the same reasons,

$$(x \cdot y) \cdot z = (\frac{3}{5} \cdot \frac{9}{8}) \cdot \frac{2}{7} = \frac{3 \cdot 9}{5 \cdot 8} \cdot \frac{2}{7} = \frac{(3 \cdot 9) \cdot 2}{(5 \cdot 8) \cdot 7}$$

In the final fraction in each set of equations, the numerators agree and the denominators agree, because multiplication of whole numbers is associative, and therefore these final fractions are equal to each other.

(4) If $x = \frac{a}{b}$, then $\frac{a}{b} \cdot 1$ is the point on the number line obtained by dividing the unit interval into b equal parts and then going a of these from 0. That is the definition of the point represented by the fraction $\frac{a}{b}$. Therefore, $\frac{a}{b} \cdot 1 = \frac{a}{b}$.

By definition of $1 \cdot x$, this is the point obtained by dividing the interval from 0 to x into 1 part and going 1 of these parts from 0. That point is x.

(5) If $x = \frac{3}{5}$, $y = \frac{9}{8}$, and $z = \frac{2}{7}$, then we can reduce $\frac{9}{8}$ and $\frac{2}{7}$ to a common denominator with $\frac{9}{8} = \frac{63}{56}$ and $\frac{2}{7} = \frac{16}{56}$. Then

$$x \cdot (y+z) = \frac{3}{5} \cdot \left(\frac{63}{56} + \frac{16}{56}\right), \text{ by Definition 5.82}$$
$$= \frac{3}{5} \cdot \frac{63+16}{56}, \text{ Theorem 5.47}$$
$$= \frac{3 \cdot (63+16)}{5 \cdot 56}, \text{ Theorem 5.78}$$
$$= \frac{3 \cdot 63 + 3 \cdot 16}{5 \cdot 56}, \text{ distributive property for whole numbers}$$
$$= \frac{3 \cdot 63}{5 \cdot 56} + \frac{3 \cdot 16}{5 \cdot 56}, \text{ Theorem 5.47}$$
$$= \frac{3}{5} \cdot \frac{63}{56} + \frac{3}{5} \cdot \frac{16}{56}, \text{ Theorem 5.78}$$
$$= x \cdot y + x \cdot z, \text{ definition of multiplication}$$

(6) If $x = \frac{3}{5}$, $y = \frac{9}{8}$, and $z = \frac{2}{7}$, then $x \neq 0$ and $z < y$. We can reduce $\frac{9}{8}$ and $\frac{2}{7}$ to a common denominator with $\frac{9}{8} = \frac{63}{56}$ and $\frac{2}{7} = \frac{16}{56}$, thus verifying that $z < y$ because $16 < 63$. Then

$$x \cdot z = \frac{3}{5} \cdot \frac{16}{56}, \text{ by definition of multiplication}$$
$$= \frac{3 \cdot 16}{5 \cdot 56}, \text{ by Theorem 5.78}$$
$$< \frac{3 \cdot 63}{5 \cdot 56}, \text{ by } 3 \cdot 16 < 3 \cdot 63 \text{ and Theorem 5.38}$$
$$= \frac{3}{5} \cdot \frac{63}{56}, \text{ by Theorem 5.78}$$
$$= x \cdot y, \text{ by definition of multiplication}$$

from which we conclude that $x \cdot z < x \cdot y$, as was to be proved.

The proof of the distributive property of multiplication over subtraction goes in the same way as that for addition. The only extra condition to be checked is that the final subtraction is defined. This follows from the order property for rational numbers. □

5.5. Multiplication of Rational Numbers

Example 5.84 (Product of mixed fractions). Use the distributive property

$$(3\frac{3}{4}) \cdot (4\frac{5}{9}) = (3 + \frac{3}{4}) \cdot (4 + \frac{5}{9})$$
$$= 3 \cdot (4 + \frac{5}{9}) + \frac{3}{4} \cdot (4 + \frac{5}{9}), \text{ distributive property}$$
$$= 3 \cdot 4 + 3 \cdot \frac{5}{9} + \frac{3}{4} \cdot 4 + \frac{3}{4} \cdot \frac{5}{9}, \text{ distributive property}$$
$$= 12 + \frac{5}{3} + 3 + \frac{5}{12}, \text{ cancellation}$$
$$= 15 + \frac{25}{12} = 15 + 2\frac{1}{12} = 17\frac{1}{12}$$

or, convert to improper fractions

$$(3\frac{3}{4}) \cdot (4\frac{5}{9}) = \frac{15}{4} \cdot \frac{41}{9} = \frac{\overset{5}{\cancel{15}} \cdot 41}{\cancel{4} \cdot \cancel{9}} = \frac{205}{12} = 17\frac{1}{12}$$

Oral exercise 5.85. Compute and explain.

(1) What is 4 times $\frac{5}{8}$? $\frac{5}{6}$? $\frac{3}{14}$? $\frac{10}{16}$?
(2) What is 8 of $\frac{10}{12}$? $\frac{3}{16}$? $5\frac{1}{2}$? $\frac{9}{12}$?
(3) What is 10 times $1\frac{3}{20}$? $2\frac{5}{24}$? $6\frac{1}{5}$? $3\frac{7}{170}$?
(4) Explain the equality of $\frac{3}{5}$ and $\frac{1}{5}$ of 3. *Ans.* $\frac{3}{5}$ is 3 $\frac{1}{5}$'s and $\frac{1}{5}$ of 3 is $\frac{1}{5}$ of $5 \cdot 3$ $\frac{1}{5}$'s, which is $5 \cdot 3 \div 5 = 3$ $\frac{1}{5}$'s.
(5) Explain the equality of $\frac{7}{4}$ and $\frac{1}{4}$ of 7.
(6) Explain the equality of $\frac{1}{12}$ and $\frac{1}{3}$ of $\frac{3}{12}$.
(7) Explain the equality of $\frac{1}{12}$ and $\frac{1}{3}$ of $\frac{1}{4}$.

5.5.1. Averages and partitions.

Definition 5.86. The *average* (or *mean*) of two rational numbers x and y is their sum divided by 2, which is $\frac{1}{2} \cdot (x+y) = \frac{1}{2} \cdot x + \frac{1}{2} \cdot y$. The average of n rational numbers is their sum divided by n, which is $\frac{1}{n}$ times their sum.

Example 5.87. The average of $\frac{3}{8}$ and $\frac{4}{5}$ is $\frac{1}{2} \cdot (\frac{3}{8} + \frac{4}{5}) = \frac{1}{2} \cdot (\frac{15+32}{40}) = \frac{47}{80}$, which lies halfway from $\frac{3}{8}$ to $\frac{4}{5}$ as shown on the following number line.

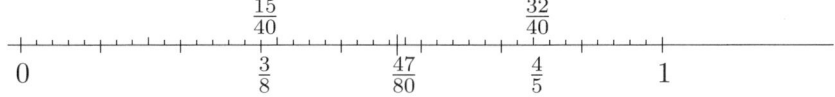

Theorem 5.88. *The average of rational numbers x and y is a rational number lying halfway between x and y.*

Proof. Suppose that $x < y$. This means that $y - x$ is defined and a rational number. The average $\frac{1}{2} \cdot (x + y)$ is a rational number, because sums and products of rational numbers are rational numbers. The number halfway between x and y is $x + \frac{1}{2} \cdot (y - x)$, which is

$$\begin{aligned}
x + \frac{1}{2} \cdot (y - x) &= x + (\frac{1}{2} \cdot y - \frac{1}{2} \cdot x), \text{ dist. prop. (5) of Theorem 5.83} \\
&= (x + \frac{1}{2} \cdot y) - \frac{1}{2} \cdot x, \text{ (4) of Theorem 5.65} \\
&= (\frac{1}{2} \cdot y + x) - \frac{1}{2} \cdot x, \text{ commutative property} \\
&= \frac{1}{2} \cdot y + (x - \frac{1}{2} \cdot x), \text{ (4) of Theorem 5.65} \\
&= \frac{1}{2} \cdot y + (1 - \frac{1}{2}) \cdot x, \ x = 1 \cdot x \text{ and (5) of Theorem 5.83} \\
&= \frac{1}{2} \cdot y + \frac{1}{2} \cdot x = \frac{1}{2} \cdot (x + y), \text{ (5) of Theorem 5.83}
\end{aligned}$$

the average of x and y. □

Oral exercise 5.89. (1) Find
  (a) $3 - \frac{1}{2} \cdot 3$ and $\frac{3}{4} - \frac{1}{2} \cdot \frac{3}{4}$.
  (b) $\frac{1}{2} \cdot 7 + \frac{1}{2} \cdot 3$ and $\frac{1}{2} \cdot \frac{3}{5} + \frac{1}{2} \cdot \frac{7}{5}$.

(2) Find the average of 3 and 4. Explain why the average lies halfway between the two given numbers.

(3) Find the rational number halfway between $\frac{1}{2}$ and $\frac{3}{4}$.

(4) Find the rational number halfway between $\frac{7}{11}$ and $\frac{8}{11}$.

In Theorem 5.88 we saw that the average of two numbers is the number halfway between them. For example, the average of 5 and 8 is $\frac{1}{2} \cdot (5 + 8) = 5 + \frac{1}{2} \cdot (8 - 5) = 6\frac{1}{2}$, and that is the number halfway from 5 to 8.

In the same way $5 + \frac{1}{4} \cdot (8 - 5) = 5\frac{3}{4}$ is the number $\frac{1}{4}$ of the way from 5 to 8.

The number which is $\frac{7}{5}$ of the way from 5 to 8 is $5 + \frac{7}{5} \cdot (8 - 5) = 5 + \frac{21}{5} = 9\frac{1}{5}$.

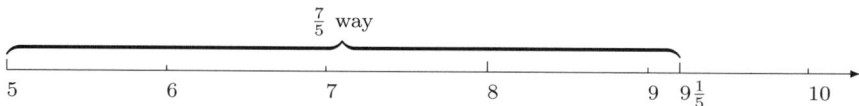

The general procedure is expressed as follows.

Theorem 5.90. *If x, y, and z are rational numbers, and if $x < y$, then the rational number that lies z of the way from x to y is*
$$x + z \cdot (y - x)$$
This is equal to $(1 - z) \cdot x + z \cdot y$. This number lies between x and y if and only if z is between 0 and 1. If $z = \frac{1}{2}$, then $x + z \cdot (y - x)$ is the average of x and y.

Oral exercise 5.91. Find the number $\frac{1}{2}$ of the way from 2 to 8; $\frac{2}{3}$ of the way; $\frac{5}{3}$ of the way.

§5.5 Exercises

Explain your solutions. Use a line segment diagram whenever possible.

(1) (a) Explain the meaning of $\frac{3}{5}$ of a line segment? Draw a picture.
(b) What is the definition of $\frac{3}{5} \cdot y$, where y is any rational number? Draw a picture.
(c) From the definition find $\frac{3}{5} \cdot \frac{10}{13}$. Draw a picture.

(2) As in the examples preceding Theorem 5.78, explain the following.
(1) $\frac{1}{4} \cdot 3$. (2) $\frac{3}{4} \cdot 8$. (3) $\frac{3}{4} \cdot \frac{12}{5}$. (4) $\frac{3}{4} \cdot \frac{7}{5}$.

(3) Explain why $\frac{3}{5} \cdot \frac{1}{2}$ is equal to $\frac{6}{10} \cdot \frac{1}{2}$, as in the proof of Theorem 5.79.

(4) Use the associative and commutative properties of multiplication of rational numbers to explain why $\frac{5}{8} \cdot \frac{7}{11} = 5 \cdot 7 \cdot \frac{1}{8} \cdot \frac{1}{11}$.

(5) Explain how to find $\frac{2}{3}$ of $5\frac{1}{2}$.

(6) Multiply –
(a) $\frac{9}{10}$ by $\frac{5}{21}$ by $\frac{7}{5}$. Ans. $\frac{3}{10}$.
(b) $\frac{6}{7}$ by $\frac{10}{12} \cdot 6\frac{3}{5}$. Ans. $4\frac{5}{7}$.
(c) $\frac{13}{14}$ of $\frac{20}{21}$ by $\frac{15}{18}$ of $\frac{7}{39}$. Ans. $\frac{25}{189}$.

(7) Find the average of 1, 2, and 4.

(8) Find the rational number halfway between $\frac{11}{10}$ and $\frac{17}{15}$.

(9) Explain why $3\frac{1}{4} \cdot 8 = 3 \cdot 8 + \frac{1}{4} \cdot 8 = 26$.

(10) Explain why $5 \cdot 7\frac{3}{5} = 5 \cdot 7 + 5 \cdot \frac{3}{5} = 38$.

(11) Explain why $3\frac{1}{4} \cdot 8\frac{5}{6} = 3 \cdot 8 + 3 \cdot \frac{5}{6} + \frac{1}{4} \cdot 8 + \frac{1}{4} \cdot \frac{5}{6} = 36\frac{5}{24}$.

(12) Use the Order Property of Multiplication to explain why $\frac{13}{14} \cdot \frac{17}{19} < \frac{13}{14}$. Hint: Multiply both sides of $\frac{17}{19} < 1$ by $\frac{13}{14}$.

(13) Find the number which is $\frac{3}{8}$ of the way from $\frac{11}{13}$ to $\frac{12}{13}$.

(14) Find the number which is $\frac{7}{5}$ of the way from $7\frac{1}{2}$ to $13\frac{7}{8}$.

(15) If a recipe calls for $\frac{3}{4}$ cup of butter for a batch of cookies, how much butter is needed for $2\frac{1}{2}$ batches?

(16) If a fish weighs $\frac{3}{8}$ of a pound, how much will 10 weigh if they are of the same size?

(17) If your car uses $\frac{4}{7}$ of a tank of gas in a week, how much gas does it use in 14 weeks?

(18) A store had 18 chickens and sold $\frac{1}{3}$ of them: how many were sold?

(19) Joshua has 60 cents and Caitlin has $\frac{5}{4}$ as many: how many cents has Caitlin?

(20) If 2 hats cost $14, what will $\frac{3}{7}$ of 42 hats cost?

(21) Alexi had $35. She used $\frac{2}{7}$ of it to buy a CD. How much money did she have left? Explain how to solve this problem by the following three methods.
  (a) Use the definition of a fraction of something to find how much she spent on the CD, then subtract this from the total to find how much is left.
  (b) Find the fraction of money she has left and then use the definition of fraction of something to find how much money she has left.
  (c) Unitary method: Divide a line segment into 7 equal parts and indicate on it which parts correspond to what she spent and which parts correspond to what she has left. Decide on a unit. Find the value of one unit, and use this to solve the problem.

(22) Ralph had 600 bushels of wheat. He sold $\frac{3}{10}$ of it to Bob and $\frac{1}{4}$ of it to Dick. What fraction of it remained? How many bushels remained?

(23) If Suella has $4\frac{3}{5}$ bushels of wheat, and sells $\frac{2}{3}$ of it, what fraction remains? How many bushels remain?

(24) Bo divided $\frac{2}{3}$ of his pizza equally among Stephanie, Cameron, David, and Bridget. What fraction of the pizza did Stephanie receive?

(25) Find the area in square feet of a rectangle which is $\frac{1}{3}$ ft. wide and $\frac{5}{6}$ ft. long. What is the area in square inches? Answer this last question in two ways: (i) By first finding the width and length in inches. (ii) By determining the number of square inches in 1 square foot and then converting the area in square feet to square inches.

(26) Three-fourths of a wall is painted white. One-third of the remaining part is painted green. What fraction of the wall is painted green? Illustrate.

5.6. Division of Rational Numbers

Division of rational numbers is defined as the inverse operation to multiplication. Division is to multiplication as subtraction is to addition. As for subtraction and for division of whole numbers, there are two kinds of division, partitive and measurement. If the divisor specifies the size of the group, it is measurement division. If the divisor specifies the number of groups, it is partitive division.

If the divisor is not a whole number, then partitive division seems confusing, because the divisor is specifying a fractional number of groups. If we rephrase the partitive division question to be "b of what makes a?", then it makes sense even when b and a are positive rational numbers. The measurement division problem can be rephrased as "How many of b in a?", with the understanding that the answer can be a fraction.

In contrast to the situation with whole numbers, the division $x \div y$ is always defined for rational numbers if $y \neq 0$.

Definition 5.92 (Partitive division). If x and $y > 0$ are rational numbers, then the partitive quotient $x \div y$ is the rational number z such that $y \cdot z = x$.

In words, y of z is x, so z is the answer to the question "y of what is x?".

Definition 5.93 (Measurement division). If x and $y > 0$ are rational numbers, then the measurement quotient $x \div y$ is the rational number z such that $z \cdot y = x$.

In words, z of y is x, so z is the answer to the question "How many of y is x?". We understand that the answer can be a fraction, or mixed fraction, expressing the fact that there are a certain whole number of y's, plus a part of y, in x.

In the notation $x \div y = z$, the number x is called the *dividend*, y is called the *divisor*, and z is called the *quotient*. We use the same notation for both kinds of division, because they both produce the same quotient.

Theorem 5.94. *If x and $y > 0$ are rational numbers, then $x \div y$ in partitive division is equal to $x \div y$ in measurement division.*

Proof. If $z = x \div y$ in partitive division, then $x = y \cdot z$. By the commutative property of multiplication, $y \cdot z = z \cdot y$, which implies that $z = x \div y$ in measurement division. Conversely, starting with the measurement division quotient, we see in the same way that it is also the partitive division quotient. □

We shall see that for numbers that are not whole, measurement division seems conceptually simpler than partitive division. Partitive division problems can usually be done by the unitary method, in which the only division being done is by a whole number.

Example 5.95. Measurement division $1 \div 5 = \frac{1}{5}$ because $\frac{1}{5} \cdot 5 = 1$. This solves the problem: how many intervals of length 5 are in the unit interval? The answer is $\frac{1}{5}$ of an interval of length 5 is in a unit interval. The answer to "How many?" can be a fraction.

Partitive division $1 \div 5 = \frac{1}{5}$ because $5 \cdot \frac{1}{5} = 1$. This solves the problem: 5 of what is 1? The answer of $\frac{1}{5}$ is what we get when we divide the unit interval into 5 equal subintervals. This is the problem we solve in our definition of the value of the fraction $\frac{1}{5}$.

Example 5.96. Measurement division $\frac{2}{3} \div \frac{5}{8} = \frac{16}{15}$ because $\frac{16}{15} \cdot \frac{5}{8} = \frac{2}{3}$, as can be computed quickly using cancellation. This solves the problem: How many intervals of length $\frac{5}{8}$ in an interval of length $\frac{2}{3}$? The answer of $\frac{16}{15} = 1\frac{1}{15}$ means that there is one such interval and $\frac{1}{15}$ of such an interval more. The solution can be illustrated as follows, where the smallest subintervals shown are twenty-fourths. In the picture we can see 1 interval of length $\frac{5}{8}$ within the interval from 0 to $\frac{2}{3}$, and the small interval from $\frac{5}{8}$ to $\frac{2}{3}$ is $\frac{1}{15}$ of $\frac{5}{8}$, which is $\frac{1}{15} \cdot \frac{5}{8} = \frac{1}{24}$.

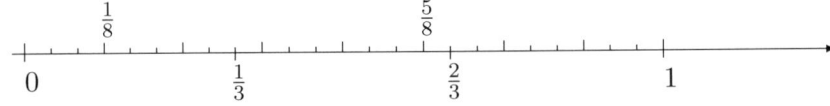

Partitive division $\frac{2}{3} \div \frac{5}{8} = \frac{16}{15}$ because $\frac{5}{8} \cdot \frac{16}{15} = \frac{2}{3}$, as can be calculated using cancellation. This solves the problem: $\frac{5}{8}$ of what is $\frac{2}{3}$? The solution can be illustrated as follows. The smallest subintervals are fifteenths. These need to be counted to see for certain that $\frac{2}{3}$, which is 10 fifteenths, is $\frac{5}{8}$ of 16 fifteenths.

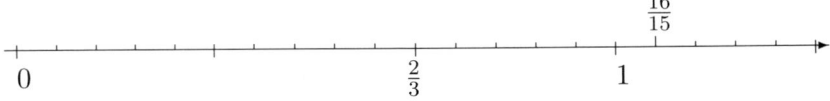

Theorem 5.97 (Invert and Multiply). *For rational numbers $x \geq 0$ and $y \neq 0$, represented by fractions $x = \frac{a}{b}$ and $y = \frac{p}{q}$, the quotient $x \div y$ is*

5.6. Division of Rational Numbers

represented by the fraction $\frac{q}{p} \cdot \frac{a}{b} = \frac{q \cdot a}{p \cdot b}$. In terms of fractions, this is

$$\frac{a}{b} \div \frac{p}{q} = \frac{q}{p} \cdot \frac{a}{b} = \frac{q \cdot a}{p \cdot b}$$

In particular, division of a rational number by a nonzero rational number always exists and the quotient is a rational number.

Proof. For example, the theorem says that $\frac{2}{3} \div \frac{5}{7} = \frac{7}{5} \cdot \frac{2}{3}$. In order to verify the truth of this, the definition of partitive division requires that $\frac{5}{7} \cdot (\frac{7}{5} \cdot \frac{2}{3})$ must equal $\frac{2}{3}$, which is the case by the associative property of multiplication. In fact,

$$\frac{5}{7} \cdot (\frac{7}{5} \cdot \frac{2}{3}) = (\frac{5}{7} \cdot \frac{7}{5}) \cdot \frac{2}{3} = 1 \cdot \frac{2}{3} = \frac{2}{3}$$

The general argument goes in the same way. Notice that the quotient calculated in this way does not depend on the choice of fraction representing the given rational numbers, because multiplication does not depend on this choice. For example, $\frac{2}{5} = \frac{12}{30}$ and

$$\frac{3}{4} \div \frac{2}{5} = \frac{5}{2} \cdot \frac{3}{4} = \frac{30}{12} \cdot \frac{3}{4} = \frac{3}{4} \div \frac{12}{30}$$

$\square$

According to this theorem, if a and $b \neq 0$ are whole numbers, then regarded as rational numbers their quotient is

(5.13) $$a \div b = \frac{a}{1} \div \frac{b}{1} = \frac{1}{b} \cdot \frac{a}{1} = \frac{a}{b}$$

which completes the identification of a/b as either a quotient or a fraction.

Definition 5.98. The *reciprocal* of a nonzero rational number x is the rational number $1 \div x$. Therefore, x times its reciprocal is 1.

The reciprocal of a whole number, such as 7, is the fraction 1 over this number, $\frac{1}{7}$ in this case. In fact, the reciprocal is the quotient $1 \div 7 = \frac{1}{7}$, as remarked above. The reciprocal of a fraction $\frac{a}{b}$ is the fraction $\frac{b}{a}$, since $1 \div \frac{a}{b} = \frac{b}{a} \cdot 1 = \frac{b}{a}$ by the above theorem.

Taking the reciprocal of a fraction is also called *inverting* the fraction. According to Theorem 5.97, to divide by a fraction you multiply by its reciprocal. This is also expressed by saying that to divide by a fraction, invert the fraction and multiply.

Oral exercise 5.99. Follow the model of Examples 5.95 and 5.96: find the quotient and verify that it satisfies the definitions of measurement and partitive division. Pose the measurement division problem and interpret the quotient. Pose the partitive division problem and interpret the quotient.

(1) $1 \div 4$. *Answer:* $1 \div 4 = \frac{1}{4} \cdot 1 = \frac{1}{4}$.

Partitive question: 4 of what is 1? Answer is $1 \div 4 = \frac{1}{4}$, because $4 \cdot \frac{1}{4} = 1$, so 4 of $\frac{1}{4}$ is 1.

Measurement question: How many 4's in 1? Answer is $1 \div 4 = \frac{1}{4}$, because $\frac{1}{4} \cdot 4 = 1$, so there is $\frac{1}{4}$ of 4 in 1.

(2) $\frac{17}{3} \div 4$

(3) $1 \div \frac{4}{5}$

(4) $\frac{7}{3} \div \frac{4}{5}$

Oral exercise 5.100. Using cancellation when possible, calculate the quotient of:

(1) $\frac{15}{17}$ by 5; $\frac{35}{40}$ by 7; $\frac{7}{12}$ by 5; $7\frac{1}{5}$ by 6.

(2) 10 by $\frac{5}{7}$; 12 by $\frac{3}{8}$; 60 by $\frac{7}{9}$; 125 by $3\frac{4}{7}$.

(3) $\frac{7}{8}$ by $\frac{2}{5}$; $\frac{9}{20}$ by $\frac{3}{4}$.

(4) $\frac{1}{3}$ of $\frac{1}{2}$ by $\frac{3}{4}$; $\frac{1}{4}$ of $\frac{1}{5}$ by $\frac{1}{3}$.

Oral exercise 5.101. Find the reciprocal of

(1) 1; 3; 10.

(2) $\frac{1}{5}$; $\frac{1}{10}$; $\frac{1}{7}$.

(3) $\frac{2}{3}$; $\frac{23}{7}$; $3\frac{1}{2}$.

Oral exercise 5.102. Identify each problem as partitive or measurement division and solve it, as shown in the sample problem (a).

(1) If 4 of a given interval makes an interval of $\frac{17}{3}$ units, how many units in the given interval? *Ans.* Partitive division problem: 4 of what is $\frac{17}{3}$? The solution is $\frac{17}{3} \div 4 = \frac{1}{4} \cdot \frac{17}{3} = \frac{17}{12}$.

(2) If $\frac{4}{5}$ of a given interval makes an interval of $\frac{7}{3}$ units, how many units in the given interval?

(3) How many halves in 3 unit intervals?

(4) The unit interval is 3 of how many equal subintervals?

(5) How many 4's in $\frac{5}{3}$?

(6) The unit interval is $\frac{1}{2}$ of what interval?

(7) $\frac{5}{3}$ is 4 of what?

5.6.1. Properties of Division. Division of rational numbers has the following properties.

Theorem 5.103. *(1) If a and $d > 0$ are whole numbers and if the long division $a \div d$ has quotient q and remainder r, where $0 \leq r < d$, so that*

5.6. Division of Rational Numbers

$a = d \cdot q + r$, then in rational number division
$$a \div d = q + \frac{r}{d} = q\frac{r}{d}$$
In particular, when a is divisible by d, which means that $r = 0$, then the whole number quotient q equals the rational number quotient.

(2) Associative properties: If x, y, and z are rational numbers and if $z > 0$, then

(5.14) $$(x \cdot y) \div z = x \cdot (y \div z)$$

and if $y > 0$ as well, then

(5.15) $$(x \div y) \div z = x \div (y \cdot z)$$

(3) Identity property of 1: If x is a rational number, then
$$x \div 1 = x$$

(4) Order properties: Division by a rational number of an inequality preserves it. If $x < y$ and if $z > 0$, then

(5.16) $$x \div z < y \div z$$

Division by an inequality reverses the order. If $0 < x < y$ and $z > 0$, then

(5.17) $$z \div y < z \div x$$

In particular, for $z = 1$, this says that $1 \div y < 1 \div x$. In words, the order of the reciprocals is the reverse of the order of given rational numbers.

(5) Distributive property of division over addition: If x, y, and $z > 0$ are rational numbers, then

(5.18) $$(x + y) \div z = x \div z + y \div z$$

(6) Distributive property of division over subtraction: If x, y, and $z > 0$ are rational numbers and if $y < x$, then

(5.19) $$(x - y) \div z = x \div z - y \div z$$

Proof. Throughout the proof we shall use Theorem 5.97.

(1) If $a = d \cdot q + r$, then as rational numbers
$$\begin{aligned}
a \div d &= \frac{1}{d} \cdot a = \frac{1}{d} \cdot (d \cdot q + r) \\
&= \frac{1}{d} \cdot (d \cdot q) + \frac{1}{d} \cdot r, \text{ Distributive property} \\
&= q + \frac{r}{d}, \text{ Associative property and def. of mult.} \\
&= q\frac{r}{d}
\end{aligned}$$

For example, if $a = 7$ and $d = 2$, then whole number division $7 \div 2$ has quotient 3 and remainder 1 so that $7 = 2 \cdot 3 + 1$. In rational number division, $7 \div 2 = \frac{7}{2} = 3\frac{1}{2} = 3 + \frac{1}{2}$, as claimed.

(2) To prove the associative property expressed by equation (5.14), consider the case $x = \frac{2}{3}$, $y = \frac{5}{8}$, and $z = \frac{7}{4}$. Then the left side of equation (5.14) is

$$(\frac{2}{3} \cdot \frac{5}{8}) \div \frac{7}{4} = \frac{4}{7} \cdot (\frac{2}{3} \cdot \frac{5}{8})$$

and the right side is

$$\frac{2}{3} \cdot (\frac{5}{8} \div \frac{7}{4}) = \frac{2}{3} \cdot (\frac{4}{7} \cdot \frac{5}{8})$$

The two sides are equal by the associative and commutative properties of multiplication of rational numbers.

Use the same numbers to see how to prove the associative property expressed by equation (5.15). Then the left side of equation (5.15) is

$$(\frac{2}{3} \div \frac{5}{8}) \div \frac{7}{4} = \frac{4}{7} \cdot (\frac{8}{5} \cdot \frac{2}{3}) = \frac{4 \cdot (8 \cdot 2)}{7 \cdot (5 \cdot 3)}$$

and the right side is

$$\frac{2}{3} \div (\frac{5}{8} \cdot \frac{7}{4}) = \frac{2}{3} \div \frac{5 \cdot 7}{8 \cdot 4} = \frac{8 \cdot 4}{5 \cdot 7} \cdot \frac{2}{3} = \frac{(8 \cdot 4) \cdot 2}{(5 \cdot 7) \cdot 3}$$

The two sides are equal by the associative and commutative properties of multiplication of whole numbers.

(3) The identity property $x \div 1 = \frac{1}{1} \cdot x = x$, by the identity property of 1 in multiplication.

(4) Illustrate the proof of the order property expressed by equation (5.16) with $x = \frac{2}{3}$, $y = \frac{7}{8}$, and $z = \frac{4}{5}$. Then $\frac{2}{3} < \frac{7}{8}$ and $\frac{4}{5} > 0$ and

$$\frac{2}{3} \div \frac{4}{5} = \frac{5}{4} \cdot \frac{2}{3} \quad \text{and} \quad \frac{7}{8} \div \frac{4}{5} = \frac{5}{4} \cdot \frac{7}{8}$$

That is, division by $\frac{4}{5}$ is multiplication by its reciprocal. By the order property of multiplication, order is preserved when an inequality is multiplied by the reciprocal of $\frac{4}{5}$; namely, $\frac{5}{4} \cdot \frac{2}{3} < \frac{5}{4} \cdot \frac{7}{8}$. Therefore, $\frac{2}{3} \div \frac{4}{5} < \frac{7}{8} \div \frac{4}{5}$, as we wished to prove.

Illustrate the proof of the order property expressed by equation (5.17) with the same numbers. Multiply $\frac{2}{3} < \frac{7}{8}$ by the reciprocal of $\frac{2}{3}$, to conclude by the order property of multiplication that

$$\frac{3}{2} \cdot \frac{2}{3} < \frac{3}{2} \cdot \frac{7}{8}$$

5.6. Division of Rational Numbers

which is $1 < \frac{3}{2} \cdot \frac{7}{8}$. Now multiply this inequality by the reciprocal of $\frac{7}{8}$, to conclude again by the order property of multiplication that
$$1 \cdot \frac{8}{7} < (\frac{3}{2} \cdot \frac{7}{8}) \cdot \frac{8}{7}$$
which is $\frac{8}{7} < \frac{3}{2}$. This shows that the reciprocals of $\frac{2}{3}$ and $\frac{7}{8}$ are in the opposite order from the order of $\frac{2}{3}$ and $\frac{7}{8}$. Multiply the order of the reciprocals by $\frac{4}{5}$ and again apply the order property of multiplication to conclude that
$$\frac{8}{7} \cdot \frac{4}{5} < \frac{3}{2} \cdot \frac{4}{5}$$
which is $\frac{4}{5} \div \frac{7}{8} < \frac{4}{5} \div \frac{2}{3}$ as desired.

(5) Illustrate the proof with $x = \frac{2}{3}$, $y = \frac{7}{8}$, and $z = \frac{4}{5}$. The distributive property over addition follows from the distributive properties of multiplication over addition, as follows.
$$(\frac{2}{3} + \frac{7}{8}) \div \frac{4}{5} = \frac{5}{4} \cdot (\frac{2}{3} + \frac{7}{8})$$
$$= \frac{5}{4} \cdot \frac{2}{3} + \frac{5}{4} \cdot \frac{7}{8}, \text{ dist. prop. of mult.}$$
$$= \frac{2}{3} \div \frac{4}{5} + \frac{7}{8} \div \frac{4}{5}$$

(6) The proof is similar to the proof of (5). □

Oral exercise 5.104. (1) Divide $\frac{11}{12}$ of $\frac{7}{8}$ by $\frac{1}{8}$.

(2) Divide $\frac{12}{13} \cdot 25$ by 5.

(3) Divide $\frac{27}{23}$ of 11 by $\frac{9}{23}$.

(4) Divide 3 by $\frac{4}{5}$ and the result by $\frac{5}{7}$.

(5) Divide $\frac{25}{23}$ by $\frac{5}{23}$ of $\frac{1}{2}$.

(6) Divide $\frac{3}{17}$ by $\frac{2}{3}$ and $\frac{14}{17}$ by $\frac{2}{3}$ and add the results.

(7) Divide $\frac{25}{23}$ and $\frac{2}{23}$ by $\frac{7}{9}$ and subtract the smaller from the larger.

Corollary 5.105. *For nonzero rational numbers x and z,*

(1) *if $x < 1$, then $z < z \div x$;*

(2) *if $1 < x$, then $z \div x < z$.*

Proof. For the first statement, consider $x = \frac{2}{3}$ and $z = \frac{5}{8}$. Then $\frac{2}{3} < 1$, so the order property 4(b) says that $\frac{5}{8} \div 1 < \frac{5}{8} \div \frac{2}{3}$; that is, $\frac{5}{8} < \frac{5}{8} \div \frac{2}{3}$.

For the second statement, consider the case $x = \frac{4}{3}$ and $z = \frac{5}{8}$. Then $1 < \frac{4}{3}$ and the order property 4(b) says that $\frac{5}{8} \div \frac{4}{3} < \frac{5}{8} \div 1 = \frac{5}{8}$. □

Oral exercise 5.106. Which is larger:

(1) 31 or $31 \div \frac{2}{3}$?

(2) 27 or $27 \div \frac{13}{4}$?
(3) $\frac{5}{8}$ or $\frac{5}{8} \div \frac{3}{4}$?
(4) $\frac{11}{4}$ or $\frac{11}{4} \div \frac{21}{9}$?

5.6.2. Notation. It is convenient to extend the notation, $x \div y = \frac{x}{y}$, to rational numbers. Many of the properties of division have simple expressions in this notation. For example, the distributive properties of division expressed in equations (5.18) and (5.19) become

$$\frac{x+y}{z} = \frac{x}{z} + \frac{y}{z} \quad \text{and} \quad \frac{x-y}{z} = \frac{x}{z} - \frac{y}{z}$$

respectively. Expressions obtained with this notation are sometimes called *compound fractions*. For example,

$$\frac{2}{3} \div \frac{7}{4} = \frac{\frac{2}{3}}{\frac{7}{4}} = \frac{4}{7} \cdot \frac{2}{3}$$

Learning to work with expressions such as these provides excellent preparation for algebra.

Oral exercise 5.107. Find

(5.20) $\quad \dfrac{\frac{11}{12} \cdot \frac{7}{8}}{\frac{1}{8}}$

(5.21) $\quad \dfrac{\frac{3}{17}}{\frac{2}{3}} + \dfrac{\frac{14}{17}}{\frac{2}{3}}$

(5.22) $\quad \dfrac{\frac{3}{4} + \frac{9}{8}}{\frac{5}{4}}$

5.6.3. Applications. To use division of rational numbers, rephrase a problem in the standard measurement or partitive division statement, then calculate.

Example 5.108. How many three-quarter hour appointments can a physician schedule in a $3\frac{1}{2}$ hour period? This problem can be rephrased as: How many $\frac{3}{4}$'s in $3\frac{1}{2}$? We recognize this as a measurement division problem whose solution is

$$3\frac{1}{2} \div \frac{3}{4} = \frac{4}{3} \cdot 3\frac{1}{2} = \frac{4}{3} \cdot (3 + \frac{1}{2}) = \frac{4}{3} \cdot 3 + \frac{4}{3} \cdot \frac{1}{2} = 4 + \frac{2}{3} = 4\frac{2}{3}$$

where we have used the distributive property and cancellation in this calculation. The physician can schedule 4 appointments, but there will remain only $\frac{2}{3}$ of the time needed for another appointment.

Example 5.109. After travelling $2\frac{7}{8}$ miles, Ashley has completed $\frac{2}{3}$ of the way from her house to Forsyth. How far from her house is Forsyth?

5.6. Division of Rational Numbers

Partitive division: $\frac{2}{3}$ of what is $2\frac{7}{8}$? This is a partitive division problem whose solution is

$$2\frac{7}{8} \div \frac{2}{3} = \frac{3}{2} \cdot 2\frac{7}{8} = \frac{3}{2} \cdot (2 + \frac{7}{8}) = \frac{3}{2} \cdot 2 + \frac{3}{2} \cdot \frac{7}{8} = 3 + \frac{21}{16} = 3\frac{21}{16}$$

which is $4\frac{5}{16}$ miles from Ashley's house to Forsyth.

Unitary method:

Home to Forsyth $\underbrace{\overbrace{\rule{3cm}{0pt}}^{2\frac{7}{8} \text{ miles}}\rule{1.5cm}{0pt}}_{1 \text{ unit}}$

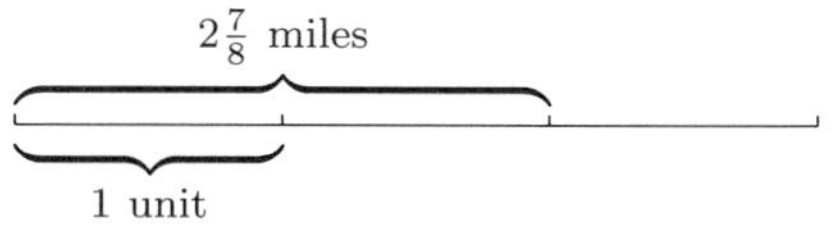

Home to Forsyth is 3 units.
2 units $= 2\frac{7}{8}$ miles.
1 unit $= (2\frac{7}{8}) \div 2 = \frac{1}{2} \cdot (2\frac{7}{8}) = 1\frac{7}{16}$ miles
3 units $= 3 \cdot (1\frac{7}{16}) = 3\frac{21}{16} = 4\frac{5}{16}$ miles.

Example 5.110. At 70 mph, how many seconds does it take to go 1 mile?
Solution: 70 miles in 1 hour
   1 mile in $1 \div 70 = \frac{1}{70}$ hour
   $\frac{1}{70}$ of 1 hour is $\frac{1}{70}$ of 60 minutes $= \frac{1}{70} \cdot 60 = \frac{6}{7}$ minute
   $\frac{6}{7}$ minute $= \frac{6}{7}$ of 60 seconds $= \frac{6}{7} \cdot 60 = \frac{360}{7} = 51\frac{3}{7}$ seconds
At 70 mph it takes $51\frac{3}{7}$ seconds to go 1 mile.

Example 5.111 (Area of a Circle)**.** Divide a circle of radius r into n equal parts (called *sectors*). Cut the parts and rearrange them as shown below for $n = 8$. Divide one end sector into two equal parts by cutting along a radius and move one of these parts to the other end. The resulting figure is approximately a rectangle with height r and length $C/2$, where C is the circumference of the circle. We conclude that the area A of the circle is approximately the area of this rectangle,

$$A = r \cdot C/2$$

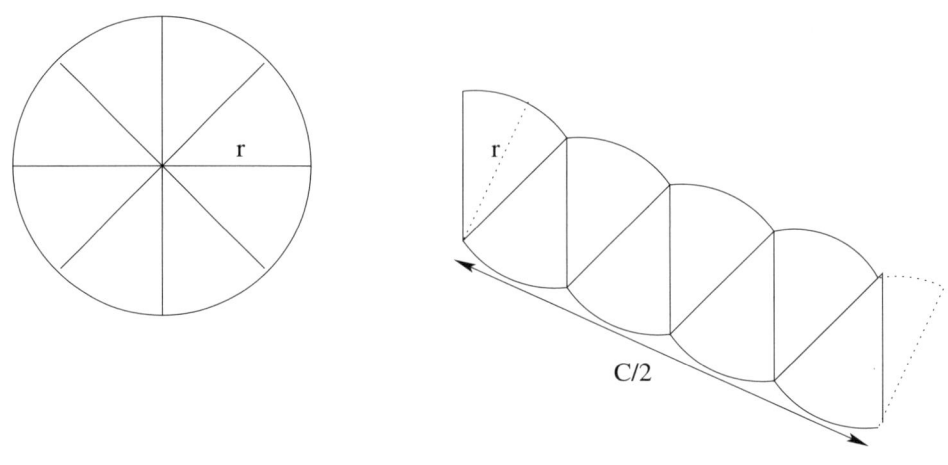

§5.6 Exercises

In answering each problem, state whether it is a measurement or partitive division. Illustrate your solution with a line segment diagram.

(1) How many $\frac{2}{3}$'s in 2? How many $\frac{2}{3}$'s in $3\frac{1}{2}$?

(2) 2 is $\frac{2}{3}$'s of what? $3\frac{1}{2}$ is $\frac{2}{3}$ of what?

(3) The interval from 0 to 3 is 5 of what length equal segments?

(4) How many 7's in $\frac{3}{2}$?

(5) $\frac{5}{2}$ of what is 1?

(6) $\frac{3}{4}$ is $\frac{2}{7}$ of what?

(7) If $\frac{2}{3}$ of Anthony's money is added to $\frac{1}{6}$ of his money, the sum will equal $125: how much money has Anthony?

(8) If Joshua spent $\frac{4}{7}$ of his allowance and had $21 left, what was his allowance? Explain the solution in two ways: as an application of subtraction and division; and by the unitary method with a line segment diagram.

(9) Rita sold $\frac{1}{3}$ of her corn to Ruth, $\frac{3}{6}$ to Nettie, and had 50 bushels remaining. (a) What fraction of her corn did she sell? (b) What fraction remains? (c) How much corn had she at first?

(10) Gavin had 600 bushels of wheat. He sold $\frac{3}{10}$ of it to Dale. He sold $\frac{3}{4}$ of what remained to Newell. (a) What fraction of his wheat did he sell all together? (b) What fraction remained? (c) How many bushels remained?

§5.6 Exercises 239

(11) If 8 women can do $\frac{3}{5}$ of a piece of work, how much can 1 woman do (assuming each does an equal amount)? *Hint:*

(12) Robert shares $\frac{8}{9}$ of a ton of coal among 5 poor neighbors: how much does each get?

(13) If 7 horses can eat $\frac{2}{5}$ of a load of hay in a day, how much can 1 horse eat in the same time? How much can 10 horses eat in the same time? *Hint:*

(14) After traveling 18 miles from home Carol is $\frac{3}{5}$ of the way from her house to Pete's house: how far from her house to Pete's house?

(15) How many baskets holding $\frac{3}{4}$ of a bushel each will be required to hold 15 bushels of peaches?

(16) If $\frac{7}{8}$ yards of gingham make 1 apron, how many can be made out of 14 yards?

(17) When a bushel of apples can be traded for $\frac{3}{4}$ pound of butter, how many bushels of apples can be obtained for $\frac{3}{5}$ pound of butter?

(18) At 75 mph, how many seconds does it take to go 1 mile?

(19) (a) Make up a measurement division word problem whose solution is $(1\frac{3}{4}) \div \frac{2}{3}$.
 (b) Make up a partitive division word problem whose solution is $\frac{7}{8} \div \frac{3}{5}$.

(20) Morris can unload a truck load of sugar in $\frac{3}{4}$ hours. Morris and Tony together can unload it in $\frac{1}{2}$ hour. How long would it take Tony alone to unload the truck? Assume that he unloads at the same constant rate alone as when he works with Morris. *Hint:*

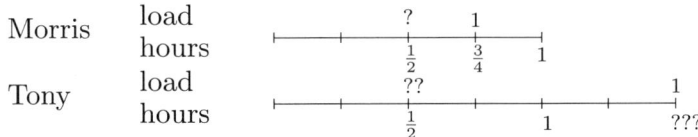

(21) On a piece of card stock draw as large a circle as you can. Use a protractor to measure angles to divide this circle into 18 equal parts. With scissors, cut out the 18 sectors and arrange them as described in Example 5.111. Cut one sector into two equal pieces and rearrange one of these pieces to make a figure which is very nearly a rectangle. Explain why the area of this figure is approximately $r \cdot C/2$, where r is the radius and C is the circumference of the circle.

Review Exercises

For each exercise explain how you decide on which computations to make, then show how to make them. Conclude with a statement of your answer and what it means. Here is a sample problem with solutions.

Example 5.112. Ella received $1200 last week, but that was $\frac{4}{5}$ of what she should have received. How much more did she have coming to her?

Solution 1: What she should have received is the answer to the partitive division problem: 1200 is $\frac{4}{5}$ of what? Therefore, she should have received $1200 \div \frac{4}{5} = \frac{5}{4} \cdot 1200 = 1500$ dollars. How much more she had coming to her is the answer to the comparative subtraction question: how much more than 1200 is 1500? The answer to this question is $1500 - 1200 = 300$ dollars. She has $300 coming to her.

Solution 2: Use the unitary method.

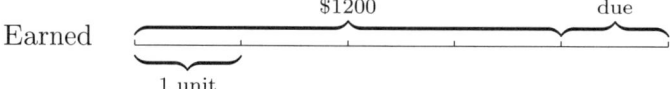

4 units = 1200 dollars, 1 unit due.

1 unit = 1200/4 = 300 dollars.

Ella had $300 more coming to her.

(1) After working $5\frac{1}{2}$ days on a job, Suella is $\frac{3}{4}$ done: how many days total needed for the job?

(2) If a serving is $\frac{1}{6}$ of a candy bar, how many servings can be obtained from $15\frac{1}{3}$ candy bars?

(3) How many pieces $\frac{2}{3}$ foot long can be obtained from a board $8\frac{1}{2}$ feet long?

(4) Morris can unload a truck load of sugar in $1\frac{3}{4}$ hours. He asks Tony to help him unload the truck so that he can unload in 1 hour. How long must it take Tony to unload the truck alone in order for the two of them together to unload it in 1 hour? (Assume they work together at their normal speed, without interfering with each other). *Hint:*

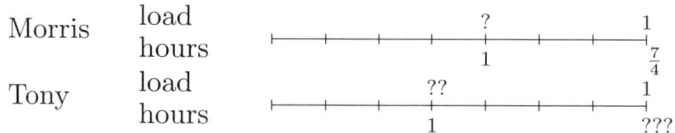

(5) Patti bartered $3\frac{3}{4}$ yards of cloth for $17\frac{1}{8}$ pounds of cheese, $5\frac{1}{2}$ yards for $23\frac{1}{4}$ pounds, and $8\frac{1}{8}$ yards for $27\frac{1}{5}$ pounds: how many yards did she trade, and how much cheese did she receive?

(6) Viola has 3 fields; the first contains $16\frac{4}{5}$ acres; the second, $19\frac{1}{4}$ acres; and the third, $21\frac{2}{3}$ acres: how many acres in the 3 fields?

(7) Niels received \$1500 last week, but that was $\frac{4}{3}$ of what he should have received. How much did he have to return?

(8) If $\frac{2}{3}$ of your money is increased by $\frac{1}{7}$ of your money, the sum will equal \$136: how much money have you?

(9) A piece of carpet contains $65\frac{1}{2}$ yards: if Ralph sells to Richard $22\frac{1}{4}$ yards and to Robert $27\frac{7}{8}$ yards, how much remains?

(10) You have two rolls of carpet, each containing $125\frac{1}{4}$ yards; from one piece you sell $26\frac{7}{8}$ yards, and from the other $47\frac{1}{2}$ yards: how much remains? *Ans.* $176\frac{1}{8}$ yds.

(11) Jen spent $\frac{2}{3}$ of her paycheck and found that \$600 was $\frac{3}{7}$ of the remainder: how big was her paycheck? *Ans.* \$4200.

(12) You had $\frac{3}{4}$ of $\frac{1}{2}$ of \$1000 and lost $\frac{5}{6}$ of $\frac{3}{7}$ of \$630: how much have you remaining? *Ans.* \$150.

(13) Gavin had 600 bushels of wheat; at one time he sold $\frac{3}{10}$ of it and at another $\frac{1}{4}$ of what remained: how many bushels of the wheat were sold?

(14) If $\frac{3}{4}$ of $18\frac{3}{4}$ years is $\frac{5}{16}$ of my age, how old am I? *Ans.* 45 years.

(15) Nicholas has $\frac{3}{4}$ of \$375, Joshua has $\frac{4}{5}$ as much as Nicholas, and Caitlin has $\frac{3}{5}$ as much as both: how many dollars has each, and how many have they all?

5.7. Ratio and Proportion

A ratio is a fraction formed by the number of one group of objects divided by the number of another group. Ratio is another word for fraction. The name *rational number* comes from ratio. The theory of ratios is the subject of Book V of Euclid's *Elements* [**Euc56**, Vol. 2, pp. 112-186]. Over many centuries the material of Book V evolved into our modern notation of fractions. Common discourse continues to use the language of ratio. Ratios provide a simple way to express quantitative relationships. The *unitary method* will be our standard procedure for solving many problems involving ratios. Keep in mind that any ratio can be expressed as a fraction and conversely, any fraction can be regarded as a ratio.

Euclid's Book V tells us that ratios can be formed between rational numbers, not just whole numbers. This is important for applications in geometry, such as for the result that for similar triangles the ratios of corresponding sides are in proportion. Even more, Euclid saw the need to consider ratios involving irrational numbers, such as in $\frac{1}{a} = \frac{a}{2}$, or in the ratio of the circumference of a circle to its diameter.

5.7.1. Ratios.

Definition 5.113. The *ratio* of a number a to a number b is the fraction $\frac{a}{b}$. It is read a to b and is denoted $a : b$. The ratio $c : d$ is *in proportion to* the ratio $a : b$ if $\frac{c}{d} = \frac{a}{b}$, that is, if the corresponding fractions have the same value. A ratio $a : b$ is expressed in *simplest form* if the fraction $\frac{a}{b}$ is expressed in lowest terms.

Modern texts sometimes use the term *equivalent to* or *equal to* in place of *in proportion to*. To say that the ratio $a : b$ is in proportion to the ratio $c : d$ means that $a : b$ is equivalent to $c : d$ or that $a : b$ is equal to $c : d$. This can be written
$$a : b :: c : d \text{ or } a : b = c : d$$
which are read "a is to b as c is to d". Both mean $\frac{a}{b} = \frac{c}{d}$. Theorem 5.17 tells us that for any natural number n, the ratio $a : b$ is in proportion to $n \cdot a : n \cdot b$, because $\frac{a}{b} = \frac{n \cdot a}{n \cdot b}$.

Example 5.114. In the following set of circles and squares
$$\{\bigcirc \ \bigcirc \ \bigcirc \ \square \ \square \ \square \ \square \ \square\}$$
the ratio of the number of circles to the number of squares is $3 : 5$. Here the unit is each individual item. The ratio of the number of squares to the number of circles is $5 : 3$.

In terms of fractions, the number of circles is $\frac{3}{5}$ of the number of squares. The number of squares is $\frac{5}{3}$ times the number of circles.

The ratio of the number of circles to the total number of figures in the set is $3 : 8$, since $3 + 5 = 8$ is the total number of figures in the set. The number of circles is $\frac{3}{8}$ of the total number of figures in the set.

In a ratio, the unit can be individual elements of the set, as in the preceding example, or they can be groups of elements, as in the next example.

Example 5.115. In the set of bowties and stars
$$\left\{ \begin{array}{cccccccc} \bowtie & \bowtie & \star & \star & \star & \star & \star \\ \bowtie & \bowtie & \star & \star & \star & \star & \star \\ \bowtie & \bowtie & \star & \star & \star & \star & \star \\ \bowtie & \bowtie & \star & \star & \star & \star & \star \\ \bowtie & \bowtie & \star & \star & \star & \star & \star \end{array} \right\}$$
the ratio of the number of bowties to the number of stars is $10 : 25$, if we take the unit to be each individual bowtie and star. If we take the unit to be each column of bowties and stars, then the ratio of the number of bowties to the number of stars is $2 : 5$. This ratio is in proportion to $10 : 25$ because $\frac{2}{5} = \frac{10}{25}$. The equivalence of these ratios required that each unit contain the same number of bowties or stars.

5.7. Ratio and Proportion

The ratio of stars to bowties is 5 : 2. Line segments can picture this ratio of 5 units to 2 units as follows.

```
         1 unit
        ⌢⎯⎯⎯⎯
bowties ├────┤
stars   ├────┼────┼────┼────┼────┤
```

The number of bowties is $\frac{2}{5}$ of the number of stars. The number of stars is $\frac{5}{2}$ of the number of bowties. The number of bowties is $\frac{2}{7}$ of the total number of bowties and stars, since $2 + 5 = 7$ is the total number of units of bowties and stars.

Example 5.116. Many measurements are expressed as ratios.

(1) A rectangle has height 4 inches and length 7 inches. The ratio of its height to its length is 4 : 7. The ratio of its length to its height is 7 : 4. The height is $\frac{4}{7}$ times the length. The length is $\frac{7}{4}$ times the height. The ratio of its length to its perimeter is 7 : 22.

(2) A recipe calls for 4 cups of flour and 2 cups of sugar. The ratio of flour to sugar in this recipe is 4 : 2. Expressed in simplest form, the ratio of flour to sugar in this recipe is 2 : 1. The amount of flour used in this recipe is 2 times the amount of sugar used. The amount of sugar used is $\frac{1}{2}$ times the amount of flour used.

Oral exercise 5.117. Peggy has $70 and Diane has $40.

(1). What is the ratio of Diane's money to Peggy's money? What is this ratio in simplest form?

(2). Peggy's money is what fraction of Diane's money?

(3). Diane's money is what fraction of Peggy's money?

(4). Diane's money is what fraction of the money they have altogether?

(5). What is the ratio of Diane's money to the money they have altogether?

Oral exercise 5.118. Of the 30 students in a math class, 20 of them are women.

(1). What is the ratio of the number of men in the class to the number of women in the class? What is this ratio in simplest form?

(2). The number of men is what fraction of the number of women?

(3). The number of women is what fraction of the total number of students in the class?

(4). What is the ratio of the number of men to the total number of students in the class?

Oral exercise 5.119. In the set of circles, squares, and triangles

$$\{\bigcirc \bigcirc \bigcirc \bigcirc \square\square\square \triangle \triangle \triangle \triangle \triangle\}$$

(1). What is the ratio of the number of circles to the number of squares?

(2). What is the ratio of the number of circles to the number of triangles?

(3). What is the ratio of the number of triangles to the number of squares?

Example 5.120. In a math class of 32 students the ratio of the number of men to the number of women is 3 : 5. How many men are there?

Unitary method: Use line segments to picture the ratio as well as the unit.

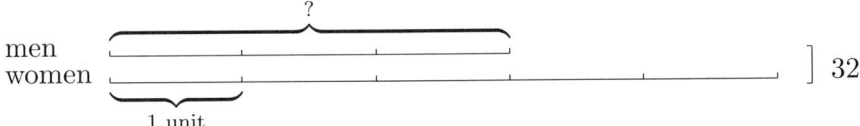

8 units = 32 students.

1 unit = 32 ÷ 8 = 4 students.

3 units = 3 · 4 = 12 students.

There are 12 men in the class.

Fractions: Of the 3 + 5 = 8 units making up the 32 people in the class, there are 3 units of men. The number of men is $\frac{3}{8}$ of 32, which is

$$\frac{3}{8} \cdot 32 = 3 \cdot 4 = 12 \text{ men}$$

Oral exercise 5.121. The ratio of Pete's weight to Carol's weight is 5 : 3. If Pete weighs 60 pounds, what does Carol weigh?

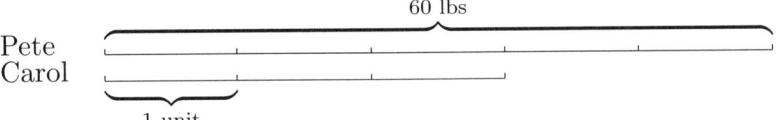

Unitary method:

5 units equals how many pounds?

1 unit equals how many pounds?

How many units does Carol weigh? How many pounds is that?

Fractions:

Carol's weight is what fraction of Pete's weight?

How many pounds is that?

Oral exercise 5.122. Rosemary and Diane bought their mother a present which cost $80. Rosemary being older and having more money decided that the ratio of the amount she paid to the amount that Diane paid should be 3 : 2. How much did Diane pay?

5.7. Ratio and Proportion

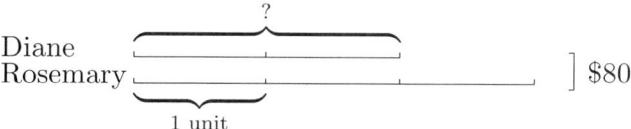

How many units make $80?

1 unit equals how many dollars?

How many units did Diane pay? How much did Diane pay?

What fraction of the cost did Diane pay?

What fraction calculation gives what Diane paid?

We may consider the ratios of pairs of numbers from any sequence of numbers. For example, in Oral Exercise 5.119, the ratio of the number of circles to the number of squares to the number of triangles is 4 : 3 : 5. This means that the ratio of the number of circles to the number of squares is 4 : 3 and the ratio of the number of squares to the number of triangles is 3 : 5. It also means that the ratio of the number of circles to the number of triangles is 4 : 5.

Example 5.123. David and Ella have $20 each week to distribute as allowance to their three daughters, Sigrid, Sue, and Jen, whose ages are 10, 6, and 4 years, respectively. David and Ella distribute the money in the ratio of the children's ages. How much weekly allowance does Jen receive?

Solution: The ratio of Sigrid's age to Sue's age to Jen's age is 10 : 6 : 4 = 5 : 3 : 2 in simplest form.

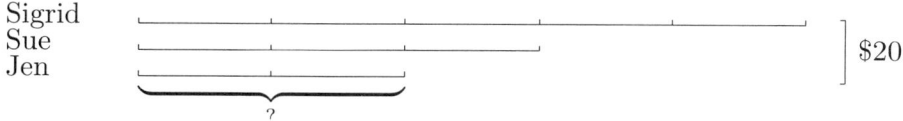

10 units = 20 dollars

1 unit = 20 ÷ 10 = 2 dollars

2 units = 2 · 2 = 4 dollars

Jen receives $4 each week.

What fraction of the $20 does Jen receive?

Oral exercise 5.124. There are 13 girls and 17 boys in a class. Find the ratio of the number of girls to the number of boys to the total number of students in the class.

The number of girls is what fraction of the total number of students in the class?

The number of girls is what fraction of the number of boys in the class?

The term proportion is often used to express a known ratio. For example, a civil engineer would say that in making concrete the sand and cement is mixed in some proportion. In recipes, the ingredients are mixed according to some proportion. In tinting paint, the color and base are mixed in some proportion.

Example 5.125. In a recipe for Scotch shortbread, the proportion of flour to sugar is 5 to 3. This means that the ratio of the amount of flour to the amount of sugar is 5 : 3. Every 5 units of flour require 3 units of sugar. The ratio of the amount of sugar to the amount of flour is 3 : 5. Every 3 units of sugar require 5 units of flour. In the recipe, the proportion of sugar to flour is 3 to 5.

Using this proportion of flour to sugar, we must mix how many cups of flour with 9 cups of sugar?

The question can be phrased: 5 is to 3 as **what** is to 9? In brief, this is written $5 : 3 :: x : 9$; what is x?

Fractions: The ratios are $\frac{5}{3} = \frac{x}{9}$. By the arithmetic of fractions,

$$x = 9 \cdot \frac{x}{9} = 9 \cdot \frac{5}{3} = 3 \cdot 5 = 15$$

Unitary Method: The proportion of flour to sugar is 5 to 3 means that the ratio of the amount of flour to the amount of sugar is 5 : 3. Picture this ratio as

flour
sugar

9 cups

3 units = 9 cups.

1 unit = $9 \div 3 = 3$ cups.

5 units = $5 \cdot 3 = 15$ cups.

15 cups of flour are needed with 9 cups of sugar.

Example 5.126. In a math class, the ratio of the number of sophomores to the number of juniors is 3 : 2. The ratio of the number of juniors to the number of seniors is 5 : 7. What is the ratio of the number of sophomores to the number of seniors?

Solution: For juniors the units are different in the two ratios.

sophs
juniors

juniors
seniors

In the ratio of sophomores to juniors, the unit is $\frac{1}{2}$ the number of juniors. In the ratio of juniors to seniors, the unit is $\frac{1}{5}$ the number of juniors. A common denominator for these two fractions is 10, so the unit should be $\frac{1}{10}$

5.7. Ratio and Proportion

the number of juniors. To use this unit, each $\frac{1}{2}$ must be divided into 5 equal parts and each $\frac{1}{5}$ must be divided into 2 equal parts. Then the ratios look like this:

sophs
juniors
seniors

Therefore, the ratio of the number of sophomores to the number of juniors to the number of seniors is $15 : 10 : 14$, and the ratio of the number of sophomores to the number of seniors is $15 : 14$, as pictured.

This result is obtained algebraically by multiplying the ratio $3 : 2$ by 5 and the ratio $5 : 7$ by 2 to get

$$3 : 2 = 15 : 10 \text{ and } 5 : 7 = 10 : 14$$

Oral exercise 5.127. In a box of candy, the ratio of the number of reds to the number of yellows is $3 : 2$. The ratio of the number of yellows to the number of greens is $4 : 5$. What is the ratio of the number of reds to the number of greens?

reds
yellows

yellows
greens

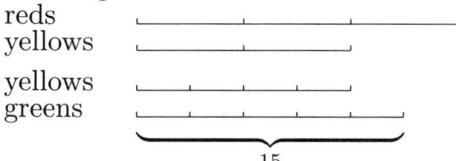

15

There are 15 green candies in the box. How many red candies are there in the box?

What is the total number of candies in the box?

Example 5.128. Prices are ratios. A price is the ratio of the number of objects to the number of dollars required to buy that number of objects. If 6 pairs of shoes cost $180, what will 8 pairs cost? This question can be posed in terms of the ratio of the price to the number of pairs of shoes, where the given price of 6 pairs of shoes determines the proportion

$$180 : 6 :: x : 8$$

where x is the price of 8 pairs of shoes. Rewritten in terms of fractions, this is

$$\frac{180}{6} = \frac{x}{8}$$

Method 1: Fraction arithmetic leads to the solution

$$x = 8 \cdot \frac{x}{8} = 8 \cdot \frac{180}{6} = 8 \cdot 30 = 240$$

Thus, 8 pairs of shoes cost $240.

Method 2: The unitary method is the most sensible way to solve this problem. Notice that the given proportion of 180 dollars for 6 pairs of shoes reduces to 30 : 1 in simplest form, which says that $30 is the price of 1 pair of shoes.

pairs of shoes

$180

6 pairs cost 180 dollars.

1 pair costs $180/6 = 30$ dollars.

8 pairs cost $8 \cdot 30 = 240$ dollars.

5.7.2. Ratio Changes. How does the ratio of one number to another change if one of the numbers is changed?

Example 5.129. Ragna set up her family's household budget so that the ratio of monthly expenditures to monthly savings is 7 : 2. Her monthly income is $2,700. If her monthly income increases by $100 and if she saves this whole amount, what then will be the ratio of monthly expenditures to monthly savings?

Solution: The initial situation can be pictured as follows:

spends
saves   } $2,700

1 unit

The initial amount saved each month is 2 units and spent each month is 7 units.

9 units = 2700 dollars.

1 unit = $2700/9 = 300$ dollars.

Initial monthly savings = 2 units = $2 \cdot 300 = 600$ dollars.

Monthly expenditures = 7 units = $7 \cdot 300 = 2100$ dollars.

Method 1: After her raise, Ragna saves $600 + 100 = 700$ dollars each month and the amount she spends does not change. Now the ratio of monthly expenditures to monthly savings is 2100 : 700, which is 3 : 1 in simplest form.

Method 2: After her raise, her monthly saving is increased by $\frac{1}{3}$ of a unit.

spends
saves

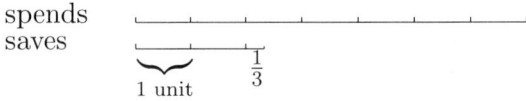

1 unit $\frac{1}{3}$

Divide all units into 3 equal parts, then monthly expenditures = $7 \cdot 3 = 21$ units and monthly savings = $2 \cdot 3 + 1 = 7$ units and the ratio of monthly expenditures to monthly savings has become 21 : 7 = 3 : 1 in simplest form.

Notice that we may consider the ratio of rational numbers. Her monthly expenditure is 7 units and the new monthly saving is $2\frac{1}{3}$ units, so the ratio of monthly expenditure to monthly saving has become

$$7 : 2\frac{1}{3} :: 7 : \frac{7}{3} :: 3 \cdot 7 : 3 \cdot \frac{7}{3} :: 21 : 7 :: 3 : 1$$

Example 5.130. To fertilize his fields, Newell plants peas and wheat together. The amount of peas to the amount of wheat is in the ratio of 2 : 3. This year, having an extra 7 bushels of seed peas, Newell decides to plant them with the usual amount of wheat. The ratio then becomes 5 : 6. How many bushels of peas did he usually plant? How many did he plant this year?

Before:

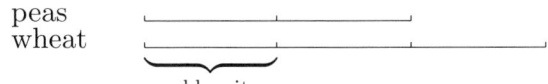

This year, the amount of wheat doesn't change:

The new unit is $\frac{1}{2}$ of the old unit, since the fixed amount of wheat is divided into 6 equal parts by dividing the old units into 2 equal parts.

1 new unit = 7 bushels, since the amount added to the peas is 1 additional new unit.

5 new units = $5 \cdot 7 = 35$ bushels.

This year Newell planted 35 bushels of peas. In years preceding, he planted $35 - 7 = 28$ bushels of peas.

The geometry is showing us the following. Express the before and after ratio of peas to wheat with the wheat expressed by the same number of units in both.

Before: ratio of peas to wheat is 2 : 3 :: 4 : 6.

After: ratio of peas to wheat is 5 : 6, so peas increased by 1 unit.

Peas increased by 7 bushels, so 7 bushels of peas = 1 unit.

Newell usually planted 4 units of peas, which is $4 \cdot 7 = 28$ bushels of peas.

Oral exercise 5.131. Janet set up her family's household budget so that their monthly income was divided between expenditures and savings. The ratio of monthly expenditures to monthly savings is 7 : 2. An increase in her monthly rent requires her to shift $\frac{1}{4}$ of her monthly savings over to monthly expenditures.

Before:

spends
saves

After:

spends
saves

What is the new ratio of monthly expenditures to savings?

What fraction of their monthly income is Janet saving now?

Explain how the following calculation finds the new ratio of monthly expenditures to monthly savings.

$$(7 + \frac{1}{4} \cdot 2) : (2 - \frac{1}{4} \cdot 2) = \frac{15}{2} : \frac{3}{2} = 2 \cdot \frac{15}{2} : 2 \cdot \frac{3}{2} = 15 : 3 = 5 : 1$$

Beware of forming the wrong ratios.

Example 5.132. Joshua and Caitlin were given the job of removing the nails from a large pile of used lumber. They both remove nails at the same rate. Together they can do the job in 5 hours. If Nicholas can remove nails at the same rate as they can, how long will it take all three of them to remove the nails from the pile of lumber?

Unitary method: What fraction of the job can be done by one child in one hour?

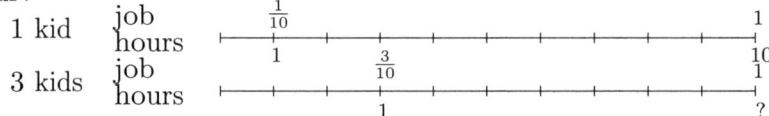

2 children do 1 job in 5 hours.

1 child does $\frac{1}{2}$ job in 5 hours.

1 child does $\frac{1}{5} \cdot \frac{1}{2} = \frac{1}{10}$ job in 1 hour.

3 children do $3 \cdot \frac{1}{10} = \frac{3}{10}$ job in 1 hour.

3 children do whole job in $\frac{10}{3} = 3\frac{1}{3}$ hours, which is 3 hours and 20 minutes.

A variation: Instead of Nicholas coming to help them, Caitlin and Joshua's dad does and he can pull nails three times as fast as they can. How long does it take the three of them to do the job?

Each child can do $\frac{1}{10}$ of the job in 1 hour.

Dad can do $3 \cdot \frac{1}{10} = \frac{3}{10}$ of the job in 1 hour.

Joshua, Caitlin, and Dad can do $\frac{1}{10} + \frac{1}{10} + \frac{3}{10} = \frac{1}{2}$ of the job in 1 hour.

Together they can do the whole job in $2 \cdot 1 = 2$ hours.

5.7.3. Ratios in Geometry.

5.7.3.1. Similar Triangles.

Definition 5.133. A triangle with sides of length a, b, and c units is *similar* to a triangle with sides of length A, B, and C units if

$$a : A = b : B = c : C$$

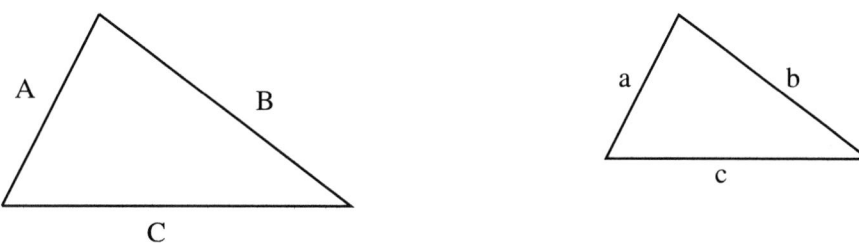

Theorem 5.134. *If two triangles have corresponding angles equal, then the two triangles are similar, with sides opposite the equal angles having the same ratios.*

A frequent source of similar triangles comes from cutting a given triangle by a line parallel to one of its sides. Corresponding angles are equal.

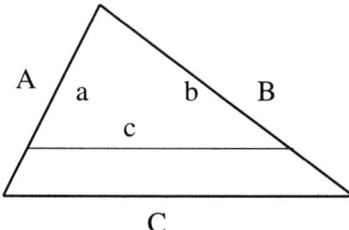

Example 5.135. A vertical pole 6 feet long casts a shadow 4 feet long. At the same time a vertical tree casts a shadow 12 feet long. How tall is the tree?

Solution: Draw the right triangles formed by the pole and its shadow and the tree and its shadow.

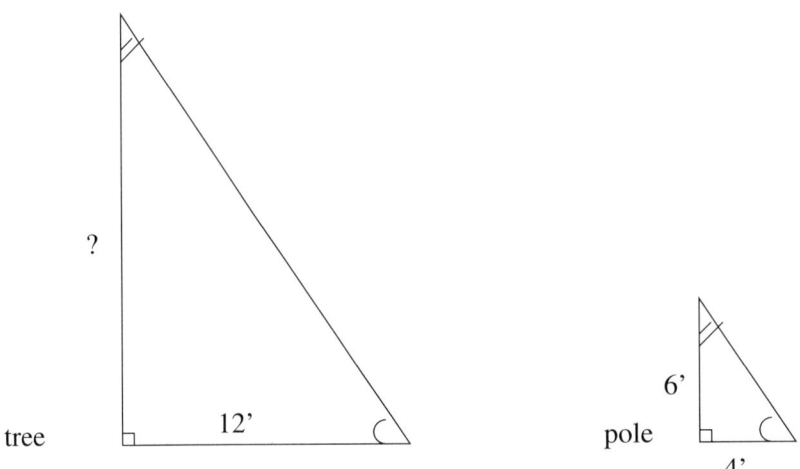

These triangles are similar, since their corresponding angles are equal. Therefore, the ratio of the height of the tree to the length of its shadow is in proportion to the ratio of the height of the pole to the length of its shadow, that is, $x : 12 :: 6 : 4$. Therefore,

$$x = 12 \cdot \frac{x}{12} = 12 \cdot \frac{6}{4} = 18$$

The tree is 18 feet tall.

5.7.3.2. *Circumference of Circles.* The ratio of the circumference of a circle to its diameter is in proportion for all circles. This can be seen as follows. Draw a circle of diameter d on a sheet of rubber. Now stretch the rubber by the same amount in all directions from the center of the circle until the radius of the drawn circle has doubled. It seems evident that all lengths on the rubber sheet have doubled. In particular, the length of the circumference has doubled. Consequently, the ratio of circumference to diameter for the initial circle and the stretched circle is in proportion, because $\frac{C}{d} = \frac{2 \cdot C}{2 \cdot d}$.

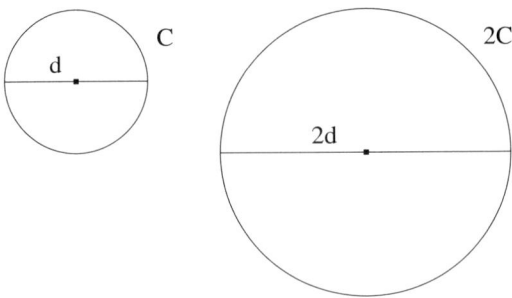

Definition 5.136. The number π is defined to be the ratio of the circumference of a circle to its diameter.

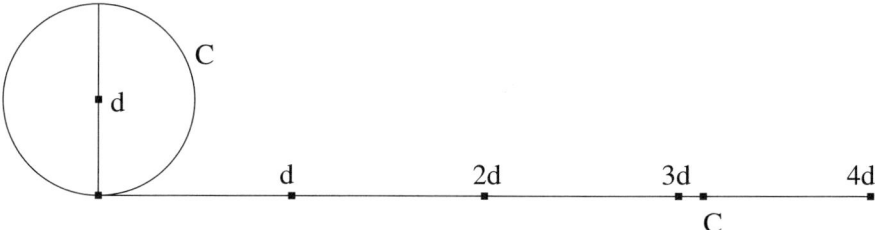

C is at the point πd on this line.

Theorem 5.137. *The ratio of the area A of a circle of radius r to the area of a square of side length r is π. That is, $A = \pi r^2$.*

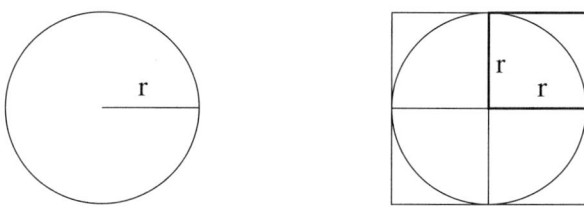

§5.7 Exercises

Explain how to solve the following problems. Use line segment diagrams and the unitary method whenever possible.

(1) This year the ratio of Joshua's age to his grandfather's age is 6 : 31 and his grandfather is 50 years older than he is. How old is Joshua? What will be the ratio 8 years from now?

(2) The length of a rectangle is 16 inches and its width is 12 inches. Find the ratio of its length to its perimeter, in simplest form.

(3) Two-thirds of JB's money is equal to $\frac{3}{5}$ of Alexi's money. What is the ratio of JB's money to Alexi's money? Use a line segment diagram to illustrate your solution.

(4) Ashley and Austin had an equal amount of money each. After Ashley spent $35 and Austin spent $22, the ratio of Ashley's money to Austin's money was 3 : 5. How much money did each girl have at first? Use a line segment diagram to illustrate your solution.

(5) Rochelle, Denise, and Kristi shared a sum of money in the ratio 2 : 3 : 9. If Kristi received $42 more than Rochelle, how much did Denise receive?

(6) The sides of a triangle are in the ratio 4 : 5 : 6. If the perimeter of the triangle is 60 inches, find the length of the shortest side.

(7) If water is pumped into a tank at the rate of 24 gallons per minute, it takes 3 hours to fill it. At what rate must water be pumped into the tank in order to fill it in 2 hours?

(8) Morris can unload a truck load of sugar in $\frac{3}{4}$ hours. Morris and Tony together can unload it in $\frac{1}{2}$ hour. How long would it take Tony alone to unload the truck? Assume that he unloads at the same constant rate alone as when he works with Morris. *Hint:* What fraction of the truck can Morris unload in $\frac{1}{2}$ hour? Draw a line segment diagram.

(9) If a train travels 300 mi in 9 hr 40 min, how long will it take to travel 223 mi at the same rate? *Ans.* 7 hr 11 min 8 sec.

(10) If 16 men can build a house in 20 days, how long would it take 12 men to build it (all men working at the same rate)? *Ans.* $26\frac{2}{3}$ days.

(11) If $6\frac{1}{2}$ cords of wood cost $140, how many can be bought for $210?

(12) When 180 qt of milk produce 24 lb of butter, how many pounds will 480 qt of the same milk produce? Explain how to solve this problem by the unitary method.

(13) If 360 qt of milk produce 48 lb of butter, how much of the same milk will be required to produce 50 lb of butter?

(14) If 12 men can build a house in 30 days, how many men can build it in 24 days (all working at the same rate)?

(15) A bankrupt merchant paid 33 cents on the dollar. If he owed Niels $6742.50, how much did Niels receive?

(16) When the tax on a property valued at $6000 is $532.50, how much is the tax, at the same rate, on a property valued at $14,500?

(17) A vertical pole 50 ft high casts a shadow 27 ft long. How long at the same time of day will be the shadow of a vertical pole 70 ft high?

(18) A pole 25 ft high casts a shadow 18 ft long. How high is the pole which at the same time casts a shadow 39 ft long?

(19) Joshua has corn at $.75 a bushel to exchange for Anthony's wheat at $1.20 a bushel. If Anthony asks $1.40 a bushel for his wheat, how much a bushel should Joshua charge for his corn to maintain the same exchange rate?

(20) It costs $1680 in wages to employ 3 men at 8 hours a day for 10 days. At the same rate, what would it cost to employ 5 men for 12 days at 10 hours a day? *Hint:* Find the cost of 1 man-hour.

(21) Nicholas has 6 horses, Joshua has 4 cows, and Caitlin has 10 sheep. During the time each is away at camp, their parents feed the horses for 3 weeks, the cows for 2 weeks, and the sheep for 3 weeks. The parents pay $60 for all this feed. If a cow eats 5 times as much as a sheep and a

horse eats twice as much as a cow, what is each child's share of this feed bill? Hint: let a sheep-week of feed be the amount of feed consumed by 1 sheep in 1 week. How many sheep-weeks of feed are consumed during the times the parents are feeding? What is the cost of a sheep-week of feed?

(22) Jen has $2500 more than Sue and they together have $7000. How much has each? *Hint:*

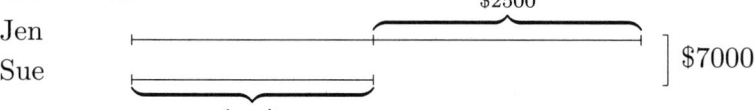

(23) A coat, a pair of boots, and a hat cost $330. The boots cost $100 less than the coat and the hat $40 less than the boots. How much did each cost?

(24) Andre and Newell together have 600 cows. Andre has twice as many as Newell. How many cows has each? *Hint:*

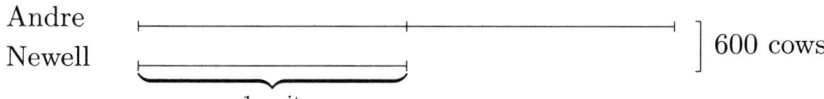

(25) The ratio of the number of Newell's cows to the number of Andre's cows is 2 : 3. After selling $\frac{1}{2}$ of her cows, Andre has 60 fewer cows than Newell. How many cows does Newell have? Use line segment diagrams to illustrate the original ratio, and then the situation after the sale of cows. Use the diagrams to explain the solution.

(26) The sum of Jen's age and Ella's age is 120 years. If Jen's age is $\frac{3}{5}$ of Ella's age, how old is Ella? Explain two methods of solution: as an application of addition and division; and by the unitary method with a line segment diagram.

Chapter 6

Decimals

Review our base ten numeration system as defined in Section 3.1 on page 102.

6.1. Decimal Expansions

6.1.1. Decimal numeration. There is a natural way to extend our numeration system beyond the set of whole numbers. Recall that our base 10 numeration system begins with the one's place and in each place to the left an element is worth ten times an element in the adjacent place to the right. Following this logic, we could consider the place to the right of the one's place. An element in it would have to be $\frac{1}{10}$ in order that an element in the adjacent place to the left, the one's place, have ten times its value, because $10 \cdot \frac{1}{10} = 1$. Place a period, called the decimal point, to the right of the one's place in order to keep track of the one's place. For example, 12.3 is then 1 ten + 2 ones + 3 tenths, pronounced twelve and three-tenths (the decimal point is designated by saying *and*).

Definition 6.1. The Hindu-Arabic numeration system to base ten is extended by using places to the right of the one's place. An element in each place is ten times the value of an element in the place to its immediate right. The one's place is identified by placing a period, called the decimal point, between its place and the next place to the right. The names of the places to the right of the one's place are tenths, hundredths, thousandths, ten-thousandths, etc., corresponding to the value of an element in each place. Numerals of this extended numeration system are called decimal expansions.

Decimal is the Latin word for tenth and this is the value of the unit in the first place to the right of the one's place. A tenth of a tenth is a hundredth, since $\frac{1}{10} \cdot \frac{1}{10} = \frac{1}{100}$. A tenth of a tenth of a tenth is a thousandth, and so on.

Example 6.2. (1) $.6 = \frac{6}{10} = \frac{3}{5}$.

(2) $.75 = \frac{7}{10} + \frac{5}{100} = \frac{75}{100} = \frac{3}{4}$.

(3) $32.57 = 3$ tens $+ 2 + 5$ tenths $+ 7$ hundreths $= 32 + \frac{5}{10} + \frac{7}{100} = 32 + \frac{57}{100} = 32\frac{57}{100}$, pronounced thirty-two and fifty-seven hundredths.

(4) $5.03 = 5 + 3$ hundredths $= 5\frac{3}{100}$, pronounced five and three hundredths.

(5) $7.392 = 7 + 3$ tenths $+ 9$ hundredths $+ 2$ thousandths $= 7\frac{392}{1000}$, pronounced seven and three-hundred-ninety-two thousandths.

(6) $.7 = 7$ tenths $= \frac{7}{10}$, pronounced seven tenths. Adjoining a zero in the place to the right of .7 gives $.70 = 7$ tenths $+ 0$ hundredths $= \frac{7}{10} = \frac{70}{100}$, from multiplying both numerator and denominator by 10.

We call the decimal numeral of a number its *decimal expansion*. We shall always mean finite decimal expansion when we say decimal expansion. Infinite decimal expansions will always have the modifier infinite in their name. It is important to realize that a decimal expansion of a number is just a new notation, not a new concept. The decimal expansion .4 means $\frac{4}{10}$. To take .4 of an object, divide the object into 10 equal parts and take 4 of them, as pictured here:

In a decimal expansion, the digits to the left of the decimal point represent a whole number, while the digits to the right of the decimal point, decimal point included, represent a rational number between 0 and 1 (possibly 0), whose denominator is the value of the unit of the last place to the right. For example, in the decimal expansion 27.392 the digits to the left of the decimal point represent the whole number 27 while the digits to the right of the decimal point, decimal point included, represent the rational number $.392 = \frac{392}{1000}$. The decimal expansion is the sum of these two numbers,

$$27.392 = 27 + .392 = 27 + \frac{392}{1000}$$

Adjoining a zero at the end of a decimal fraction, in a place to the right of the decimal point, does not change the value of the number. It has the effect of multiplying the numerator and denominator by ten. For example,

6.1. Decimal Expansions

$.3 = \frac{3}{10}$ is equal to

$$.30 = \frac{30}{100} = \frac{3 \cdot 10}{10 \cdot 10} = \frac{3}{10} = .3$$

6.1.2. Decimal Inequalities. The rule for determining an inequality between two positive decimal expansions is the same as the rule for whole numbers. To determine which of two decimal expansions is smaller, find the first place from the left in which the digits differ. The smaller number is the one with the smaller digit in this place. For example, this rule says that $1.349 < 1.358$ because the first place from the left where these numbers have different digits is the hundredths place, where $4 < 5$. This is the correct answer, because $\frac{49}{1000} < \frac{58}{1000}$, which in turn is true because $49 < 58$, which we determine by the usual rule for the order of whole numbers.

Oral exercise 6.3. (1) Which is larger, 4.789 or 4.798?

(2) Which is larger, 51.701 or 51.071?

Where on the number line is a decimal expansion such as 3.1415? From what we know about decimal inequalities, $3.1 < 3.1415 < 3.2$, which says that 3.1415 lies on the number line between 3.1 and 3.2

and $3.14 < 3.1415 < 3.15$ says that 3.1415 lies between 3.14 and 3.15

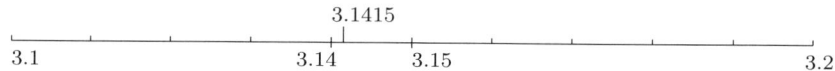

and $3.141 < 3.1415 < 3.142$ says that 3.1415 lies between 3.141 and 3.142

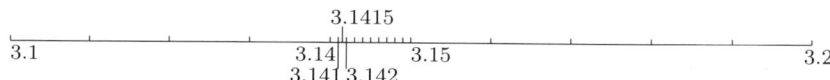

Definition 6.4 (Rounding Off). A decimal expansion is *rounded* to the nearest designated place, if all places to the right of that place are set equal to zero. The place itself remains the same if the next place to the right contains a digit less than 5, but is increased by 1 otherwise.

For example,

3.1415 rounded to the nearest tenth is 3.1

3.1415 rounded to the nearest hundredth is 3.14

3.1415 rounded to the nearest thousandth is 3.142

Decimal expansions may be rounded off at any place, not just at places to the right of the decimal point. Whole numbers can be rounded off. For example, 7,895,201 rounded to the nearest million is 8,000,000 and 581.76 rounded to the nearest ten is 580.

A rounded-off number is an estimate of the number. As with all estimates, we want to know if it is too big or too small, and by about how much.

Proposition 6.5. *If a decimal expansion is rounded to the nearest designated place, then the round-off is smaller than the number if the digit in the place remains the same. It is larger otherwise. The difference between the number and the round-off is less than 5 of the next place to the right.*

Proof. The statement and proof are clarified by some examples. The number 3.14159 rounded to the nearest hundredth is 3.14, which is smaller than 3.14159. As we observed above, 3.14 < 3.14159 < 3.15. The midpoint between 3.14 and 3.15 is 3.145, which is larger than 3.14159. Therefore,

$$3.14159 - 3.14 < 3.145 - 3.14 = .005$$

as we can see in the picture

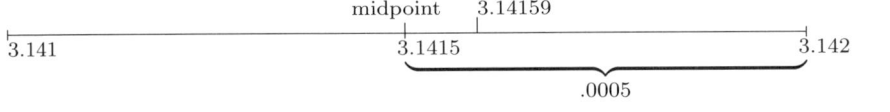

Rounded to the nearest thousandth, 3.14159 is 3.142, which is larger than 3.14159, because the digit in the thousandths place was increased by 1 in the round off. Therefore, 3.14159 is larger than the midpoint between 3.141 and 3.142, which is 3.1415, and

$$3.142 - 3.14159 < 3.142 - 3.1415 = .0005$$

as we can see in the picture

□

6.1.3. Decimal Fractions. Many rational numbers have no (finite) decimal expansion. We consider now how to determine which rational numbers have a decimal expansion.

Definition 6.6. A *power* of ten is a product of ten with itself a certain number of times, this number being the power. Ten itself is considered to be ten to the power one. Ten to the power two is $10 \cdot 10 = 100$, and so on.

This idea of power is related to exponents, which we will consider in a subsequent section.

6.1. Decimal Expansions

Definition 6.7. A *decimal fraction* is any rational number which can be represented by a fraction whose denominator is a power of ten.

Example 6.8. The fraction $\frac{37}{100}$ is a decimal fraction because its denominator is a power of ten. The fraction $\frac{3}{5}$ is a decimal fraction because it has the same value as $\frac{3 \cdot 2}{5 \cdot 2} = \frac{6}{10}$, which is a fraction whose denominator is a power of ten. The fraction $\frac{1}{3}$ is not a decimal fraction because any fraction of the same value must be of the form $\frac{1 \cdot n}{3 \cdot n}$ (by Theorem 5.30). Since 3 is always a prime factor of the denominator, the denominator can never be a power of ten.

Oral exercise 6.9.   (1) Give examples of decimal fractions. Does $\frac{3}{15}$ represent a decimal fraction?

(2) Give examples of rational numbers which are not decimal fractions. Explain why each is not a decimal fraction.

Theorem 6.10. *(1) If a decimal fraction is reduced to lowest terms, then the denominator has no prime factors other than twos and fives.*

(2) If a rational number is represented by a fraction in lowest terms whose denominator has no prime factors other than twos and fives, then the rational number is a decimal fraction.

Proof. (1) Consider a number that is a decimal fraction. By Definition 6.7, it can be represented by a fraction whose denominator is a power of ten. Since $10 = 2 \cdot 5$, any fraction whose denominator is a power of 10 must have the property that reduced to lowest terms its denominator has no prime factors other than twos and fives, because reducing it to lowest terms will never introduce any new prime factors into the denominator.

(2) Consider any number that can be represented by a fraction which, reduced to lowest terms, has a denominator containing no prime factors other than twos and fives. To be specific, suppose we have the fraction $\frac{21}{60}$, which reduced to lowest terms is $\frac{7}{20}$, whose denominator has no prime factors other than twos and fives. In fact, $20 = 2 \cdot 2 \cdot 5$, so 2 is a prime factor whose multiplicity is two while 5 is a prime factor whose multiplicity is one. If the multiplicities of twos and fives are unequal, as in this example, multiply numerator and denominator by enough of the one with smaller multiplicity so that the multiplicities of twos and fives in the denominator become equal. In the present example, multiply numerator and denominator by 5, which gives us the fraction
$$\frac{7 \cdot 5}{2 \cdot 2 \cdot 5 \cdot 5} = \frac{35}{100}$$
which is a decimal fraction. Once the denominator has the same number of twos and fives, these twos and fives can be paired off so that the product of each pair is 10, and therefore the denominator is a power of 10. $\square$

Example 6.11. (1) The number $\frac{8}{12}$ reduced to lowest terms is $\frac{2}{3}$, whose denominator contains the prime factor 3. Therefore, $\frac{8}{12}$ is not a decimal fraction.

(2) The numbers $\frac{1}{3}$, $\frac{5}{7}$, and $\frac{3}{22}$ are not decimal fractions. Each is already in lowest terms, and the denominator of each contains a prime factor other than 2 or 5.

(3) The numbers $\frac{3}{5}$, $\frac{7}{10}$, and $\frac{39}{600}$ are decimal fractions. The first two are already in lowest terms and the third reduces to $\frac{39}{600} = \frac{13}{200}$. The denominator of each reduced fraction contains no prime factor other than 2 or 5.

Theorem 6.12. *(1) If a rational number has a decimal expansion, then it is a decimal fraction.*

(2) If a rational number is a decimal fraction, then it has a decimal expansion.

Proof. (1) An example can show us how the general argument goes. Consider the decimal expansion

$$42.6931 = 42 + \frac{6}{10} + \frac{9}{100} + \frac{3}{1000} + \frac{1}{10000}$$

As with any decimal expansion, the fractional part is a sum of fractions each denominator of which is a power of 10. The greatest power of 10 occuring, which is 10000 in this case, is the least common multiple of these denominators, and thus is the least common denominator. In our example, $42.6931 = 42 + \frac{6931}{10000} = \frac{426931}{10000}$, which is a decimal fraction. We conclude that a decimal expansion always represents a decimal fraction.

(2) We want to prove that any decimal fraction has a decimal expansion. As an example, consider the decimal fraction $\frac{394}{100}$, which is equal to the mixed fraction $3\frac{94}{100}$ (see Section 5.1.1), which has the decimal expansion 3.94.

As another example, consider the case of a decimal fraction that is being represented by a fraction whose denominator is not a power of ten, say $\frac{81}{60}$. As a mixed fraction this is $1\frac{21}{60}$, and reduced to lowest terms

$$\frac{21}{60} = \frac{7}{20} = \frac{7 \cdot 5}{20 \cdot 5} = \frac{35}{100}$$

Therefore, $\frac{81}{60}$ has the decimal expansion 1.35.

These examples illustrate the general argument for proving that any decimal fraction has a decimal expansion. □

Oral exercise 6.13. (1) Explain why $\frac{2}{3}$ has no (finite) decimal expansion.

(2) Explain why $\frac{6}{15}$ has a decimal expansion. Find it.

6.1. Decimal Expansions

(3) Read .8, .16, .2910, 38.806, and 16.00043.

(4) How many fifths in .6?

(5) How many fourths in .75?

(6) How many twentieths in .15? In .35?

(7) Reduce the following decimals to common fractions in their lowest terms: .125, .025, 3.25, 56.375. *Ans.*: $\frac{1}{8}$, $\frac{1}{40}$, $3\frac{1}{4}$, $56\frac{3}{8}$.

(8) Express as decimals: $\frac{1}{5}$, $\frac{3}{4}$, $\frac{1}{20}$, $\frac{3}{15}$.

The proofs of Theorems 6.10 and 6.12 contain a procedure for finding a decimal representation for any decimal fraction. It is: convert to a mixed fraction, reduce the fraction to lowest terms, multiply its numerator and denominator by enough twos or fives so that both of these primes occur with equal multiplicity in the denominator, and express it as a decimal in the evident way. For example, to find a decimal expansion of $\frac{147}{84}$, first divide to convert it to a mixed fraction $\frac{147}{84} = 1\frac{63}{84}$, then reduce the proper fraction to lowest terms: $\frac{63}{84} = \frac{3 \cdot 3 \cdot 7}{3 \cdot 7 \cdot 2 \cdot 2} = \frac{3}{4}$, then multiply numerator and denominator by $5 \cdot 5$ to get $\frac{3}{4} = \frac{75}{100} = .75$. Thus, $\frac{147}{84} = 1.75$.

6.1.4. Long Division Extended. Long division can be used to find the decimal expansion of a decimal fraction. For example, the decimal expansion of $\frac{147}{84}$ is found by carrying out the long division $147 \div 84$ as follows.

```
           1.75    quotient
   84 ) 147.00    Begin with 84 into 147 ones
        −84        Subtract 84 · 1 ones from 147 ones
          63 0    63 ones are 630 tenths, 630 ÷ 84 = 7 tenths
         −58 8    Subtract 84 · 7 tenths from 630 tenths
           4 20   42 tenths are 420 hundredths, 420 ÷ 84 = 5 hundredths
          −4 20   Subtract 84 · 5 = 420 hundredths from 420 hundredths
              0   No remainder
```

6.1.5. Quintal Fractions. How would this extension of the numeration system look in base five? As an analog to the Latin expression decimal, for tenth, we can use quintal, for fifth.

Definition 6.14. A *quintal fraction* is any rational number that can be represented by a fraction whose denominator is a power of five.

Oral exercise 6.15. Explain why a rational number is a quintal fraction if and only if its fraction in lowest terms has denominator that is a power of five.

Definition 6.16. The Hindu-Arabic numeration system to base 5 is extended by using places to the right of the one's place. An element in a place is five times the value of an element in the place to its immediate right. The one's place is identified by placing a period, called the quintal point, between its place and the next place to the right. The names of the places to the right of the one's place are fifths, twenty-fifths, hundred-twenty-fifths, six-hundred-twenty-fifths, etc., corresponding to the value of an element in each place. Numerals of this extended numeration system are called *quintal expansions*.

Quintal means fifth and this is the value of an element in the first place to the right of the one's place. A fifth of a fifth is a twenty-fifth, since $\frac{1}{5} \cdot \frac{1}{5} = \frac{1}{25}$, using base ten numerals here. In base five numerals, five is 10_5 and a fifth is $\frac{1}{10_5}$, so that a fifth of a fifth is $\frac{1}{10_5} \cdot \frac{1}{10_5} = \frac{1}{100_5}$.

Example 6.17. (1) Three-fifths is a quintal fraction, but one-half is not, because reduced to lowest terms the denominator of one-half is two, which has a prime factor other than five.

(2) The quintal expansion $4.3 = 4 + 3 \cdot \frac{1}{10_5} = \frac{43_5}{10_5}$, where numerator and denominator are base five numerals, denotes the number four and three-fifths.

(3) The quintal expansion $1.203 = 1 + 2$ fifths $+ 3$ hundred-twenty-fifths is $1 + 2 \cdot \frac{1}{10_5} + 3 \cdot \frac{1}{100_5} = 1 + \frac{203_5}{1000_5}$, where all numerators and denominators are base five numerals. The English name for this number is one and fifty-three hundred-twenty-fifths.

6.1.6. Dimidial Fractions. How would this extension of the numeration system look in base two? As an analog to the Latin expression decimal, for tenth, we can use dimidial, for half.

Oral exercise 6.18. (1) What is the correct version of Theorem 6.12 for quintal expansions?

(2) What is the correct definition of dimidial fraction?

(3) Give an example of a rational number which is a dimidial fraction. Give an example of a rational number which is not a dimidial fraction.

(4) Give the correct definition of dimidial expansion, as an extension of the base 2 numeration system.

(5) What is the correct version of Theorem 6.12 for dimidial expansions?

(6) What is the logical name and the usual English name of the number whose dimidial expansion is 1011.011_2? *Ans*: logical: 1 two of two of two 1 two one and 1 half of half 1 half of half of half; English: eleven and three eighths.

(7) What is the logical name and the usual English name of the number whose quintal expansion is 2.14_5? *Ans*: logical: 2 and 1 fifth 4 fifth of fifths; English: two and nine twenty-fifths.

§6.1 Exercises

(1) Find the decimal expansion of the following by first reducing each to a decimal fraction: (a) $\frac{1}{8}$. (b) $\frac{5}{80}$. (c) $\frac{21}{40}$. (d) $\frac{9}{120}$. (e) $\frac{24}{15}$.

(2) Use long division to find the decimal expansion of the following:
(a) $\frac{141}{400}$. (b) $3\frac{21}{40}$. (c) $\frac{11}{16}$. (d) $24\frac{1}{125}$.

(3) Find the quintal expansion of the following by first writing each as a quintal fraction using base five numerals for numerator and denominator:
(a) seventeen twenty-fifths. (b) seven and three-fifths.

(4) Find the dimidial expansion of the following by first writing each as a dimidial fraction using base two numerals for numerator and denominator:
(a) seven-eighths. (b) five-fourths.

6.2. Arithmetic of Decimals

In this section we develop the rules of arithmetic with decimals.

Oral exercise 6.19. (1) Give an example of two decimal fractions and demonstrate that their sum is again a decimal fraction.

(2) Give an example of two decimal fractions and demonstrate that their difference is again a decimal fraction.

(3) Give an example of two decimal fractions and demonstrate that their product is again a decimal fraction.

(4) Give an example of two decimal fractions whose quotient is not a decimal fraction.

Theorem 6.20. *If two decimal fractions are added, subtracted, or multiplied, then the sum, difference, or product is again a decimal fraction. If two decimal fractions are divided, then the quotient need not be a decimal fraction.*

Proof. Any decimal fraction can be reduced to a fraction whose denominator is a power of 10. Given two such fractions, then their least common denominator would be the greater power of 10 appearing, and therefore the sum or difference would have a denominator which is a power of 10. The denominator of their product would be the product of their denominators, and therefore is a power of 10, since the product of two powers of 10 is a power of 10. In each case, then, the sum, difference, or product would be a fraction whose denominator is a power of 10, which means it is a decimal fraction.

For division, $\frac{3}{10}$ and $\frac{1}{5}$ are decimal fractions, but $\frac{1}{5} \div \frac{3}{10} = \frac{1}{5} \cdot \frac{10}{3} = \frac{2}{3}$ is not a decimal fraction. (Explain why $\frac{2}{3}$ is not a decimal fraction). □

6.2.1. Addition. The addition algorithm for positive base ten numerals carries over to decimal expansions without change. Add the digits of each place, then compose tens and carry. It is important to line up the decimal points in each number to insure that places of the same value are being added. When more than two numbers are being added, scratch addition can be used, so that only digits are being added at each step. The following examples illustrate the method. The method works for the same reason that it works for whole numbers.

Example 6.21. Find $25.75 + 36.84$. We begin with a formal explanation of the calculation. Commutative and associative properties are used without comment. The first step is numeration.

$$25.75 + 36.84 = (2 \cdot 10 + 5 + \frac{7}{10} + \frac{5}{100}) + (3 \cdot 10 + 6 + \frac{8}{10} + \frac{4}{100})$$
$$= (2+3) \cdot 10 + (5+6) + \frac{7+8}{10} + (\frac{5+4}{100}), \text{ add each place}$$
$$= 5 \cdot 10 + (1 \cdot 10 + 1) + \frac{1 \cdot 10 + 5}{10} + \frac{9}{100}, \text{ compose tens}$$
$$= (5+1) \cdot 10 + (1+1) + \frac{5}{10} + \frac{9}{100}, \text{ carry}$$
$$= 62.59, \text{ numeration}$$

As with whole numbers, this formal explanation can be summarized in the following algorithm, in which we add the numbers in each place, then compose tens and then carry.

```
            25.75
           +36.84
   (2+3)(5+6).(7+8)(5+4) add each place
       5(10+1).(10+5)9 compose tens
       (5+1)(1+1).59 carry
            62.59
```

All of these steps are combined as follows, where we begin adding in the place farthest to the right, compose tens and carry before moving to the next place to the left. The ten tenths carried to the one's place is indicated

6.2. Arithmetic of Decimals

with a 1 over the 5, and the ten ones carried to the ten's place is indicated with a 1 over the 2.

$$\begin{array}{r} \overset{1\,1}{25}.75 \\ +36.84 \\ \hline 62.59 \end{array}$$

Example 6.22. Use scratch addition to find the sum of 25.002, 206.31, 505.05, 1.015, 8.33. In this illustration, columns are added from top down. Remember that an empty place means zero is the digit in that place. Each scratch indicates one group of ten to be carried to the next place to the left. See Subsection 3.2.2 to review details of scratch addition.

$$\begin{array}{r} 25.00\,2 \\ 20\cancel{6}_1.31 \\ 505.05 \\ 1.01\,5 \\ +\ \cancel{8}_5.3\cancel{3}_0 \\ \hline 745.707 \end{array}$$

Oral exercise 6.23.   (1) Add 1.8 and 4.3.

(2) Add 1.8 and 4.03.

(3) Add 1.8 and 40.003.

(4) In each of these problems, convert the decimal expansion to a common fraction, do the addition and verify that the answers agree.

6.2.2. Subtraction. The subtraction algorithm for positive base ten numerals carries over to decimal expansions without change. To subtract a smaller number from a larger one, each given as a decimal expansion, line up the decimal points in a column and subtract place by place in the same way used for whole numbers. In case the subtrahend contains more digits to the right of the decimal point than does the minuend, annex zeros to the right of the decimal point of the minuend. Recall that the *subtrahend* is that which is being taken from the *minuend*. The method works for the same reasons that it works for whole numbers.

Example 6.24. From 8.7 take 3.1.

$$\begin{array}{r} 8.7 \\ -3.1 \\ \hline (8-3).(7-1) \text{ subtract by place} \\ 5.6 \end{array}$$

A formal explanation of this algorithm is

$$8.7 - 3.1 = 8 + \frac{7}{10} - (3 + \frac{1}{10}) = (8-3) + \frac{7-1}{10} = 5 + \frac{6}{10} = 5.6$$

in which we have used subtraction associative properties from Theorem 5.65.

Example 6.25. What must be added to 23.125 to make 78.75? The answer is $78.75 - 23.125$, which we calculate as follows.

$$\begin{array}{r} 78.750 \\ -23.125 \\ \hline \end{array} \quad \text{annex a 0 on the right}$$

The first step requires taking 5 thousandths from 0 thousandths, which cannot be done in whole numbers. In the minuend we decompose the 5 hundredths in the adjacent place to the left into 4 hundredths and 10 thousandths. Now the first step becomes taking 5 thousandths from 10 thousandths, a difference of 5 thousandths. The problem now looks like this.

$$\begin{array}{r} 78.7\,\overset{4}{\cancel{5}}\,{}^{1}0 \\ -23.1\,2\,5 \\ \hline 55.6\,2\,5 \end{array}$$

since 4 hundredths minus 2 hundredths is 2 hundredths, 7 tenths minus 1 tenth is 6 tenths, 8 ones minus 3 ones is 5 ones, and 7 tens minus 2 tens is 5 tens.

Oral exercise 6.26. (1) Explain how to compute $2 - .7$. Explain why your method gives the correct answer.

(2) Explain how to check your answer in the preceding problem by adding together the answer and the subtrahend. Explain why this is a check.

6.2.3. Multiplication. The multiplication algorithm for positive base ten numerals carries over to decimal expansions with one additional step: the placement of the decimal point in the resulting product. The rule is to multiply the decimal expansions as if there were no decimal point present, and then point off from the right as many decimal places in the product as there are in the multiplier and the multiplicand combined. For example

$$\begin{array}{r} 6.13 \\ \times 7.24 \\ \hline 2452 \\ 1226 \\ 4291 \\ \hline 44.3812 \end{array}$$

$4 \cdot 613$
$2 \cdot 613$ tens
$7 \cdot 613$ hundreds
Add. Decimal point is 4 places from the right

One way to see why this rule gives the correct answer is to convert the decimal expansions to common fractions and then multiply, then convert back to a decimal expansion. In the present example, $6.13 = \frac{613}{100}$ and $7.24 = \frac{724}{100}$ and

$$\frac{613}{100} \cdot \frac{724}{100} = \frac{613 \cdot 724}{100 \cdot 100} = \frac{443812}{10000}$$

6.2. Arithmetic of Decimals

The final step is to understand why $\frac{443812}{10000} = 44.3812$. Using long division to find the decimal expansion of this common fraction, we see that $443812 \div 10000 = 44 + \frac{3812}{10000}$ and then $\frac{3812}{10000} = \frac{3000}{10000} + \frac{800}{10000} + \frac{10}{10000} + \frac{2}{10000} = \frac{3}{10} + \frac{8}{100} + \frac{1}{1000} + \frac{2}{10000} = .3812$. Therefore, $\frac{443812}{10000} = 44.3812$. There is a general rule here which we state as a theorem.

Recall that a power of ten is 10 times 10 as many times as the power. Thus 10, 100, 1000, 10000 are powers of ten. Also recall that the decimal point of a whole number is at the right of the one's place, even though it is usually not written.

Theorem 6.27. *To multiply a decimal expansion by a power of ten, move the decimal point as many places to the right as the power. Annex zeros on the right if necessary. To divide a decimal expansion by a power of ten, move the decimal point as many places to the left as the power. Annex zeros on the left if necessary.*

Proof. For whole numbers, the multiplication rule is a direct consequence of our numeration system, as we observed back in the section on computing products. For example, $10 \cdot 7 = 7$ tens $= 70$ and $100 \cdot 813 = 81300$. It is easy to see why the rule works for a general decimal expansion if we first convert it to a decimal fraction. For example,
$$100 \cdot 12.675 = 100 \cdot \frac{12675}{1000} = \frac{12675}{10} = 1267.5$$
so the decimal point moved 2 places to the right, precisely corresponding to the cancellation of two zeros in the denominator of the fraction $\frac{100}{1000} = \frac{1}{10}$.

The division rule works because the product of two powers of ten is a power of ten whose power is the sum of the powers of the two given numbers. For example, 100 is 10 to the power 2 and 1000 is 10 to the power 3, and we know that $100 \cdot 1000 = 100000$ is 10 to the power $2 + 3 = 5$. Notice that the power is the number of zeros in the numeral. As an example of a whole number divided by a power of 10, consider $392 \div 100$, which is
$$(3 \cdot 100 + 9 \cdot 10 + 2) \cdot \frac{1}{100} = 3 \cdot \frac{100}{100} + 9 \cdot \frac{10}{100} + \frac{2}{100} = 3 + \frac{9}{10} + \frac{2}{100} = 3.92$$
so the decimal point of 392 moved 2 places to the left, because the division of each place by 100 reduces the value of an element of that place to that of an element two places to its right. As an example of a decimal expansion divided by a power of 10, we have
$$13.68 \div 1000 = \frac{1368}{100} \cdot \frac{1}{1000} = \frac{1368}{100000} = .01368$$

$\square$

Oral exercise 6.28. (1) Compute $10 \cdot 37.81$ and explain why your method is correct.

(2) Compute $.024 \cdot 100$ and explain why your method is correct.

(3) Compute $8.02 \div 1000$ and explain why your method is correct.

(4) Express the quotient $\frac{3}{4} \div 100$ as a decimal expansion. Explain your method.

(5) Compute $.3 \cdot .05$.

(6) Compute $.07 \cdot .8$.

6.2.4. Division. Long division for base ten numerals carries over to decimal expansions with one additional step: the placement of the decimal point in the resulting quotient. The rule is to divide the decimal expansions as if there were no decimal point present, and then point off from the right as many decimal places in the quotient as the number of decimal places in the dividend exceeds the number of those in the divisor. In practice, this is accomplished by moving the decimal point to the right in both divisor and dividend by as many places as it has in the divisor. As a consequence of this move, the divisor is a whole number. In the following example we use margin notes to indicate the procedure at each step. Justification of the procedure is identical to the justification of long division of whole numbers expressed in base ten numerals.

Consider 578.5728 divided by 8.32. We begin by shifting the decimal point two places (the number of places it has in the divisor) to the right in both divisor and dividend. The result of this is the problem 57857.28 divided by 832. But shifting the decimal point two places to the right is the same as multiplication by 100, so this new problem is $578.5728 \cdot 100$ divided by $8.32 \cdot 100$, and this problem is the same as the original problem, because $(578.5728 \cdot 100) \div (8.32 \cdot 100) = (578.5728 \div 8.32) \cdot (100 \div 100)$ by the associative properties of division given in Theorem 5.103. Now

```
              69.54    Decimal point above point in dividend
     832 ) 57857.28    Begin with 832 into 5785 tens
          −4992        Subtract 832 · 6 tens from 5785 tens
           7937        793 tens are 7930 ones, bring down 7 ones
          −7488        Subtract 832 · 9 ones from 7937 ones
            449 2      449 ones are 4490 tenths, bring down 2 tenths
           −416 0      Subtract 832 · 5 = 4160 tenths from 4492 tenths
             33 28     332 tenths = 3320 hundredths, bring down 8 100ths
             33 28     Subtract 832 · 4 hundredths from 3328 hundredths
                 0     No remainder
```

We conclude that $578.5728 \div 8.32 = 69.54$. This can be checked by verifying that $8.32 \cdot 69.54 = 578.5728$.

Oral exercise 6.29.   (1) Divide 1846 by 100.

6.2. Arithmetic of Decimals

(2) Divide 739 by .01.

(3) Divide 37 by .002.

6.2.5. Applications of Decimals. The most familiar use of decimals is in denoting United States currency. The unit is the dollar and a cent is a hundredth of a dollar. We denote four dollars and 35 cents by the decimal expansion $4.35, which is completely consistent with our definition of decimal expansion, since 35 cents is 35 hundredths of a dollar.

As with all application problems in these notes, the idea is to be able to solve the problem and then be able to articulate the reasoning used to solve the problem. The articulation process should strive to explain how we decide which arithmetic operations to use. Here are a few example problems with their solutions and explanations. The explanations should be informative, but brief.

Example 6.30. What will 675 ft of boards cost at $18.50 per thousand ft?

By unitary method:

1000 ft = 18.50 dollars

1 ft = $\frac{18.50}{1000}$ dollars

675 ft = $675 \cdot \frac{18.50}{1000}$ = 12.4875 dollars which the lumber yard will no doubt round off to $12.49.

By ratios: The ratio of cost to feet is always in proportion to the given price of $18.50 per 1000 ft. Thus,

$$\frac{\text{cost}}{675} = \frac{18.50}{1000}$$

$$\text{cost} = 675 \cdot \frac{18.50}{1000} = 12.4875$$

as shown in the following line segment diagram

Example 6.31. A store was valued at $290,000. It was insured for .625 of its value. If a woman owns .45 of the stock, what is the value of her insured part?

Solution: The value of the insured stock is $.625 \cdot 290000$ dollars. The woman owns .45 of this, which is

$$.45 \cdot (.625 \cdot 290000) = 81562.50 \text{ dollars}$$

as pictured in the line segment diagram

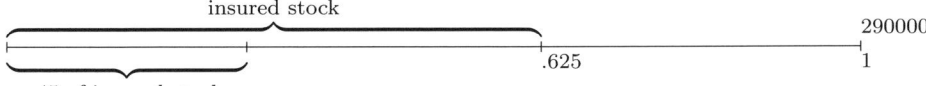

§6.2 Exercises

(1) Add 89.07 and 76.79. Use notes and margin notes to explain what you are doing.

(2) Use scratch addition to find the sum of 34.883, 13.4072, 12.0734, 110.642, and 8.0008. Use notes and margin notes to explain what you are doing.

(3) Five pieces of silver weigh as follows: .33 pounds, 1.275 pounds, .127 pounds, .5 pounds, and 1.324 pounds. How much do the five pieces weigh? Explain how you decide which arithmetic operation to use. Show the calculation, with notes.

(4) What added to 3.47 gives 6.21? Explain your procedure and do the computation with margin notes. Explain how to check that your answer is correct.

(5) Do the following computations with margin notes and other comments to explain your procedure.
 (a) From 274.684 take 217.423. *Ans.* 57.261
 (b) 186.004 − 87.621. *Ans.* 98.383
 (c) What added to 3.17 gives 621? *Ans.* 617.83
 (d) Use decimals to take $3\frac{15}{16}$ from 7.642. *Ans.* 3.7045

(6) Compute the following products, with notes describing your method. Explain why the method is correct.
 (a) $25.75 \cdot 5.6$ *Ans.* 144.2
 (b) $16.004 \cdot 1.68$ *Ans.* 26.88672
 (c) $448.9 \cdot .008$ *Ans.* 3.5912

(7) Gavin exchanged 28 bushels of wheat at $1.25 a bushel for 200 pounds of sugar at $13\frac{3}{4}$ cents a pound, and the balance in cash. How much is due him? *Ans.* $7.50

(8) Compute the following quotients, with notes describing your method. Explain why the method is correct.
 (a) Divide 12.5111 by 30.5. *Ans.* .4102
 (b) Divide .0189 by .025. *Ans.* .756
 (c) Divide 9.455 by $12\frac{1}{2}$. *Ans.* .7564

In the following exercises, write out the solution with each operation indicated, then carry out the calculations, either by hand or with a calculator. Follow this with a brief explanation for how you solved the problem.

(8) John walks 3.745 miles and Sadie 3.8005 miles. Which walks the farthest and by how much?

6.3. Repeating Decimals

(9) Clarence owned 64.803 acres of land. He bought 10.7045 acres and sold all but 16.455 acres. How many acres did he sell?

(10) What is the value of 7.5 tons of hay at $18.75 a ton?

(11) Mortie exchanged 12.55 tons of hay at $49.95 per ton for 2 head of cattle at $460.25 a head, the balance to be paid in cash. How much cash did he have to pay?

(12) Addison purchased oil at $29.43 per barrel and sold it at $32.48 a barrel, for a profit of $671. How many barrels did he have?

(13) A store containing $88,000 worth of stock was damaged by fire, but $\frac{3}{5}$ of its goods were saved. If Ragna owns .40, Leah .35, and Niels the remainder of the stock, what is the value of Niels's share of goods saved?

6.3. Repeating Decimals

We have seen that only decimal fractions have a decimal expansion. The decimal fractions were those rational numbers which can be represented by a common fraction in lowest terms whose denominator contains no prime factor other than 2 or 5. Many useful fractions, such as $\frac{1}{3}, \frac{2}{3}, \frac{3}{11}$, and $\frac{1}{7}$, are not decimal fractions. We have proved that these numbers do not have a (finite) decimal expansion, so we ask about the possibility of approximating these numbers by decimal expansions. One approach to answering this question is to use the long division method for finding a decimal expansion.

Example 6.32. Consider the case of $\frac{1}{3}$. Using the rules of long division with decimal expansions, we begin the calculation of $1 \div 3$.

$$
\begin{array}{r}
0.3 \\
3 \overline{\smash{)}1.00} \\
-9 \\
\hline
10
\end{array}
$$

- Begin with 3 into 10 tenths
- Subtract $3 \cdot 3$ tenths from 10 tenths
- Now have 3 into 10 hundredths

Except for the change of value of an element from the hundredth's place as opposed to the tenth's place, the next step again requires us to find $10 \div 3$, which again, as always, is 3 (hundredths now) with remainder 1 hundredth, which, for the next step, would be decomposed into 10 thousandths. The next step would then again be $10 \div 3$. This process goes on forever, with the 3 being repeated forever in the quotient. We call the result a repeating decimal, and write it $.\overline{3}$, where the bar over the 3 indicates that the 3 repeats forever. This phrase *repeats forever* means that no matter how many threes are written, you can annex yet another. The idea is similar to what we mean by the counting numbers going on forever: no matter where you stop, you can always add one more.

Example 6.33. Consider what decimal expansion we obtain for $\frac{3}{11}$ when we calculate $3 \div 11$.

```
           0.27
    11 | 3.00     Begin with 11 into 30 tenths
        −22       Subtract 11 · 2 tenths from 30 tenths
          80      Now have 11 into 80 hundredths
        − 77      Subtract 11 · 7 hundredths from 80 hundredths
           3      We are back to the first step
```

The next step will be $30 \div 11$, which is the **same as the first step**. The next two steps in the long division will yield 27 ten-thousandths in the quotient, at which point 3 will again be the remainder. The 27 will repeat forever. We write the result as the repeating decimal $.\overline{27}$, where the bar over the 27 means that the two-digit block, 27, repeats forever.

Example 6.34. Consider what decimal expansion we obtain for $\frac{1}{7}$ when we calculate $1 \div 7$.

```
         0.142857
    7 | 1.000000    Begin with 7 into 10 tenths
       − 7          Subtract 7 · 1 tenths from 10 tenths
         30         Now have 7 into 30 hundredths
       − 28         Subtract 7 · 4 hundredths from 30 hundredths
         20         Now have 7 into 20 thousandths
       − 14         Subtract 7 · 2 thousandths from 20 thousandths
         60         Now have 7 into 60 ten-thousandths
       − 56         Subtract 7 · 8 ten-thousandths from 60 ten-thousandths
         40         Now have 7 into 40 hundred-thousandths
       − 35         Subtract 7 · 5 1/100000's from 40 1/100000's
         50         Now have 7 into 50 millionths
       − 49         Subtract 7 · 7 millionths from 50 millionths
          1         We are back to the first step
```

The next step is $10 \div 7$, which is the same as the first step. As only zeros followed the 1 in the dividend, and that is again the case, each successive step will be repeated until the remainder becomes 1 again. We conclude that the repeating decimal expansion of $\frac{1}{7}$ is $.\overline{142857}$, where the whole six digit block 142857 repeats forever.

Definition 6.35. A *repeating decimal* is a decimal expansion for which at some place to the right of the decimal point a finite block of digits repeats forever. The block of repeating digits is called its *repetend* or *repeating part*.

Examples of repeating decimals are given above. A repeating decimal expansion of a rational number is a representation of that number, which

6.3. Repeating Decimals

means that we can write these examples as equalities:

$$\frac{1}{3} = .\overline{3}, \quad \frac{3}{11} = .\overline{27}, \quad \frac{1}{7} = .\overline{142857}$$

Their repetends are 3, 27, and 142857, respectively. Some more examples are $1.2\overline{3}$, whose repetend is 3, and $128.7\overline{09}$, whose repetend is 09. By this definition of repeating decimal, the decimal expansion of a decimal fraction is a repeating decimal in which 0 is the repetend. For example, .5 is the same as $.5\overline{0}$, because $.5 = .50 = .500$, etc.

Theorem 6.36. *If x is a rational number, then x has a repeating decimal expansion.*

Proof. Any positive rational number can be represented by a common fraction, $\frac{m}{n}$, say, where m and n are whole numbers. At each step of the long division $m \div n$, the remainder must be a whole number less than n, which means it must be one of the numbers in the set $\{0, 1, \ldots, n-1\}$. There are n numbers in this set, and so there are at most n different possible remainders.

If the remainder is ever 0, then the division stops and it must be the case that $\frac{m}{n}$ is a decimal fraction.

Otherwise, the remainder is never 0. At each step it must be a number in the set $\{1, \ldots, n-1\}$. In the long division, from any given place, the remainder must repeat in at most $n - 1$ steps. In particular, once past the decimal point, the remainder must repeat after at most $n - 1$ steps. The quotients at each step will then repeat, because each partial division is the same. The digits in the quotient between the repeated remainders is the repetend. Here is an illustration of what this argument is saying.

If the number is $\frac{23}{15}$, then we do the long division $23 \div 15$ as

```
         1.53
   15 ) 23.00
        15
        ‾‾
         8     Now past the decimal point
         75
         ‾‾
          5    First remainder is 5
          45
          ‾‾
           5   Remainder is 5 again
```

The repetend is thus 3 in this example. The repeating decimal expansion of $\frac{23}{15}$ is $1.5\overline{3}$.

□

Oral exercise 6.37. (1) Find the repeating decimal expansion of $\frac{2}{3}$. Of $\frac{20}{3}$.

(2) What is the repeating decimal expansion of $6+\frac{2}{3}$? How is it related to the repeating decimal fraction of 6 and of $\frac{2}{3}$?

(3) Find the repeating decimal expansion of $\frac{1}{11}$. Of $\frac{10}{11}$. Of $\frac{100}{11}$.

(4) What is the repeating decimal expansion of $9+\frac{1}{11}$? How is it related to the repeating decimal fraction of 9 and of $\frac{1}{11}$?

(5) How is the repeating decimal expansion of $\frac{m}{n}$ related to that of $\frac{10 \cdot m}{n}$?

(6) How is the repeating decimal expansion of $\frac{m}{n}$ related to that of $\frac{100 \cdot m}{n}$?

Proposition 6.38. *If x is a rational number with a repeating decimal expansion, then a power of 10 times x has the same repeating decimal expansion but with the decimal point moved to the right as many places as the power. In other words, the rule for multiplying a repeating decimal by a power of 10 is the same as the rule for multiplying a decimal by a power of 10.*

Proof. For example, if the number $\frac{m}{n}$ is multiplied by 1000 (which is 10 to the third power), then it is represented by the common fraction $\frac{1000 \cdot m}{n}$. But the base ten numeral of $1000 \cdot m$ is that of m with the decimal point moved 3 places to the right. It follows that the decimal point in dividend and quotient of $(1000 \cdot m) \div n$ is 3 places to the right of its position in $m \div n$.

For example, as shown above, $\frac{2}{11} = .\overline{18}$. The theorem at hand claims that $10 \cdot \frac{2}{11} = 1.\overline{81}$. This is true, because, from the long division $2 \div 11$ shown above, we see that the long division of $20 \div 11$ is

$$\begin{array}{r} 1.818 \\ 11 \overline{\smash{\big)}\ 20.000} \end{array}$$

from which we see that the quotient is the repeating decimal $1.\overline{81}$, which is the same as $.\overline{18}$ with the decimal point moved 1 place to the right. In the same way, the long division of $200 \div 11$,

$$\begin{array}{r} 18.18 \\ 11 \overline{\smash{\big)}\ 200.00} \end{array}$$

shows that the repeating decimal of $\frac{200}{11}$ is $18.\overline{18}$, that is, $100 \cdot .\overline{18} = 18.\overline{18}$. In the same way, multiplication of a repeating decimal by any power of 10 just moves the decimal point to the right as many places as the power. □

Proposition 6.39. *If x is a rational number with a repeating decimal expansion, then x is the sum of a finite decimal and a rational number whose infinite repeating decimal has zeros in every place to the left of the repetend, and whose repetend is the same as that in the decimal expansion of x.*

6.3. Repeating Decimals

Proof. It is best to see what this Proposition is saying for a specific example. Consider the rational number $\frac{47}{198}$. The long division

$$
\begin{array}{r}
.237 \\
198\overline{\smash{)}47.000} \\
\underline{-39\,6} \\
7\,40 \\
\underline{-5\,94} \\
1\,460 \\
\underline{-1\,386} \\
74
\end{array}
$$

shows that $\frac{47}{198} = .2\overline{37}$. After the first step of the long division we see that the quotient is .2 with remainder 7.4, which we can write as

$$\frac{47}{198} = .2 + \frac{7.4}{198}$$

The rest of the long division amounts to the long division $7.4 \div 198$ and shows that

$$\frac{7.4}{198} = .0\overline{37}$$

Therefore we may conclude that

$$\frac{47}{198} = .2 + .0\overline{37}$$

as claimed by the Proposition. $\square$

Theorem 6.40. *Any repeating decimal is the repeating decimal expansion of some rational number.*

Proof. The proof gives us an effective way to find the rational number represented by a given repeating decimal. The method is best illustrated with examples.

What rational number is represented by the repeating decimal $.\overline{21}$? Suppose that $\frac{m}{n} = .\overline{21}$, where m and n are whole numbers. Then Proposition 6.38 says that the repeating decimal expansion of $\frac{100 \cdot m}{n}$ is $21.\overline{21}$, and Proposition 6.39 says that this repeating decimal expansion can be written as $21 + .\overline{21}$. From this we subtract $\frac{m}{n}$ and proceed as follows.

$$
\begin{align}
(6.1) \qquad 100 \cdot \frac{m}{n} &= 21.\overline{21} = 21 + .\overline{21} \\
(6.2) \qquad \frac{m}{n} &= .\overline{21} \\
\hline
(6.3) \qquad 99 \cdot \frac{m}{n} &= 21
\end{align}
$$

which implies that $\frac{m}{n} = 21 \div 99 = \frac{7}{33}$. By long division we can verify that $\frac{7}{33} = .\overline{21}$.

Why did we multiply the given repeating decimal $.\overline{21}$ by 100? The answer is that we multiplied by a power of 10 for which the power is equal to the length of the repetend. By Proposition 6.38, each multiplication by 10 shifts each digit over one place to the left, so multiplying by 10 as many times as the length of the repetend moves each digit to the left by the length of the repetend. In particular, it moves the second occurrence of the repetend into the position of the first occurrence. From the second occurrence of the repetend this product is identical to the given decimal from its first occurrence of the repetend. Splitting the numbers into a sum and subtracting we get the infinite repeating decimals to cancel out.

As another example, find the rational number whose repeating decimal expansion is $.47\overline{135}$. Suppose that $\frac{m}{n} = .47\overline{135}$, for some natural numbers m and n. The length of the repetend is 3, so multiply $\frac{m}{n}$ by $10 \cdot 10 \cdot 10 = 1000$, using Proposition 6.38, decompose the result into a sum as described in Proposition 6.39 and then subtract the original number from this.

$$(6.4) \qquad 1000 \cdot \frac{m}{n} = 471.35\overline{135} = 471.35 + .00\overline{135}$$

$$(6.5) \qquad \frac{m}{n} = .47\overline{135} = .47 + .00\overline{135}$$

$$(6.6) \qquad (1000 - 1) \cdot \frac{m}{n} = 471.35 - .47 = \frac{47088}{100}$$

we conclude that

$$\frac{m}{n} = \frac{47088}{100} \div 999 = \frac{47088}{99900} = \frac{436}{925}$$

in lowest terms. By long division we verify that $\frac{436}{925} = .47\overline{135}$.

In summary, the method is to suppose the given repeating decimal is a rational number. Multiply it by the power of ten for which the power is equal to the length of the repetend and write the product as the sum of a finite decimal and an infinite repeating decimal which will be identical to that obtained from the original number by starting from the first occurrence of the repetend. Subtract the original number, thus cancelling out the infinite repeating parts and solve for the rational number. Long division verifies that the repeating decimal expansion of the fraction obtained is the given decimal. □

Example 6.41. Find the rational number whose repeating decimal expansion is $2.\overline{3}$. Begin by supposing that $\frac{m}{n} = 2.\overline{3}$. The length of the repetend is 1, so we multiply by 10, decompose into a sum and subtract as described

6.3. Repeating Decimals

in the general method:

$$(6.7) \qquad 10 \cdot \frac{m}{n} = 23.\overline{3} = 23 + .\overline{3}$$

$$(6.8) \qquad \frac{m}{n} = 2 + .\overline{3}$$

$$(6.9) \qquad 9 \cdot \frac{m}{n} = 23 - 2 = 21$$

from which we conclude that $2.\overline{3} = \frac{21}{9} = \frac{7}{3} = 2\frac{1}{3}$. Notice that if we already knew that $.\overline{3} = \frac{1}{3}$, then we have a briefer calculation:

$$2.\overline{3} = 2 + .\overline{3} = 2 + \frac{1}{3}$$

Oral exercise 6.42. Find the rational number whose repeating decimal expansion is $7.\overline{3}$.

Example 6.43. Find the rational number whose repeating decimal expansion is $.0\overline{3}$. Suppose that $.0\overline{3} = \frac{m}{n}$. The length of the repetend is 1, so we multiply by 10 and subtract as in the general method

$$(6.10) \qquad 10 \cdot \frac{m}{n} = .3\overline{3} = .3 + .0\overline{3}$$

$$(6.11) \qquad \frac{m}{n} = .0\overline{3}$$

$$(6.12) \qquad 9 \cdot \frac{m}{n} = .3 = \frac{3}{10}$$

from which we conclude that

$$.0\overline{3} = \frac{1}{30} = \frac{1}{10} \cdot \frac{1}{3}$$

A briefer calculation is $10 \cdot .0\overline{3} = .\overline{3} = \frac{1}{3}$, so $.0\overline{3} = \frac{1}{10} \cdot \frac{1}{3} = \frac{1}{30}$.

Oral exercise 6.44. In two ways find the rational number whose repeating decimal is $.7\overline{3}$.

Example 6.45. Find the rational number whose repeating decimal expansion is $.\overline{9}$. Suppose $.\overline{9} = \frac{m}{n}$. Then

$$(6.13) \qquad 10 \cdot \frac{m}{n} = 9.\overline{9} = 9 + .\overline{9}$$

$$(6.14) \qquad \frac{m}{n} = .\overline{9}$$

$$(6.15) \qquad 9 \cdot \frac{m}{n} = 9$$

from which we conclude that $.\overline{9} = \frac{m}{n} = 1$.

Oral exercise 6.46. Demonstrate two ways to find the rational number whose repeating decimal is $.2\overline{9}$.

Can two different repeating decimals be the expansion of the same rational number? Give an example or explain why not.

§6.3 Exercises

In the first ten problems, find the rational number whose repeating decimal expansion is given. In more old fashioned terms, reduce the following repeating decimals to common fractions. In the last five problems, use long division to find the repeating decimal expansion of the given fraction.

Show your method. Do the long divisions by hand, showing all steps.

(1) $.\overline{6}$ Ans: $\frac{2}{3}$

(2) $.\overline{2}$

(3) $.\overline{42}$ Ans: $\frac{14}{33}$

(4) $.\overline{24}$

(5) $.\overline{852}$ Ans: $\frac{284}{333}$

(6) $.\overline{743}$

(7) $.25\overline{64}$ Ans: $\frac{2539}{9900}$

(8) $.46\overline{52}$

(9) $9.3\overline{209}$ Ans: $9\frac{353}{1100}$

(10) $3.2\overline{708}$

(11) $\frac{5}{11}$ Ans: $.\overline{45}$

(12) $\frac{6}{7}$

(13) $\frac{3}{13}$ Ans: $.\overline{230769}$

(14) $\frac{5}{6}$

(15) $\frac{7}{22}$ Ans: $.3\overline{18}$

6.4. Arithmetic of Infinite Decimals

In applications, infinite decimals are approximated by the finite decimal obtained by truncating it at some point and adjusting the last digit in this truncation by the rules of rounding off. The definition of how to round an infinite decimal to the nearest designated place is the same as for (finite) decimal expansions given in Definition 6.4.

6.4. Arithmetic of Infinite Decimals

Example 6.47. (1) $.\overline{6}$ rounded to the nearest tenth is .7; rounded to the nearest hundredth is .67; rounded to the nearest thousandth is .667.

(2) $.\overline{09}$ rounded to the nearest tenth is .1; rounded to the nearest hundredth is .09, rounded to the nearest thousandth is .091.

Oral exercise 6.48. (1) Round $.\overline{18}$ to the nearest tenth. To the nearest hundredth. To the nearest thousandth.

(2) Explain why $\frac{.2}{7} = \frac{2 \cdot .1}{7} = \frac{2}{7} \cdot .1$.

(3) Explain why $\frac{.06}{7} = \frac{6 \cdot .01}{7} = \frac{6}{7} \cdot .01$.

How well does the rounded-off decimal approximate the repeating decimal? The answer is the same as it was for the case of finite decimal expansions and is given in Proposition 6.5: the difference between the number and its round-off to a place is less than five elements of the next place to the right.

The proof is not as elementary as it was for the case of a finite decimal expansion, because the part discarded is now itself an infinite repeating decimal. Here is a proof using the long division process we followed to find the infinite repeating decimal.

Proof. We want to prove the round-off rule of Proposition 6.5 for the case of repeating decimal expansions. We shall illustrate the proof with the example of the repeating decimal expansion $\frac{3}{7} = .\overline{428571}$.

Rounded to the nearest tenth, $.\overline{428571}$ is .4. We want to prove that they differ by less than .05. From the long division $3 \div 7$

$$\begin{array}{r} .4 \\ 7 \overline{\smash{)}3.0} \\ -2\,8 \\ \hline 2 \end{array} \text{ Remainder of 2 tenths}$$

This means that $3 \div 7 = .4$ plus a remainder of 2 tenths, which we write as $3 = 7 \cdot .4 + .2$. Divide through by 7 to get

$$\frac{3}{7} = .4 + \frac{.2}{7} = .4 + \frac{2}{7} \cdot .1$$

because $\frac{.2}{7} = \frac{2 \cdot .1}{7} = \frac{2}{7} \cdot .1$, where the last step uses the division associative property expressed in equation (5.14) of Theorem 5.103. The difference between $\frac{3}{7}$ and its round-off to the nearest tenth is then

$$\frac{3}{7} - .4 = \frac{2}{7} \cdot (.1) < \frac{1}{2} \cdot (.1) = .05,$$

where the inequality comes from $\frac{2}{7} < \frac{1}{2}$ and the order property of multiplication in Theorem 5.83. Thus, .4 is smaller than $.\overline{428571}$ by less than .05.

Rounded to the nearest hundredth, $.\overline{428571}$ is .43. We want to prove that they differ by less than .005. The long division carried another step,

$$
\begin{array}{r}
.42 \\
7 \overline{\smash{)}3.00} \\
-2\,8 \\
\hline
2\,0 \\
1\,4 \\
\hline
6 \quad \text{Remainder of 6 hundredths}
\end{array}
$$

means that $3 = 7 \cdot .42 + .06$. Divide through by 7 to get

$$\frac{3}{7} = .42 + \frac{.06}{7} = .42 + \frac{6}{7} \cdot (.01)$$

because $\frac{.06}{7} = \frac{6 \cdot .01}{7} = \frac{6}{7} \cdot .01$. The difference between $\frac{3}{7}$ and its repeating decimal rounded to the nearest hundredth is then

$$.43 - \frac{3}{7} = .43 - (\frac{6}{7} \cdot (.01) + .42) = .01 - \frac{6}{7} \cdot (.01) = \frac{1}{7} \cdot (.01) < \frac{1}{2} \cdot (.01) = .005$$

Thus, .43 is bigger than $.\overline{428571}$ by an amount less than .005, which is what we wanted to prove.

Let's do one more step just to be sure we see the pattern to the proof. The long division carried another step,

$$
\begin{array}{r}
.428 \\
7 \overline{\smash{)}3.000} \\
-2\,8 \\
\hline
20 \\
14 \\
\hline
60 \\
56 \\
\hline
4 \quad \text{Remainder of 4 thousandths}
\end{array}
$$

means that $3 = 7 \cdot .428 + .004$. Divide through by 7 to get

$$\frac{3}{7} = .428 + \frac{.004}{7} = .428 + \frac{4}{7} \cdot (.001)$$

because $\frac{.004}{7} = \frac{4 \cdot .001}{7} = \frac{4}{7} \cdot .001$. The repeating decimal $.\overline{428571}$ rounded to the nearest thousandth is .429. It is larger than $.\overline{428571}$ by

$$.429 - (.428 + \frac{4}{7} \cdot (.001)) = .001 - \frac{4}{7} \cdot (.001) = \frac{3}{7} \cdot .001 < \frac{1}{2} \cdot (.001) = .0005$$

which is what we wanted to prove. $\square$

6.4. Arithmetic of Infinite Decimals

The preceding proof gives us the following number line interpretation of an infinite repeating decimal. We illustrate it with $\frac{3}{7} = .\overline{428571}$. Since $.4 < .\overline{428571} < .5$, we know that on the number line it lies between .4 and .5

```
                          ↓ 3/7 in here
├─────────────────────────────┼──┼──────────────────────────┤
0                            .4  .5                          1
```

Then $.42 < .\overline{428571} < .43$ means that

```
              ↓ 3/7 in here
├─────────────┼──┼────────────────────────────────────────┤
.4           .42 .43                                      .5
```

and $.428 < .\overline{428571} < .429$ means that

```
                                          ↓ 3/7 in here
├──────────────────────────────────────────┼──┼───────────┤
.42                                      .428 .429       .43
```

and $.4285 < .\overline{428571} < .4286$ means that

```
                           ↓ 3/7 in here
├──────────────────────────┼──┼───────────────────────────┤
.428                     .4285 .4286                    .429
```

and so forth.

Oral exercise 6.49. Explain why $.\overline{3} > .33$ and $.\overline{3} - .33 < .005$. What is the exact value of this difference? (Hint: It is $\frac{1}{3} - \frac{33}{100}$). What is the repeating decimal expansion of the difference? (Hint: Do the long division $1 \div 300$).

Oral exercise 6.50. The repeating decimal expansion of $\frac{7}{11}$ is $.\overline{63}$. Explain why $.64 > .\overline{63}$ and $.64 - .\overline{63} < .005$. What is the exact value of this difference? (Hint: It is $\frac{64}{100} - \frac{7}{11}$. Compare this to $.005 = \frac{1}{200}$).

If a repeating decimal is not a finite decimal, that is, if its repetend is not 0, then any round-off of it has a value different from the value of the given repeating decimal. This fact has important implications for calculators and computers, all of which hold numbers as decimals to a fixed number of places. Actually, calculators and computers use base 2 numeration, which means they use dimidial fractions to a given number of places. The following discussion applies to any calculator or computer.

Suppose you decide that you prefer working with decimals, not too long, of course, say out to four places to the right of the decimal point. Actually, calculators and computers are more restricted, since they work with a fixed total number of places. The sum of the number of places to the left with the number of places to the right, of the decimal point, is fixed. This is called floating point arithmetic. In any case, if you work with fixed length decimals, then any rational number which is not a decimal fraction will be represented by an approximation which is not exact. Even decimal fractions whose decimal is too long will be approximated by a round-off of its decimal. When these approximations are multiplied or divided, the approximation often becomes worse. As a consequence, one must avoid too many computations with a computer or calculator in solving a given problem.

For example, suppose all numbers must be written with no more than three places to the right of the decimal point. Then $\frac{7}{11}$ would be represented by .636 so that $\frac{7}{11} \cdot \frac{7}{11}$ would be represented by $.636 \cdot .636 = .404$, when rounded off to the nearest thousandth. The actual value $\frac{7}{11} \cdot \frac{7}{11} = \frac{49}{121} = .405$ when rounded off to the nearest thousandth. This kind of error is called round-off error. It accumulates with each operation, but in a random way that is difficult to estimate.

To minimize round-off error, avoid rounding-off until you reach the final answer.

Example 6.51. At 3 for a dollar, how much do 25 oranges cost, to the nearest cent?

Solution: $(1 \cdot 25) \div 3 = 8\frac{1}{3}$ dollar, which is $8.33, to the nearest cent.

Explanation 1: If 3 oranges cost 1 dollar, then 1 orange costs $1 \div 3$ dollars, so 25 oranges cost $(1 \div 3) \cdot 25$ dollars, which is $8\frac{1}{3} = 8.\overline{3}$ dollars. At this stage, and not before, we round off to the nearest cent (which means the nearest hundredth), to get $8.33. If we were to round off at an earlier step, for example round off the price of one orange to the nearest cent is .33 dollar, then we would find that 25 oranges cost $25 \cdot .33 = 8.25$ dollars, an 8 cent error.

Explanation 2: Each group of 3 oranges costs 1 dollar, and there are $25/3 = 8\frac{1}{3}$ such groups in a set of 25 oranges. The cost of $8\frac{1}{3}$ groups is therefore $1 \cdot (8\frac{1}{3}) = 8\frac{1}{3}$ dollars, as obtained above.

Example 6.52. Suppose you own a million shares of Acme Gold Inc. stock valued at $6\frac{3}{8}$ dollars per share. The stock exchange decides to switch to the use of decimals in the quotation of all stock prices. If this is done by rounding off each share price to the nearest cent, how much money will you lose or gain by this change?

Solution: Before the conversion your million shares are worth

$$1,000,000 \cdot (6\frac{3}{8}) = 6,000,000 + (3,000,000 \div 8) = 6,375,000$$

dollars. The decimal expansion of $6\frac{3}{8}$ is 6.375, which rounded to the nearest hundredth is 6.38. After the conversion, your stock is worth $6.38 per share, so a million shares are worth $1,000,000 \cdot 6.38 = 6,380,000$ dollars. Therefore, you make $6,380,000 - 6,375,000 = 5,000$ dollars from the conversion.

6.4.1. Nonrepeating Decimals. Can you imagine a decimal which continues on forever to the right of the decimal point, without constant repetition of some repetend? How would you write this down, or prescribe how it

6.4. Arithmetic of Infinite Decimals

should be written down? Any place that you stop writing, what you have could be regarded as the repetend, so you must specify how the rest of the decimal places are filled. Here is an example of how to do that.

(6.16) .1011011101111011111...

where the next place is 0, then comes 6 ones, then 0, then 7 ones, and so on, with each block of ones separated by a 0, and the length of the block of ones increasing by one each time.

Is it possible that such an infinite nonrepeating decimal represents some number? One way to answer this question is to say that it represents the number which is approximately .1, better yet .10, even better .101, better still .1011, and so on. To the nearest millionth it is approximated by the rational number .101101. This answer begs the question, however, because it tells us better and better approximations of the number, but still doesn't tell us what the number is. Throughout this text we have identified the numbers we have been studying with points on the number line. Our philosophy has been that numbers are points on the number line. The question to ask about our infinite nonrepeating decimal is what point on the number line does it represent? The rational number approximations are then points close to this point.

Let us repeat the above description using the number line. The points of the number line constitute all possible real numbers. If formula (6.16) represents a number a, then it corresponds to some point on the number line. Where is this point? It is between .1 and .2, in which it is between .10 and .11, in which it is between .101 and .102, in which it is between .1011 and .1012, and so on. The inclusion of the next digit in the expansion locates the point in the correct subinterval of the previous interval divided into ten equal subintervals. Here is an illustration of two steps.

Assuming that you agree that we can call an infinite nonrepeating decimal, such as that in Equation (6.16), a number, is it a rational number? After all, so far in these notes, rational numbers are all the numbers we have developed. Recall that Theorem 6.36 says that any rational number has a repeating decimal expansion. But we have also seen that some rational numbers have two different decimal expansions, such as, for example, $1.\overline{0} = .\overline{9}$. Nevertheless, could there be some rational number that has both a repeating decimal expansion and an infinite nonrepeating decimal expansion? The answer is no, for the reason that we'll illustrate for the case of the number a above given by the infinite decimal (6.16).

Suppose that b is a rational number equal to a. Then b has a finite or infinite repeating decimal expansion which cannot be the same as the decimal (6.16), which is neither finite nor repeating. Starting from the decimal point, there must be some first decimal digit where they are different. For the sake of argument, suppose that the decimal expansion of b starts with $.101d$, where d is a digit unequal to 1. Then a and b have the same decimal digits up to this fourth digit. We have just observed that a is a point lying in the interval between .1011 and .1012. This means that

$$.1011 < a < .1012$$

Now if $d = 0$, then $b \leq .1011$ and therefore b cannot be equal to a. If $d \geq 2$, then $.1012 \leq b$ and again b cannot be equal to a. This argument works in general to show that an infinite nonrepeating decimal cannot represent a rational number.

There should be numbers which cannot be rational. For example, it seems natural that there should be a number which multiplied by itself gives 2. In symbols, there should be a number a such that $a \cdot a = 2$. The following theorem, due to the Pythagoreans, who preceded Euclid, is contained in [**Euc56**, page 2, Vol. 3].

Theorem 6.53. *If a is a rational number, then $a \cdot a \neq 2$.*

Proof. The proof is an example of an argument called proof by contradiction, or more historically, *reductio ad absurdum*. The idea of the proof is to assume there is a rational number a with the property that $a \cdot a = 2$, and from that deduce a statement which we know to be false, thus forcing us to conclude that our initial assumption cannot be true.

The proof goes as follows. Since a is a rational number, it can be expressed as a fraction $a = \frac{m}{n}$ in lowest terms, which means that the greatest common factor of m and n is 1. If $a \cdot a = 2$, then

$$2 = \frac{m}{n} \cdot \frac{m}{n} = \frac{m \cdot m}{n \cdot n}$$

We multiply both sides by $n \cdot n$ to get the equation

$$2 \cdot n \cdot n = m \cdot m$$

which shows that 2 is a prime factor of $m \cdot m$. But, the prime factors of $m \cdot m$ are those of m, each appearing twice. Therefore, 2 must be one of the prime factors of m, which means that $m = 2 \cdot p$, for some whole number p. Dividing both sides of the preceding equation by 2, we have

$$n \cdot n = 2 \cdot p \cdot p$$

which shows that 2 must be a prime factor of n. We made sure that $\gcf(m, n) = 1$, but now we have shown that 2 is a common prime factor of m and n. The argument has reduced to an absurdity, and we conclude that the initial assumption that $a \cdot a = 2$ is false. $\square$

Definition 6.54. The *square root* of a rational number b is a number a such that $a \cdot a = b$.

It is standard notation to designate the square root of b by $\sqrt{b}$. The symbol $\sqrt{b}$ stands for a positive number with the property that $\sqrt{b} \cdot \sqrt{b} = b$.

The above Theorem 6.53 says that the square root of 2 is not a rational number. The argument given in the proof will show that the square root of 3 is not a rational number. In fact, the argument will show that the square root of any prime number is not a rational number. This means that none of $\sqrt{2}$, $\sqrt{3}$, and $\sqrt{b}$ for any prime number b is a rational number. Following our philosophy of identifying the numbers with points on the number line, we must think of these *irrational numbers* as points on the number line.

Oral exercise 6.55. Prove that the square root of 3 is not a rational number.

Oral exercise 6.56. What is the square root of 4? Of 9? What is $\sqrt{25}$?

Oral exercise 6.57. Explain why $1.1 \cdot 1.1 < 1.2 \cdot 1.2$.

Oral exercise 6.58. Explain why if $0 < a < b$, then $a \cdot a < b \cdot b$. Ans. By the order property in Theorem 5.83, we can multiply $a < b$ by a to get $a \cdot a < a \cdot b$, and by b to get $b \cdot a < b \cdot b$. Therefore, since $a \cdot b = b \cdot a$, these inequalities say $a \cdot a < a \cdot b < b \cdot b$.

One way to locate $\sqrt{2}$ on the number line is to describe its infinite decimal expansion and then use the method described above for locating with ever greater precision the location of this infinite decimal.

Theorem 6.59. *The square root of 2 has an infinite nonrepeating decimal expansion.*

Proof. We apply the *disection method*, called that because at each step we divide some interval into ten equal subintervals. Suppose there is a number a for which $a \cdot a = 2$. Think of where the number a must lie on the number line. It must lie between 1 and 2, because $1 \cdot 1 = 1 < 2$ and $2 \cdot 2 = 4 > 2$. Thus, 1 is too small and 2 is too large.

Consider the points 1.1, 1.2, 1.3, and so on up to 1.9, that result from dividing the interval from 1 to 2 into 10 equal subintervals. Then $1.1 \cdot 1.1 = 1.21 < 2$ is too small, $1.2 \cdot 1.2 = 1.44$ is too small, $1.3 \cdot 1.3 = 1.69$ is too small, $1.4 \cdot 1.4 = 1.96$ is too small, but $1.5 \cdot 1.5 = 2.25$ is too large. We conclude that a lies between 1.4 and 1.5. Its infinite decimal expansion begins with 1.4.

```
|---|---------|---|----↓ √2 in here---|-------|---|
    1        1.4 1.5                  2
```

Next, divide the interval from 1.4 to 1.5 into 10 equal subintervals. This is done with the points 1.41, 1.42, and so forth to 1.49. Then $1.41 \cdot 1.41 = 1.9881$ is too small but $1.42 \cdot 1.42 = 2.0164$ is too large. We conclude that a lies between 1.41 and 1.42. Its infinite decimal expansion begins with 1.41.

```
|---|---↓ √2 in here---|-------------------|
    1.4 1.41 1.42                        1.5
```

Next, divide the interval from 1.41 to 1.42 into 10 equal subintervals. This is done with the points 1.411, 1.412, and so forth to 1.419. Then $1.411 \cdot 1.411 = 1.990921$ is too small, $1.412 \cdot 1.412 = 1.993744$ is too small, $1.413 \cdot 1.413 = 1.996569$ is too small, $1.414 \cdot 1.414 = 1.999396$ is too small, but $1.415 \cdot 1.415 = 2.002225$ is too large. We conclude that a lies between 1.414 and 1.415. Its infinite decimal expansion begins with 1.414. Its first three decimal places are .414.

```
|-------|---↓ √2 in here---|-------------|
    1.41    1.414 1.415                1.42
```

The procedure is hopefully clear by now. Notice that all multiplications being done here are between rational numbers and are exact. No round-off is being done in any of the products computed. This process can be continued forever, at least in principle. Each step determines one more place in the decimal expansion of a. In this sense we have determined the infinite decimal expansion of a, a number such that $a \cdot a = 2$. This decimal expansion cannot be repeating, because we know that any repeating decimal represents a rational number (Theorem 6.40), and we also know that a is not a rational number. □

Oral exercise 6.60. Explain why we could stop as soon as we found a number which was too large. For example, in the first step, 1.4 was too small, but 1.5 was too large. Is it possible that 1.6, or 1.7, etc. could again be too small? Explain.

Oral exercise 6.61. Continue the proof to find the digit in the ten-thousandth's place of the infinite decimal expansion of the square root of 2. Use a calculator only to calculate the products, not to calculate the square root. The idea here is to understand the meaning of the infinite decimal expansion of the square root of 2.

Oral exercise 6.62. Follow the method of the proof to find the first three places of the infinite decimal expansion of the square root of 3.

6.4. Arithmetic of Infinite Decimals

6.4.2. Newton's method. Isaac Newton discovered a method for finding the roots to all sorts of equations. This method produces fractions which estimate the root, rather than a decimal to so many places. Of course, the decimal expansion of each fraction can be found by long division. His method applied to finding the square root of any positive number a begins with a guess, called $x_0 > 0$, and then iterates each estimate from the previous estimate by the formula

$$(6.17) \qquad x_{n+1} = \frac{1}{2}(\frac{a}{x_n} + x_n)$$

for $n = 0, 1, 2, 3, \ldots$, as far as you wish to go with it. After the initial guess, the estimates are always too big. That is, $x_n > \sqrt{a}$ for each $n \geq 1$.

Example 6.63 ($\sqrt{2}$ by Newton's method). Here is how Newton's method is used to find estimates of $\sqrt{a}$ when $a = 2$. As an initial guess we take $x_0 = 1$. Then the estimate of the first step is

$$x_1 = \frac{1}{2}(\frac{a}{x_0} + x_0) = \frac{1}{2}(\frac{2}{1} + 1) = \frac{3}{2} = 1.5$$

The estimate of the second step is

$$x_2 = \frac{1}{2}(\frac{2}{\frac{3}{2}} + \frac{3}{2}) = \frac{2}{3} + \frac{3}{4} = \frac{17}{12} = 1.41\overline{6}$$

The estimate of the third step is

$$x_3 = \frac{1}{2}(\frac{2}{\frac{17}{12}} + \frac{17}{12}) = \frac{12}{17} + \frac{17}{24} = \frac{577}{408} = 1.414\overline{2156862745098039}$$

which is not too bad considering that

$$\left(\frac{577}{408}\right)^2 = 2\frac{1}{166464}$$

is very close to 2. This also shows that $x_3 > \sqrt{2}$.

6.4.2.1. Explanation of Newton's method. The number $\sqrt{a}$ is the length of the side of a square whose area is a. Suppose that our initial guess x_0 is too small, as shown on this square.

Let $E = \sqrt{a} - x_0$ denote the error, so that $x_0 + E = \sqrt{a}$. Squaring both sides we get

$$a = (\sqrt{a})^2 = (x_0 + E)^2 = x_0^2 + 2x_0 E + E^2$$

If we try to solve this equation for E, we will have to take a square root. If we remove the E^2 from our equation we have the inequality

$$a = x_0^2 + 2x_0 E + E^2 > x_0^2 + 2x_0 E$$

Subtract x_0^2 from both sides of this inequality, and then divide both sides by $2x_0$, to get

$$E < \frac{a - x_0^2}{2x_0} = \frac{1}{2}\left(\frac{a}{x_0} - x_0\right)$$

Then

$$\sqrt{a} = x_0 + E < x_0 + \frac{1}{2}\left(\frac{a}{x_0} - x_0\right) = \frac{1}{2}\left(\frac{a}{x_0} + x_0\right) = x_1$$

where the last step is from Newton's formula (6.17). Even though the first guess was too small, this next estimate is too big.

Let $F = x_1 - \sqrt{a}$ denote the new error, so that $x_1 + F = \sqrt{a}$, as pictured here:

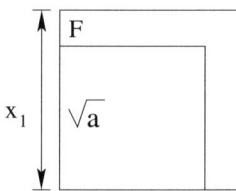

Then

$$a = (x_1 - F)^2 = x_1^2 - 2x_1 F + F^2 > x_1^2 - 2x_1 F$$

Subtract $2x_1 F - a$ from both sides of this inequality and then divide both sides by $2x_1$ to get

$$F > \frac{x_1^2 - a}{2x_1} = \frac{1}{2}\left(x_1 - \frac{a}{x_1}\right)$$

Then

$$\sqrt{a} = x_1 - F < x_1 - \frac{1}{2}\left(x_1 - \frac{a}{x_1}\right) = \frac{1}{2}\left(\frac{a}{x_1} + x_1\right) = x_2$$

where the last step is from Newton's formula (6.17). The new estimate is again too big. Each succeeding step goes in the same way. The estimates are always better, but always too big.

6.4.3. Square Roots on the Number Line. We are thinking of numbers as points on the number line. The exact location on the line of an infinite nonrepeating decimal remains elusive, because we have only been able to specify smaller and smaller subintervals in which it must lie. The location of many square roots remains elusive for the same reason, because our approximation by disection or by Newton's method keeps finding closer rational numbers, but doesn't really produce the exact point.

6.4. Arithmetic of Infinite Decimals

The Pythagorean Theorem gives us a way to use a ruler and compass to construct square roots on the number line. As a first special case, consider $\sqrt{2}$. On the number line, use a ruler and compass to construct a right triangle with the base leg being the unit interval from 0 to 1 on the number line, and the other leg erected at 1, perpendicular to the number line and of length 1.

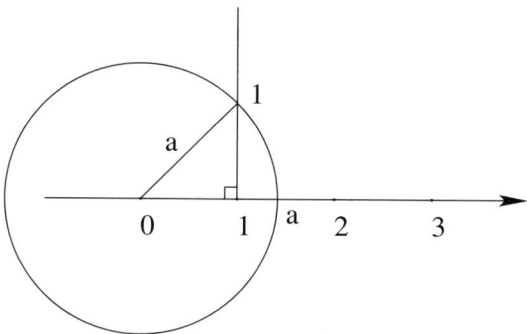

If a is the length of the hypotenuse of this isosceles right triangle, then by the Pythagorean Theorem

$$a^2 = 1^2 + 1^2 = 2$$

Therefore, $a = \sqrt{2}$, which can be marked precisely on the number line by using a compass to draw the circle with center at 0 and whose radius is the hypotenuse of this right triangle, as shown in the figure.

6.4.3.1. *Square root of a.* This construction is a special case of a construction of $\sqrt{a}$, for any positive number a. In the following figure it is assumed that $a > 1$. In the case of $0 < a < 1$, the roles of a and 1 are reversed.

Geometric Construction of a Square Root:

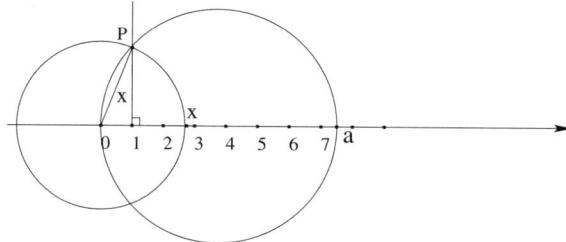

(1) Draw a horizontal number line.

(2) Draw a vertical line up from 1.

(3) Draw a circle of radius $\frac{a}{2}$ centered at $\frac{a}{2}$, so it passes through 0 and a.

(4) Let P be the point where this circle intersects the vertical line through 1.

(5) Let x denote the length of the line segment $0P$. Then $x = \sqrt{a}$.

(6) To mark x on the number line, draw a circle centered at 0 through P and mark where it intersects the number line to the right of 0.

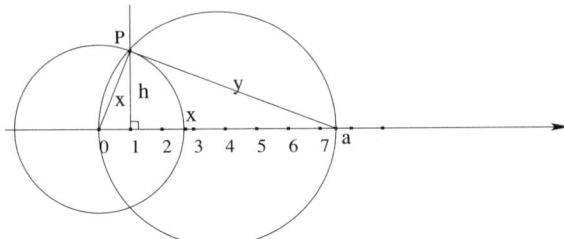

Proof that $x = \sqrt{a}$.

(1) Draw the line segment from P to a. Call its length y.

(2) Let h denote the length of the line segment from 1 to P.

(3) The triangle with vertices at 0, P, and a is inscribed in a semicircle and is therefore a right triangle, with the right angle at P.

(4) The triangles $0P1$ and $Pa1$ are also right triangles.

From the Pythagorean Theorem applied to these three right triangles we have

$$x^2 + y^2 = a^2, \quad x^2 = h^2 + 1, \quad y^2 = h^2 + (a-1)^2$$

In the first equation substitute the third equation for y^2, to get

$$x^2 + h^2 + (a-1)^2 = a^2$$

In this, substitute the second equation solved for h^2, to get

$$x^2 + x^2 - 1 + (a-1)^2 = a^2$$

Expand $(a-1)^2 = a^2 - 2a + 1$ to arrive at

$$2x^2 - 1 + a^2 - 2a + 1 = a^2$$

from which we conclude that $x^2 = a$. That is, $x = \sqrt{a}$. Put a compass at 0 and draw the circle whose radius is $0P$ in order to mark the point $\sqrt{a}$ on the number line. □

§6.4 Exercises

In each problem, present the solution and an explanation of how you reached your solution.

(1) (a) Find the infinite repeating decimal for $\frac{7}{11}$.
(b) Round it off to the nearest: (i) tenth. (ii) hundredth. (iii) thousandth.
(c) Follow the above demonstration of the proof of Proposition 6.5 for infinite repeating decimals to show that:
(i) $\frac{7}{11}$ rounded to the nearest tenth differs from $\frac{7}{11}$ by less than .05.
(ii) $\frac{7}{11}$ rounded to the nearest hundredth differs from $\frac{7}{11}$ by less than .005.
(iii) $\frac{7}{11}$ rounded to the nearest thousandth differs from $\frac{7}{11}$ by less than .0005.

(2) (a) Round $.46\overline{52}$ to the nearest thousandth.
(b) Is the round off larger or smaller than $.46\overline{52}$? Explain.
(c) Find whole numbers m and n such that $\frac{m}{n} = .46\overline{52}$ and then apply long division to $m \div n$ to prove that $.46\overline{52}$ differs from its round-off to the nearest thousandth by less than .0005.

(3) A store was valued at $70,000. It was insured for .625 of its value. Cameron owns two-thirds of the stock. (a) To the nearest cent, what is the value of the insured part of her share? (b) What is it to the nearest dollar? *Ans.* (a) $29,166.67. (b) $29,167.

(4) Niels walks 3.647 miles and Leah walks 3.8005 miles. Which walks farther, and by how much? How much to the nearest tenth of a mile? *Ans.* Leah, by .1535 mile, which is .2 mile to the nearest tenth of a mile.

(5) If apples are 7 for $2, how much for 3 dozen apples? How much is that to the nearest cent? *Ans.* $10\frac{2}{7} = 10.\overline{285714}$ dollars. $10.29.

(6) Gavin sold 2520 pounds of hay at $22 a ton, and 1860 pounds of straw at $16.50 a ton. How much did he get for both? To the nearest cent? To the nearest dollar? *Ans.* $43\frac{13}{200} = 43.065$ dollars. $43.07. $43.

(7) Give Euclid's proof of: If a is rational then $a \cdot a \neq 5$.

(8) (a) Use the disection method to find the first three decimal places (to the right of the decimal point) of the infinite decimal expansion of $\sqrt{3}$.
(b) Use this decimal approximation to indicate an interval on the number line in which $\sqrt{3}$ must lie.
(c) From your result, what is $\sqrt{3}$ rounded to the nearest hundredth?

(9) Use Newton's method to find the estimates x_1, x_2, and x_3 of $\sqrt{a}$, where $a = 4$ and the initial guess is $x_0 = 1$. Ans: $x_1 = \frac{5}{2}$, $x_2 = \frac{41}{20}$, and $x_3 = \frac{3281}{1640} = 2\frac{1}{1640}$.

(10) (a) Starting with $x_0 = 1$, find the first three estimates x_1, x_2, and x_3 of $\sqrt{3}$ given by Newton's method. Ans: $x_1 = 2$, $x_2 = \frac{7}{4}$, and $x_3 = \frac{97}{56}$.
(b) Verify that each estimate is larger than $\sqrt{3}$.
(c) By how much does the square of the third estimate exceed 3? Ans: $\frac{1}{3136}$
(d) By long division (by hand) find the infinite repeating decimal of x_3. Ans: $1.732\overline{142857}$.

(11) (a) Starting with $x_0 = 1$, find the first three estimates of $\sqrt{5}$ given by Newton's method. Ans: $x_1 = 3$, $x_2 = \frac{7}{3}$, $x_3 = \frac{47}{21}$
(b) By how much does the square of the third estimate exceed 5? Ans: $\frac{4}{441}$.

(12) Using straight edge and compass, do the geometric construction in §6.4.3.1 of $\sqrt{3}$. Prove that the number you constructed has square equal to 3.

(13) In §6.4.3.1 it is explained how to adjust the construction to find $\sqrt{a}$ when $0 < a < 1$.
(a) Do the construction to find $\sqrt{\frac{1}{2}}$.
(b) Prove that $\sqrt{\frac{1}{2}} = \frac{\sqrt{2}}{2}$ by showing that the square of the expression on the right is equal to $\frac{1}{2}$.
(c) Do the construction to find $\sqrt{2}$ and show that what you constructed for $\sqrt{\frac{1}{2}}$ is exactly half of this.

(14) For any natural number $k > 1$, Newton's method for estimating the kth root of a positive number a is

$$x_{n+1} = \frac{1}{k}\left(\frac{a}{x_n^{k-1}} + (k-1)x_n\right)$$

Use this formula for the case $k = 3$, with an initial guess of $x_0 = 1$, to find the estimates x_1 and x_2 of the cube root of 2. By how much does x_2^3 exceed 2? Ans: $x_1 = \frac{4}{3}$ and $x_2 = \frac{91}{72}$ whose cube exceeds 2 by $\frac{7075}{373248}$. For the ambitious, $x_3 = \frac{1126819}{894348}$.

6.5. Archimedes's Estimate of π

Is π rational? How do we find a decimal expansion of π? Recall the estimate of π found by Archimedes (287 - 212 BC) and cited in Equation (2.12):

$$3\frac{10}{71} < \pi < 3\frac{1}{7}$$

6.5. Archimedes's Estimate of π

It is quite remarkable that Archimedes approached the problem of calculating π by estimating below and above, rather than attempting an equality. He did not know whether π was rational. In fact, nobody knew until 1761 when the Swiss mathematician Johann Heinrich Lambert proved that π is irrational. Therefore, the decimal expansion of π is infinite nonrepeating. So far, over 208 billion places have been computed.

Here is a brief indication of how Archimedes obtained his estimates over 2200 years ago. For this I have used the article *A history of Pi* contained in the MacTutor History of Mathematics archive at the url cited in Oral Exercise 2.115. Draw a circle of radius 1 unit. Ascribe an equilateral triangle of side length $2A_1$. Inscribe an equilateral triangle of side length $2B_1$, like this:

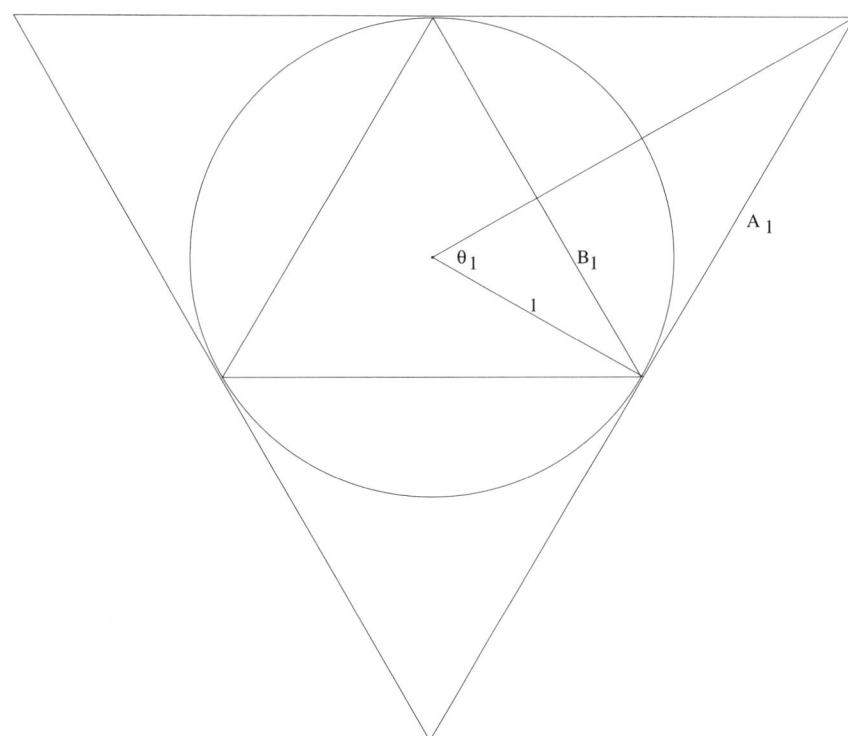

Use the Pythagorean Theorem to find $A_1 = \sqrt{3}$ and $B_1 = \frac{\sqrt{3}}{2}$. The angle $\theta_1 = 60°$, since it is the result of dividing the whole angle of $360°$ into six equal parts. For those who know trigonometry,

$$A_1 = \frac{\sin\theta_1}{\cos\theta_1}, \quad B_1 = \sin\theta_1$$

The perimeter of the inscribed triangle is $3 \cdot 2 \cdot B_1$, which is less than the circumference of the circle ($= 2 \cdot \pi$), which is less than the perimeter of the ascribed triangle, which is $3 \cdot 2 \cdot A_1$. Thus, multiplying everything by $\frac{1}{2}$ and

using the order property of multiplication, we have
$$3 \cdot \frac{\sqrt{3}}{2} < \pi < 3 \cdot \sqrt{3}$$
For $\sqrt{3}$ Archimedes used the estimates
$$\frac{265}{153} < \pi < \frac{1351}{780}$$
He left no indication of how he found these estimates. For this exposition, we shall use estimates of $\sqrt{3}$ obtained from the bisection method. For this step, we use
$$1.7 < \sqrt{3} < 1.8$$
Using the order property of multiplication, we conclude that
$$3 \times \frac{1.7}{2} < \pi < 3 \times 1.8$$
which comes out to $2.55 < \pi < 5.4$.

For the next step, inscribe a regular polygon with $2 \cdot 3$ sides, of side length $2B_2$, and ascribe a regular polygon with $2 \cdot 3$ sides, of side length $2A_2$.

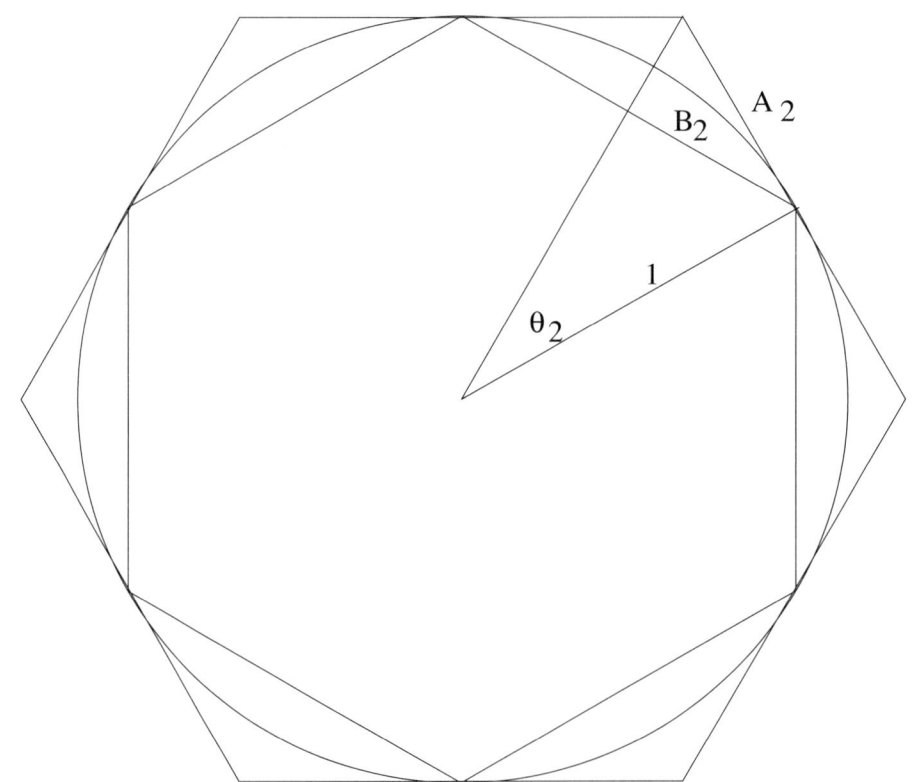

6.5. Archimedes's Estimate of π

Use the Pythagorean Theorem to find $A_2 = \frac{\sqrt{3}}{3}$ and $B_2 = \frac{1}{2}$. The angle $\theta_2 = \frac{\theta_1}{2}$. For those who know trigonometry,

$$A_2 = \frac{\sin \theta_2}{\cos \theta_2}, \quad B_2 = \sin \theta_2$$

Comparing the perimeters of these polygons to the circumference of the circle, we find the inequalities

$$6 \cdot 2 \cdot B_2 < 2 \cdot \pi < 6 \cdot 2 \cdot A_2$$

into which we put the values of A_2 and B_2 to get

$$3 < \pi < 2 \cdot \sqrt{3}$$

Using the disection method to obtain the estimate $1.73 < \sqrt{3} < 1.74$, we get

$$3 < \pi < 2 \times 1.74 = 3.48 = 3\frac{12}{25}$$

For the next step, ascribe and inscribe a regular polygon with twice as many sides, so $2 \cdot 2 \cdot 3$ sides at this step. Let $2 \cdot A_3$ be the side length of the ascribed polygon and let $2 \cdot B_3$ be the side length of the inscribed polygon.

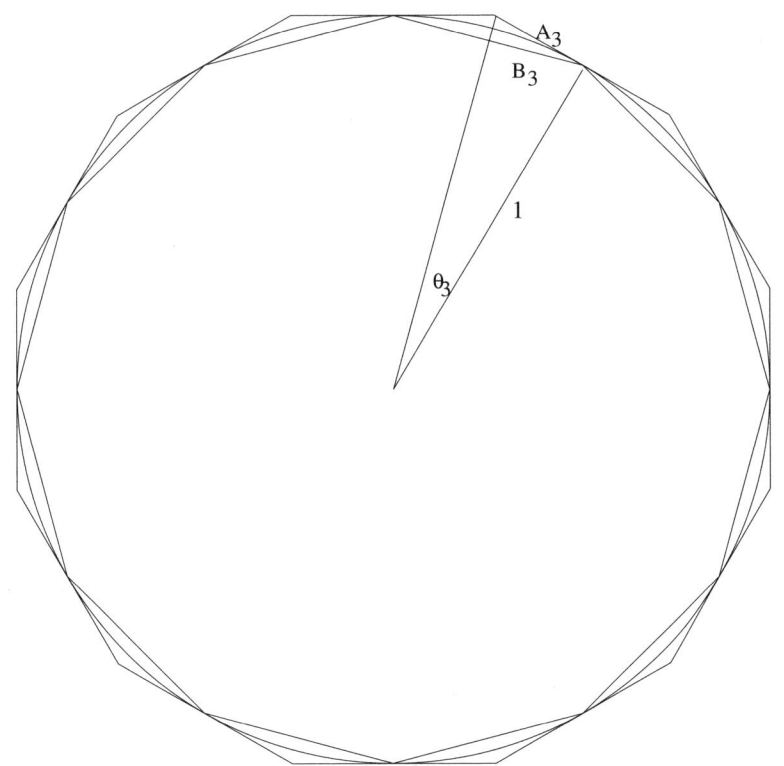

Now it is no longer easy to find A_3 and B_3. The angle $\theta_3 = \frac{\theta_2}{2}$ and, for those who know trigonometry,

$$A_3 = \frac{\sin \theta_3}{\cos \theta_3}, \qquad B_3 = \sin \theta_3$$

By using the Pythagorean Theorem and proportions coming from similar triangles, Archimedes found that

(6.18)
$$\begin{aligned}(1) & \quad \frac{1}{A_2} + \frac{1}{B_2} = \frac{1}{A_3} \\ (2) & \quad 2A_3 B_2 = 4B_3^2\end{aligned}$$

Knowing A_2 and B_2, the first equation allows us to solve for A_3. Now we know A_3 and B_2, so the second equation allows us to solve for B_3. Doing this, we find

$$\frac{1}{A_3} = \frac{1}{A_2} + \frac{1}{B_2} = \sqrt{3} + 2$$

so

$$A_3 = \frac{1}{2 + \sqrt{3}} = \frac{1 \cdot (2 - \sqrt{3})}{(2 + \sqrt{3}) \cdot (2 - \sqrt{3})} = \frac{2 - \sqrt{3}}{4 - 3} = 2 - \sqrt{3}$$

and then

$$B_3^2 = \frac{1}{4} \cdot 2 \cdot A_3 \cdot B_2 = \frac{1}{2} \cdot (2 - \sqrt{3}) \cdot \frac{1}{2} = \frac{2 - \sqrt{3}}{4}$$

so

$$B_3 = \frac{\sqrt{2 - \sqrt{3}}}{2}$$

Comparing the perimeters of the polygons with the circumference of the circle, we find that $12 \cdot 2B_3 < 2 \cdot \pi < 12 \cdot 2A_3$, so

$$6 \cdot \sqrt{2 - \sqrt{3}} < \pi < 12 \cdot (2 - \sqrt{3})$$

By the disection method

$$1.7320 < \sqrt{3} < 1.7321$$

Subtracting a larger amount from 2 leaves less, so

$$2 - \sqrt{3} < 2 - 1.732 = .268 = \frac{67}{250}$$

and

(6.19) $$\pi < 12(2 - \sqrt{3}) < 12 \cdot \frac{67}{250} = \frac{402}{125} = 3\frac{27}{125}$$

In the same way, using the upper bound for $\sqrt{3}$, we have

$$2 - \sqrt{3} > 2 - 1.7321 = .2679 = \frac{2679}{10000}$$

so, the square root of the larger number being larger than the square root of the smaller number, we have

$$\sqrt{2-\sqrt{3}} > \frac{\sqrt{2679}}{100}$$

By the disection method,

$$51.7590 < \sqrt{2679} < 51.7591$$

so

$$\sqrt{2-\sqrt{3}} > \frac{\sqrt{2679}}{100} > \frac{51.759}{100} = .51759$$

so

(6.20) $$\pi > 6\sqrt{2-\sqrt{3}} > 6 \times .51759 = \frac{155277}{50000} = 3\frac{5277}{50000}$$

Archimedes kept going, using inscribed and ascribed polygons with twice as many sides as the preceding step, through to the case of polygons with $96 = 2^5 \cdot 3$ sides. For the case of polygons with $2^n \cdot 3$ sides, if $2A_{n+1}$ is the side length of the ascribed polygon and $2B_{n+1}$ is the side length of the inscribed polygon, then the central angle $\theta_{n+1} = \frac{\theta_n}{2}$ and, for those who know trigonometry,

$$A_{n+1} = \frac{\sin \theta_{n+1}}{\cos \theta_{n+1}}, \quad B_{n+1} = \sin \theta_{n+1}$$

as can be seen on the preceding drawings. Archimedes proved the following *recursion formulas*.

Lemma 6.64 (Archimedes's Recursion Formulas).

(6.21)
$$(1) \quad \frac{1}{A_n} + \frac{1}{B_n} = \frac{1}{A_{n+1}}$$
$$(2) \quad 2A_{n+1}B_n = 4B_{n+1}^2$$

Proof. Archimedes proved this using the Pythagorean Theorem and proportions of similar triangles. Here is a simplification of his proof provided by trigonometry (developed centuries after Archimedes lived). The trigonometric identities

$$\sin^2 \theta + \cos^2 \theta = 1, \quad \sin \theta = 2 \sin \frac{\theta}{2} \cos \frac{\theta}{2}, \quad \cos \theta = \cos^2 \frac{\theta}{2} - \sin^2 \frac{\theta}{2}$$

give the following, since $\frac{\theta_n}{2} = \theta_{n+1}$,

$$\frac{1}{A_n} + \frac{1}{B_n} = \frac{1}{\frac{\sin\theta_n}{\cos\theta_n}} + \frac{1}{\sin\theta_n} = \frac{\cos\theta_n}{\sin\theta_n} + \frac{1}{\sin\theta_n}$$

$$= \frac{\cos\theta_n + 1}{\sin\theta_n} = \frac{\cos^2\theta_{n+1} - \sin^2\theta_{n+1} + 1}{2\sin\theta_{n+1}\cos\theta_{n+1}}$$

$$= \frac{2\cos^2\theta_{n+1}}{2\sin\theta_{n+1}\cos\theta_{n+1}} = \frac{\cos\theta_{n+1}}{\sin\theta_{n+1}} = \frac{1}{A_{n+1}}$$

which is (1). And

$$2A_{n+1}B_n = 2\frac{\sin\theta_{n+1}}{\cos\theta_{n+1}}\sin\theta_n = 2\frac{\sin\theta_{n+1}}{\cos\theta_{n+1}}2\sin\theta_{n+1}\cos\theta_{n+1}$$

$$= 4\sin^2\theta_{n+1} = 4B_{n+1}^2$$

proves (2). □

§6.5 Exercises

(1) On a number line, plot Archimedes's estimates in Equation (2.12) together with the estimates obtained above in Equations (6.19) and (6.20). Indicate the interval in which π lies.

(2) Find lower and upper estimates for π by going to the next step, which would be the case of inscribed and ascribed regular polygons with $2 \cdot 2 \cdot 2 \cdot 3 = 24$ sides. Use Archimedes's recursion formulas to find A_4 and B_4. Use the disection method to estimate square roots.

6.6. Percentage

The word percent means per hundred. It is a fraction whose denominator is 100. It is used in the sense of a fraction of a whole or of a given quantity. It can be an improper fraction.

Definition 6.65. If a is any number, then $a\%$ means the fraction $\frac{a}{100}$.

A percentage can be expressed as a fraction or as the decimal expansion of the fraction. For example

$$37\% = \frac{37}{100} = .37$$

37% of a whole means divide the whole into 100 equal parts and take 37 of these parts. The following square is divided into 100 equal parts and 37 of these parts are shaded. Thus, 37% of the square is shaded and 63% of the square is unshaded. There are 63 little unshaded squares.

6.6. Percentage

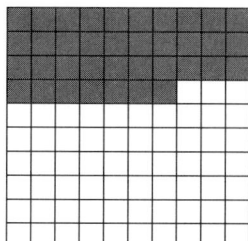

A percentage can be expressed as the proportion

part is to whole as percentage is to 100

Written as fractions, this proportion is

$$\frac{\text{part}}{\text{whole}} = \frac{\text{percentage}}{100}$$

This terminology is meant to be an aid, but it is not ideal. The part can be larger than the whole, a situation in which the percentage is greater than 100. For example, 77 is 110% of 70, since

$$\frac{77}{70} = \frac{110}{100}$$

If any two of the quantities *part*, *whole*, and *percentage* are known, then the proportion allows us to solve for the third. This gives rise to three kinds of percentage problems, depending on which of the three quantities is unknown.

6.6.1. Percentage Unknown.

Oral exercise 6.66. (1) Express as a percentage: 43 out of 100; 76 out of 100; 123 out of 100.

(2) Express as a percentage: $\frac{13}{100}$; $\frac{100}{100}$; $\frac{7}{10}$; $\frac{23}{20}$.

(3) Express as a percentage: .09; .92; .5; .6; 2.13.

(4) Express as a decimal: 35%; 77%; 1%; 100%; 120%.

(5) Express as a fraction in lowest terms: 20%; 25%; 50%; 75%; 4%; 120%.

(6) What percentage of an hour is 15 minutes? 90 minutes?

(7) Express 3 inches as a percentage of 1 foot.

(8) Express 12 ounces of water as a percentage of 1 cup of water.

In order to express a fraction as a percent, first express the fraction with denominator equal to 100. There are several ways to think about this process.

Example 6.67. Caitlin has finished stitching $\frac{3}{4}$ of her quilt. What percentage of the quilt has she stitched?

Method 1: Knowing that $100 = 4 \cdot 25$, we know that
$$\frac{3}{4} = \frac{3 \cdot 25}{4 \cdot 25} = \frac{75}{100} = 75\%$$

This method can be thought of as the ratio and proportion problem
$$3 : 4 :: \text{percentage} : 100 \quad \text{or written} \quad \frac{3}{4} = \frac{\text{percentage}}{100}$$

whose solution is
$$\text{percentage} = \frac{3}{4} \cdot 100 = 75$$

Method 2: 100% of the quilt is the whole quilt, so $\frac{3}{4}$ of the quilt is
$$\frac{3}{4} \text{ of } 100\% = \frac{3}{4} \cdot 100\% = 75\%$$

Oral exercise 6.68. Express each fraction as a percentage and explain your method:

(1) $\frac{12}{25}$; $\frac{3}{5}$; $\frac{3}{25}$; $\frac{21}{50}$; $\frac{70}{50}$.

(2) $\frac{6}{200}$; $\frac{50}{300}$; $\frac{132}{300}$; $\frac{30}{25}$.

(3) $\frac{27}{200}$; $\frac{130}{400}$; $\frac{100}{300}$; $\frac{300}{100}$.

Oral exercise 6.69. The following line segments have been divided into ten equal parts.

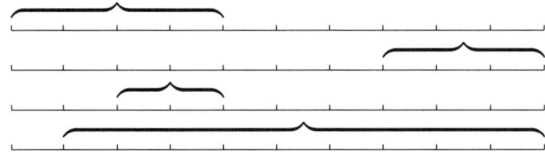

(1) What percentage of each line segment is enclosed by a brace?

(2) What percentage of each line segment is not enclosed by a brace?

The percentage depends on what is called the part and what is called the whole. This point is a source of considerable confusion. Remember that the part is a percentage of the whole.

Example 6.70. Dona has \$2,800 in her savings account and Samantha has \$2,100 in her savings account. Samantha's savings is what percentage of Dona's savings? Dona's savings is what percentage of Samantha's savings?

Solution: In the first question, Samantha's savings is the part and Dona's savings is the whole, so Dona's savings is 100%.

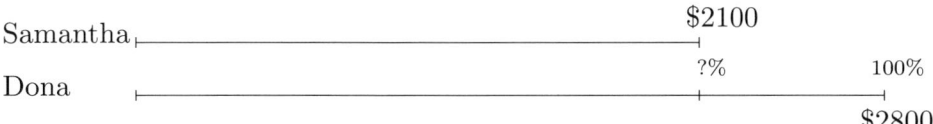

Samantha's savings is to Dona's savings as the percentage is to 100, so

$$\text{percentage} = \frac{2100}{2800} \cdot 100 = 75\%$$

Samantha's savings is 75% of Dona's savings.

In the second question, Dona's savings is the part and Samantha's savings is the whole, so Samantha's savings is 100%.

Dona's savings is to Samantha's savings as the percentage is to 100, so

$$\text{percentage} = \frac{2800}{2100} \cdot 100 = 133\frac{1}{3}\%$$

Dona's savings are $133\frac{1}{3}\%$ of Samantha's savings.

Dona's savings are what percentage more than Samantha's? This question is asking about the *difference* in their savings and what percentage of Samantha's savings is this difference. Dona's savings are $2800 - 2100 = 700$ dollars more than Samantha's. This difference is

$$\frac{700}{2100} \cdot 100 = 33\frac{1}{3}\%$$

of Samantha's savings. Dona's savings are $33\frac{1}{3}\%$ more that Samantha's savings.

Samantha's savings are what percentage less than Dona's? This question is asking what percentage of Dona's savings is the difference. It is

$$\frac{700}{2800} \cdot 100 = 25\%$$

of Dona's savings. Samantha's savings are 25% less than Dona's savings. In both of these questions, the difference in their savings was the part, but the whole was Samantha's savings in the first case and Dona's savings in the second case.

Oral exercise 6.71. Dicky has $60 and Ardis has 20% more money than Dicky. How much money does Ardis have? *Hint:* Ardis's money is what percentage of Dicky's money? Explain what the 20% means in this problem.

Dicky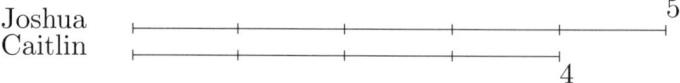

Ardis

Oral exercise 6.72. Joshua is 5 feet tall and Caitlin is 4 feet tall.

Joshua
Caitlin

What is Caitlin's height as a percentage of Joshua's? What is Joshua's height as a percentage of Caitlin's? Joshua is what percentage taller than Caitlin? Caitlin is what percentage shorter than Joshua?

Oral exercise 6.73. Nicholas's salary last summer was $800. This summer it will be $1000. Next summer it will be $900.

Last
This
Next

His salary this summer is what percentage greater than last summer's salary? His next summer's salary will be what percentage less than this summer's salary?

6.6.1.1. *Compound Interest.* Jen puts $200 into a savings account which pays 6% interest per year. Suppose that when the interest is paid it is put into the account to become the principal for the second year. How much is in the account at the end of the first year and at the end of the second year?

At the end of 1 year the account contains the original principal of $200 plus the interest it earned during that year:

$$200 + .06 \times 200 = 1 \times 200 + .06 \times 200 = (1 + .06) \times 200 = 1.06 \times 200$$

is the amount of money that will be in the account at the end of one year. This will be the principal during the second year. This principal plus the interest it earns during the second year is

$$1.06 \times 200 + .06 \times (1.06 \times 200) = 1 \times (1.06 \times 200) + .06 \times (1.06 \times 200)$$
$$= (1 + .06) \times (1.06 \times 200) = 1.06 \times (1.06 \times 200)$$

This is the amount of money that will be in the account at the end of two years.

Oral exercise 6.74. Explain in detail why $(1.06)^3 \cdot 200$ is the amount of money that will be in the account at the end of three years.

6.6. Percentage

6.6.2. Part Unknown. Problems involving a percentage of a quantity are the same as problems involving a fraction of a quantity, since a percentage is a fraction.

Example 6.75. There are 200 students in a psychology lecture. Forty-five percent of these students are women. How many women are in this psychology lecture?

Unitary method: 1% means $\frac{1}{100}$. Find $\frac{1}{100}$ of a quantity by dividing by 100.

1% of 200 is $\frac{200}{100} = 2$

45% of 200 is $45 \cdot 2 = 90$.

There are 90 women in this psychology lecture.

The unitary method is correct even when the intermediate answer is absurd. For example, if there were 240 students in the psychology lecture and 45% of these were women, then the unitary method looks like this:

1% of 240 is $\frac{240}{100} = \frac{12}{5} = 2.4$

45% of 240 is $45 \cdot \frac{12}{5} = 9 \cdot 12 = 108$.

There are 108 women in this psychology lecture.

Fraction method:

45% of 200 is $\frac{45}{100} \cdot 200 = 90$.

There were 90 women in this psychology lecture.

Arithmetic of percentages is defined since percentages are just fractions.

Example 6.76. Joshua had $70. He spent 15% of his money on lunch and 40% of his money on a new fishing pole. What percentage of his money did he have left? How much money did he have left?

Solution:

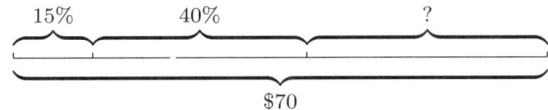

Percentage spent is $15\% + 40\% = 55\%$.

Percentage left is $100\% - 55\% = 45\%$.

Amount of money left is 45% of $70 which is $\frac{45}{100} \cdot 70 = \frac{45 \cdot 7}{10} = \$31.50$.

Example 6.77. Niels bought a TV set whose price was $480. If the sales tax was 6%, what was the total amount Niels had to pay for the TV set?

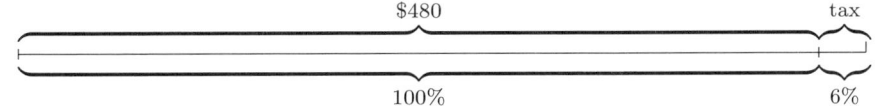

Method 1:

Sales tax is 6% of $480 is $.06 \cdot 480 = 28.80$ dollars.

He paid $480 for the TV set and $28.80 sales tax.

Total paid is $480 + 28.80 = 508.80$ dollars.

Method 2:

He paid 100% of the price for the TV set and 6% of the price for sales tax. Altogether he paid 100% + 6% of $480, which is 106% of $480, which is

$$(1 + \frac{6}{100}) \cdot 480 = \frac{106}{100} \cdot 480 = 508.80$$

Decimal fractions could be used instead. 106% is 1.06.

Oral exercise 6.78. A math class began with 20 students. After one week 15% of the students dropped the class. How many students remained in the class? Explain two ways of doing the problem (in one way you find how many students dropped the class and in the other way you find what percentage of the students remained in the class).

Example 6.79. The regular price of a shirt is $80. In a clearance sale, its price is reduced by 40%. What is the sale price? In the last week of the sale, the sale price is reduced an additional 20%. What is the new price? By what percentage has the original price been reduced?

Solution:

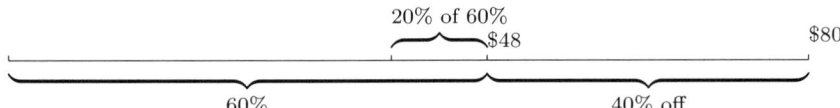

The first sale price will be $100\% - 40\% = 60\%$ of the original price, which is

$$60\% \text{ of } \$80 = .6 \cdot 80 = \$48$$

If this price is reduced by 20%, then the new sale price will be $100\% - 20\% = 80\%$ of the first sale price, which is

$$80\% \text{ of } \$48 = .8 \cdot \$48 = \$38.40$$

To find what percentage of the original price this is, calculate

$$\frac{38.4}{80} \cdot 100\% = 48\%$$

Hence, the original price has been reduced by $100\% - 48\% = 52\%$. This can also be determined as follows. The first discount is 40% of the whole $80. The second discount is 20% of 60% of the whole $80. The combined discount of the whole $80 is then

$$40\% + 20\% \text{ of } 60\% = .4 + .2 \cdot .6 = .4 + .12 = .52 = 52\%$$

6.6.3. Whole Unknown.
In these problems the part and the percentage are known, and one wants to find the whole. These are partitive division problems which can be solved directly by division or, perhaps more transparently, by the unitary method.

Example 6.80. After spending 80% of his birthday money, Anthony has $8 left. How much birthday money did he have?

Partitive division method:

$8 is $100\% - 80\% = 20\%$ of his birthday money.

8 is $\frac{20}{100}$ of what?

$8 \div \frac{20}{100} = 8 \div \frac{1}{5} = 8 \cdot 5 = 40$

Anthony had $40 birthday money.

Unitary method: Remember that 100% is the whole.

20% is 8 dollars.

1% is $\frac{8}{20} = \frac{2}{5}$ dollars.

100% is $100 \cdot \frac{2}{5} = 40$ dollars.

Oral exercise 6.81. Richard paid $24,000 for a new car. The salesman told him that he paid 20% over the list price.

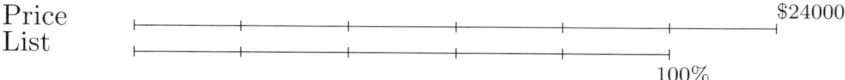

What percentage of the list price did he pay? What was the list price? Explain how to solve the latter problem by partitive division and by the unitary method.

Example 6.82. At a psychology lecture 55% of the students were women. There were 24 more women than men at the lecture. How many students were at the lecture?

Solution:

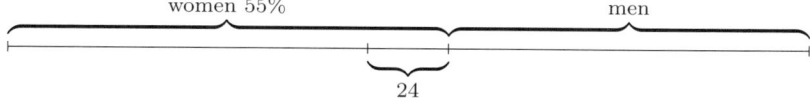

24 is what percentage of the students in the lecture?

24 is $55\% - 45\% = 10\%$ of the students at the lecture.

2.4 is 1%

$100\% = 100 \cdot 2.4 = 240$ is the number of students at the lecture.

§6.6 Exercises

Whenever possible use a line segment diagram to illustrate your explanation. In problems for which you are finding the whole, explain both the method of partitive division and the unitary method of solution.

(1) Express 2 pints as a percentage of 2 gallons.

(2) Express the length of a yard stick as a percentage of a foot ruler. How many percent longer is the yard stick than the foot ruler?

(3) From a five pound bag of flour Rita used 20 ounces to make bread. What percentage of the flour did she use?

(4) Tuition this year is $25,000. Next year it will be $26,275. By what percentage has tuition increased?

(5) Suppose that $5,000 is placed in a savings account which pays 2.4% interest per year. At the end of each year the interest is placed into the account. How much money, to the nearest cent, is in the account at the end of 3 years?

(6) A bond paid annual interest of 4.5% in 2002 and of 2.7% in 2003. Express the decrease in interest rate as a percentage of the rate in 2002.

(7) The college has 3,800 students, of whom 2,100 are women. How many percent more women than men are there (to the nearest tenth of a percent)?

(8) Laura's score on a calculus test was 63, which was 10% below the class average. What was the class average?

(9) Solveig's car gets 18 miles per gallon on the highway. This mileage is only 60% of what James gets with his car. How many miles per gallon does James get with his car?

(10) A blouse was put on sale for 40% off its original price. After one week, the sale price was reduced 30%, giving a new sale price of $84. What was the original price of the blouse? The final sale price is what percentage less than the original price?

(11) Ragna's wage was increased by 20% to $18 per hour. Find her hourly wage before the increase.

(12) Leah's monthly salary is increased by $120, which is a 10% increase. Find her monthly salary after the increase.

(13) On a trip to Switzerland, Nettie bought 40 good watches for $280 apiece. Back home, she offered to sell them to a jeweler for 30% more than she paid for them. She accepted a price which gave her a 17% profit on

her investment in the watches. What price did the jeweler pay for each watch? By what percent of her asking price did she have to reduce the price for the jeweler?

(14) Robert makes a wage of $16.20 per hour. His wage is 10% less than Ralph's hourly wage. What is Ralph's hourly wage?

(15) Ardis spends her study time on math and English. She spends 40% of her study time on math. She spends 15 minutes longer on English than on math. How long is her study time?

(16) Janet has $40 and Bud has 20% more. What percentage of his money must Bud give Janet in order that they have the same amount?

(17) Walter bought a car for $11,562.50, fixed it up and then sold it for 62.5% more. How much did he receive for it?

(18) In a town where the sales tax is 7%, Stephanie's bill for a new dishwasher was $524.30. What was the price of the dishwasher?

6.7. Exponents and Scientific Notation

6.7.1. Exponents.

Definition 6.83. If a is a positive rational number and n is a counting number, then a^n means $a \cdot \cdots \cdot a$, where there are n copies of a in this product. The number a is called the base and n is called the exponent.

Pronounce a^n as *a to the n* or as *a to the power n*. Exponents are related to our earlier terminology of power. Ten to the third power means 10^3. For any base, a to the fifth power means a^5, etc.

Example 6.84. (1) $3^2 = 3 \cdot 3 = 9$

(2) $1.2^2 = 1.2 \cdot 1.2 = 1.44$

(3) $5^3 = 5 \cdot 5 \cdot 5 = 125$

(4) $(\frac{2}{3})^2 = \frac{2}{3} \cdot \frac{2}{3} = \frac{2 \cdot 2}{3 \cdot 3} = \frac{2^2}{3^2}$.

Oral exercise 6.85. (1) In the United States, what is the name of 10^3? Of 10^6? Of 10^9? Of 10^{12}? Of 10^{15}? *Ans.* thousand, million, billion, trillion, quadrillion.

(2) Find: (a) 7^2. (b) 1.3^2. (c) 3^3. (d) 20^2. (e) $(\frac{3}{10})^3$.

Theorem 6.86 (Laws of Exponents). *(1) If a is a positive rational number and if m and n are natural numbers, then*

$$a^m \cdot a^n = a^{m+n} \quad \text{and} \quad (a^m)^n = a^{m \cdot n}$$

(2) If a and b are positive rational numbers and if m is a natural number, then

$$a^m \cdot b^m = (a \cdot b)^m$$

Oral exercise 6.87. Fill in the question marks.
- (1) (a) $10^2 \cdot 10^3 = 10^?$. (b) $(10^2)^3 = 10^?$.
- (2) (a) $2^2 \cdot 3^2 = (2 \cdot 3)^?$. (b) $(2^2 \cdot 3)^3 = 2^? \cdot 3^3$.
- (3) (a) $5^7 = 5^3 \cdot 5^?$. (b) $(5^3)^? = 5^{3 \cdot 4}$.
- (4) (a) $(\frac{2}{3})^3 \cdot 5^3 = (?)^3$. (b) $(\frac{5}{7})^2 \cdot (\frac{5}{7})^3 = (?)^{2+3}$.
- (5) What is: (a) 10^1? (b) 5^1? (c) a^1, for any base a?

Proof of Theorem. To prove the first statement in (1): a^m is the product of m a's and a^n is the product of n a's. Therefore, $a^m \cdot a^n$ is the product of $m + n$ a's, which is a^{m+n}.

To prove the second statement in (1): $(a^m)^n$ is the product of n a^m's, each of which is the product of m a's. Now n groups of m a's contain $m \cdot n$ a's. Therefore, $(a^m)^n$ is the product of $m \cdot n$ a's, which is $a^{m \cdot n}$.

To prove (2): $a^m \cdot b^m$ is a product of m a's times a product of m b's. Since there are the same number of a's and b's, namely m of each, and since multiplication of rational numbers is commutative, the order of the product can be rearranged so that each a pairs with a b and we have the product of m $a \cdot b$'s, which is $(a \cdot b)^m$. $\square$

6.7.2. Numeration and exponents. The notation of exponents allows us to express the meaning of the base 10 numeration system as follows. For example, the number 4892 means 4 thousands plus 8 hundreds plus 9 tens plus 2, which can be expressed in *expanded notation* with exponents as
$$4892 = 4 \cdot 10^3 + 8 \cdot 10^2 + 9 \cdot 10^1 + 2$$
Notice that 10^m is the value of an element m places to the left of the one's place, in terms of elements in the one's place. That is, an element one place to the left of the one's place is 10^1 elements of the one's place, and an element two places to the left of the one's place is 10^2 elements of the one's place. In particular, an element in each place is 10 times the value of an element in the adjacent place to the right, because $10 = 10 \cdot 1$, $10^2 = 10 \cdot 10^1$, $10^3 = 10 \cdot 10^2$, and in general, $10 \cdot 10^n = 10^{n+1}$, for any natural number n.

The base 5 numeral 3241_5 means 3 five of five of fives plus 2 five of fives plus 4 fives plus 1, which can be expressed with exponents as
$$3241_5 = 3 \cdot 5^3 + 2 \cdot 5^2 + 4 \cdot 5^1 + 1$$
Of course, 5 is 10_5 in base five, which means we should write this as
$$3241_5 = 3 \cdot 10_5^3 + 2 \cdot 10_5^2 + 4 \cdot 10_5^1 + 1$$
It should be crystal clear by now that there is no logical difference between numeration in one base or another.

6.7. Exponents and Scientific Notation

Oral exercise 6.88. Express the following Hindu-Arabic numerals in expanded notation with exponents.

(1) 49285

(2) 30870

(3) Base 5 numeral 34231_5

(4) Base 5 numeral 300420_5

(5) Base 2 numeral 11010111_2

Definition 6.89. For any rational number $a > 0$ and any counting number m, define a^{-m} to mean $\frac{1}{a^m}$. That is, a^{-m} is the reciprocal of a^m. Define a^0 to be 1.

This definition is made in this way so that the Laws of Exponents will continue to be valid for any integer exponents. For example, we want to define 2^{-3} in such a way that the Law of Exponents holds:

$$2^6 \cdot 2^{-3} = 2^{6-3} = 2^3$$

Dividing this equation by 2^6 gives us $2^{-3} = \frac{2^3}{2^6} = \frac{1}{2^3}$ after cancelling $2 \cdot 2 \cdot 2$ from numerator and denominator. This Law of Exponents therefore requires that $2^{-3} = \frac{1}{2^3}$, if we want the Law to hold for negative exponents as well.

Once the meaning of negative exponent is defined, then the same Law of Exponents, if it is to remain true, dictates that $2^3 \cdot 2^{-3} = 2^{3-3} = 2^0$, while $2^{-3} = \frac{1}{2^3}$ implies that $2^3 \cdot 2^{-3} = 1$. Hence, we must define $2^0 = 1$.

Theorem 6.90. *The Laws of Exponents as expressed in Theorem 6.86 are true for any positive base and any positive, negative, or zero exponents.*

Proof. We defined negative exponents and zero exponent so that Theorem 6.86 will remain true. □

Example 6.91. (1) In expanded notation

$$49.285 = 4 \cdot 10^1 + 9 \cdot 10^0 + 2 \cdot 10^{-1} + 8 \cdot 10^{-2} + 5 \cdot 10^{-3}$$

(2) The decimal expansion of

$$3 \cdot 10^2 + 8 + 7 \cdot 10^{-1} + 2 \cdot 10^{-3}$$

is 308.702.

(3) In expanded notation, the quintal expansion

$$34.231_5 = 3 \cdot 5^1 + 4 \cdot 5^0 + 2 \cdot 5^{-1} + 3 \cdot 5^{-2} + 1 \cdot 5^{-3}$$

(4) In expanded notation, the dimidial expansion

$$11010.111_2 = 2^4 + 2^3 + 2^1 + 2^{-1} + 2^{-2} + 2^{-3}$$

Oral exercise 6.92. (1) Express as a decimal expansion: 10^{-1}, 10^{-3}, 10^2.

(2) Express as a decimal expansion: $2 \cdot 10^2 + 5 \cdot 10^1 + 7 \cdot 10^0 + 3 \cdot 10^{-2} + 9 \cdot 10^{-3}$.

(3) In the decimal expansion found in the preceding problem, how is the exponent of 10 for a place related to the number of places it lies to the left or right of the one's place?

(4) Express as a quintal expansion: $2 \cdot 5^2 + 4 \cdot 5^1 + 3 \cdot 5^0 + 2 \cdot 5^{-1} + 2 \cdot 5^{-3}$.

(5) Express as a dimidial expansion: $2^5 + 2^3 + 2^2 + 2^0 + 2^{-2} + 2^{-3}$.

6.7.3. Scientific Notation. When using very large numbers, we often find that only the first few digits at the left are relevant. In a similar fashion, often only the first few nonzero digits to the right of the decimal point are relevant in very small numbers.

The federal budget in fiscal year 1999 was \$1,704,545,000,000. In a discussion, we would soon round it to the nearest hundred billion and refer to it as one trillion seven hundred billion dollars. This is written 1.7×10^{12} dollars, and pronounced 1.7 trillion dollars.

Definition 6.93. A number is expressed in *scientific notation* if it is written as $a \times 10^n$, where usually $1 \leq a < 10$, and n is positive or negative.

For convenience of reading or writing, the number a is often rounded off.

Definition 6.94. A number $a \times 10^n$, with $1 \leq a < 10$, is *expressed to k significant digits* if a is rounded off to the nearest 10^{k-1}'s place.

In scientific notation, the above federal budget figure is 1.704545×10^{12} dollars, which to two significant digits is 1.7×10^{12} dollars. One would express it to 4 significant digits, 1.705×10^{12} dollars, if that additional 5 billion dollars is important.

Example 6.95. One year is 3.1556926×10^7 seconds, or approximately, 3×10^7 seconds. Here is a case where it might be convenient to let a be greater than 10, so that the exponent part is 10^6, which is a million. To three significant places a year is 31.6×10^6 seconds, which is 31.6 million seconds.

Example 6.96. A *light-year* is the distance light travels in a vacuum in one year. It is 9.4605×10^{17} cm, which to five significant places is 946.05×10^{15} cm, which is 946 quadrillion cm to three significant places.

Example 6.97. The *rest mass of an electron* is 9.10956×10^{-28} g. This is an approximation, of course, as all measurements are. It is known that the electron mass is this number plus or minus $.00005 \times 10^{-28}$ g.

§6.7 Exercises

Oral exercise 6.98. (1) What is the maximum possible rest mass of the electron? *Ans.* 9.10961×10^{-28} g. What is the minimum possible rest mass?

(2) What is ten percent of the above federal budget figure, exactly and to two significant places?

(3) What fraction of a year is one second? Explain.

Example 6.99. *Avogadro's number* N is the number of carbon atoms in 12 grams of carbon. This number has been measured to be

$$N = 602,200,000,000,000,000,000,000 = 6.022 \times 10^{23}$$

The basic building block of a substance is the atom or molecule, depending on whether the substance is an element or a compound, respectively. A *mole* of a substance is Avogadro's number of basic building blocks of the substance. For example,

- One mole of carbon contains 6.022×10^{23} carbon atoms. The mass of a mole of carbon is 12 grams.
- A mole of oxygen contains 6.022×10^{23} oxygen atoms. The mass of a mole of oxygen is 16 grams.
- A mole of sodium contains 6.022×10^{23} sodium atoms. The mass of a mole of sodium is 23 grams.

Oral exercise 6.100. The mass of an oxygen atom is what percentage greater than the mass of a carbon atom?

§6.7 Exercises

(1) There are 100 cm in a m and 1000 m in a km. How many km in a light-year?

(2) There are 2.54 cm in an inch. How many miles in a light-year?

(3) How many miles in 1 km?

(4) (a) How many seconds in 365 days?
(b) How many more seconds in a year than in 365 days?
(c) What fraction of a day is the number of seconds calculated in part (b)? What has this fraction to do with why we have a leap year every four years?

(5) (a) What is the mass of 1 carbon atom, to three significant digits? *Ans:* 1.99×10^{-23} g.
(b) What is the mass of 1 oxygen atom, to three significant digits?

(c) The mass of 1 carbon atom is what percentage less than the mass of 1 oxygen atom?

(6) (a) What is the mass of 3 moles of sodium?

(b) The mass of the sodium atom is what percentage greater than the mass of an oxygen atom?

(c) The mass of the carbon atom is what percentage less (to the nearest tenth of a percent) than the mass of a sodium atom? *Ans:* 47.8% less.

(7) The world crude oil reserves for Jan. 1, 1998, by region, published in *World Oil*, Aug. 1998, and recorded in *The World Almanac* 2000, are (in barrels):

North America	68.8×10^9
Central & South America	63.0×10^9
Western Europe	19.7×10^9
Eastern Europe	64.3×10^9
Middle East	624.4×10^9
Africa	76.7×10^9
Far East & Oceania	58.0×10^9

Find the total number of barrels of crude oil reserves in the world on Jan 1, 1998. Explain how the distributive property allows you to ignore the 10^9 factors, add the resulting column of numbers, and then multiply the sum by 10^9. *Ans:* 974.9×10^9 barrels.

(8) The U.S. Department of Energy published the following information on the number of barrels of petroleum consumed per day in 1998:

North America	22,820,000
Central & South America	5,039,000
Western Europe	14,885,000
Eastern Europe	5,004,000
Middle East	4,328,000
Africa	2,395,000
Asia & Oceania	19,319,000

(a) Rewrite this data in scientific notation and then find the total number of barrels consumed daily in the world in 1998. Express the answer in scientific notation. *Ans:* 73.79×10^6 barrels per day.

(b) Find the total number of barrels consumed in the world during all of 1998. Express the answer in scientific notation.

(c) At the rate consumed in 1998, how many years (to the nearest tenth) will the reserves of Jan. 1, 1998, last? (See data in preceding problem).

Chapter 7

Integers

The whole numbers count "how many". Many applications involve the whole numbers in this role. Some problems involve both the concept of "how many" and of a direction, such as left-right, up-down, rising-falling, increasing-decreasing, in-out, gain-loss, debit-credit, or deposit-withdrawal. For example, to describe a car travelling along a highway running east and west, we need to give its speed, say 50 mph, and its direction, say east. Fifty dollars withdrawn from your bank account is not the same as fifty dollars deposited into your bank account. The Dow Jones industrial average changes by 60 points, and we want to know if that is up or down (a gain or a loss). The set of integers have been developed to give us numbers which simultaneously count "how many" and record the direction. They measure both quantity and direction.

The arithmetic operations of the whole numbers can be extended to the set of integers. One of the many benefits of this extension is that subtraction will be defined between any two integers.

7.1. Negative Numbers

The number line has direction built into it. To define it, we drew a line, marked 0 on it, determined a unit interval, and then began marking off the whole numbers $1, 2, 3, \ldots$ **to the right** from 0. To any left-handed person, or to someone who learned to read in Hebrew, it would be equally natural to mark $1, 2, 3, \ldots$ to the left from 0. We now do that and call the numbers to the left of 0 the *negative integers*.

Definition 7.1 (Negative integers)**.** Begin with the number line of whole numbers and extend it to the left of 0 by the same unit interval amounts.

Label these points $-1, -2, -3$, and so on. For any natural number n, the point on the line which is n unit intervals to the left of 0 is labelled $-n$ and called negative n or minus n. Call these new points the negative integers.

The set of integers, denoted **Z**, is the set of all whole numbers and all negative integers. The natural numbers are called the positive integers. Notice that 0 is an integer, neither positive nor negative. The set of integers is the union of the set of whole numbers and the set of negative integers. From now on the number line will mean the line extended indefinitely in both directions.

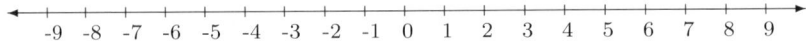

Definition 7.2 (Opposites). For any natural number n, define its opposite to be the negative integer $-n$. For any negative integer $-m$ (so here m is a natural number), define its opposite to be the natural number m. Define the opposite of 0 to be 0. For any integer a, denote its opposite by $-a$. In summary, we have defined the operation on **Z** of taking the opposite of a number. Taking the opposite sends positive to negative and negative to positive, and fixes 0.

Geometrically, taking the opposite of all integers is the same as rotating the line about 0 through an angle of 180°.

Example 7.3. (1) The opposite of 4 is -4

(2) The opposite of -7 is 7, which we can also denote as $-(-7)$. In particular, the opposite of the negative integer -7 is the positive integer 7.

(3) Notice that $-0 = 0$.

(4) Taking the opposite of a number twice always returns to that number. For example, $-(-3) = 3$ and $-(-(-5)) = -5$ (the opposite of the opposite of -5 is -5).

(5) The operation of taking the opposite defines a one-to-one correspondence between the set of negative integers and the set of positive integers.

Definition 7.4 (Absolute value). The *absolute value* of an integer x is the number of unit intervals from x to 0. Denote the absolute value of x by $|x|$. The word *magnitude* is also used for absolute value.

For example, $|8| = 8$, $|-7| = 7$ and $|0| = 0$. The absolute value of an integer is always a whole number. An integer x, other than 0, contains two pieces

7.1. Negative Numbers

of information, its magnitude $|x|$ and its sign, negative or positive, which indicates a direction of left or right, respectively, from 0 on the number line.

Theorem 7.5. *(1) If x is an integer, then either $x = a$ or $x = -a$, where a is a whole number, and $|x| = a$.*

(2) If x and y are integers and $|x| = |y|$, then $y = x$ or $y = -x$.

Proof. (1) If x is an integer, then by the definition of integers, x is either a whole number or a negative integer and these latter are the opposites of natural numbers. Therefore, $x = a$ or $x = -a$, for some whole number a. Then $|x| = |a| = |-a| = a$, which is the number of unit intervals from 0 to a or from 0 to $-a$, by our construction of the number line.

(2) If x and y are integers, then $x = a$ or $x = -a$ and $y = b$ or $y = -b$, for some whole numbers a and b. If $|x| = |y|$, then $a = b$, because $|x| = a$ and $|y| = b$. Therefore, $y = x$ or $y = -x$. □

Oral exercise 7.6. In the preceding proof, if $x = -7$ and $y = 7$, is it true that $y = -x$? Explain.

Definition 7.7 (Order). For integers x and y, define $x < y$ to mean that x lies to the left of y on the number line.

Example 7.8. $2 < 5$, $-5 < 2$, and $-7 < -5$.

x is positive means that it lies to the right of 0, which means that $x > 0$. Therefore, x is positive if and only if $x > 0$.

x is negative means that it lies to the left of 0, which means that $x < 0$. Therefore, x is negative if and only if $x < 0$.

The following theorem tells us how to use our knowledge of the order of whole numbers to decide the order of integers.

Theorem 7.9. *Suppose that x and y are integers.*

(1) If x and y are both whole numbers, then their order is the order of whole numbers.

(2) If x is negative and y is a whole number, then $x < y$.

(3) If x and y are both negative, then their order is the reverse of the order of the whole numbers $|x|$ and $|y|$.

Proof. The first statement is true by Theorem 1.38. The second statement is true because all negative integers lie to the left of all positive integers. For the third statement, if x and y are both negative, then x lies $|x|$ unit intervals to the left of 0 and y lies $|y|$ unit intervals to the left of 0. The one that lies farther to the left is the one with the larger absolute value. Therefore, the order of x and y is the reverse of the order of the whole numbers $|x|$ and $|y|$. □

Oral exercise 7.10. (1) What is a negative integer? Is $-(-6)$ negative?

(2) On the number line describe the operation of taking the opposite.

(3) On the number line, mark the integers whose distance from 0 is 5; that is, describe $\{x \in \mathbf{Z} : |x| = 5\}$.

(4) On the number line, mark the integers whose distance from 0 is no more than 5; that is, describe $\{x \in \mathbf{Z} : |x| \leq 5\}$.

§7.1 Exercises

(1) On a number line mark the integers from -10 through 10. On it indicate the absolute value of -4 and the absolute value of 4. From the definition of absolute value, explain why $|-4| = |4|$.

(2) On Monday Janet deposited $120 into her checking account. On Tuesday she deposited $76, on Wednesday she withdrew $85, on Thursday she deposited $115, and on Friday she withdrew $67. Use integers to record these transactions in a table with just one column, labelled transaction. What is the magnitude of her transaction on Friday?

7.2. Integer Addition

The goal is to define addition on the integers in such a way that it agrees with whole number addition on the subset of whole numbers and such that it has the same properties as whole number addition. We accomplish this by using the number line model of whole number addition, with the sense of direction being to the right for positive and to the left for negative.

Definition 7.11. For integers x and y, their *sum* $x+y$ is the integer arrived at by starting at x and going $|y|$ unit intervals to the right or left, according as y is positive or negative, or not going any unit intervals when $y = 0$.

Example 7.12. $2 + 5 = 7$, because to start at 2 and go 5 unit intervals to the right is to end at 7. This is the same as whole number addition. This sum is illustrated as follows.

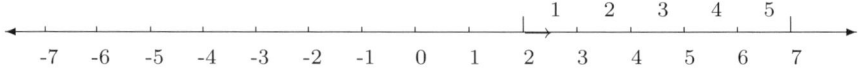

Example 7.13. $2 + (-5) = -3$, because to start at 2 and go $|-5| = 5$ unit intervals to the left is to end at -3.

7.2. Integer Addition

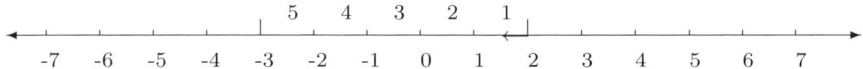

Example 7.14. $-2 + 5 = 3$, because to start at -2 and go 5 unit intervals to the right is to end at 3.

Example 7.15. $-2 + (-5) = -7$, because to start at -2 and go $|-5| = 5$ unit intervals to the left is to end at -7.

Example 7.16. $-2 + 0 = -2$, because to start at -2 and not to go any unit intervals is to end at -2.

Example 7.17. $0 + (-5) = -5$, because to start at 0 and go $|-5| = 5$ unit intervals to the left is to end at -5, by definition of -5.

Notice that when the integers a and b are both whole numbers, then the integer sum $a + b$ we have just defined agrees with the number line model of whole number addition. We have verified in Theorem 2.7 that the number line model of whole number addition agrees with the definition of whole number addition (in terms of set union). Therefore, integer addition applied to whole numbers agrees with whole number addition.

Integer addition is computed in terms of whole number addition and subtraction together with taking opposites. The next theorem gives the details for the six different cases.

Theorem 7.18. *If x and y are integers, then the sum $x + y$ falls into one of the following six cases.*

Let a and b be whole numbers such that $a \geq b$. The six cases are:

(1) $a + b$ is the whole number sum. For example, $5 + 2 = 7$.

(2) $a + (-b) = a - b$, the whole number difference. For example, $5 + (-2) = 5 - 2 = 3$.

(3) $b + (-a) = -(a - b)$, the opposite of the whole number difference. For example, $2 + (-5) = -(5 - 2) = -3$.

(4) $(-a) + b = -(a - b)$, the opposite of the whole number difference. For example, $(-5) + 2 = -(5 - 2) = -3$.

(5) $(-b) + a = a - b$, the whole number difference. For example, $(-2) + 5 = 5 - 2 = 3$.

(6) $(-a) + (-b) = -(a+b)$, the opposite of the whole number sum. For example, $(-5) + (-2) = -(5+2) = -7$.

Proof. To see that $x + y$ always falls into one of these cases, we can make a systematic list of all possibilities:

$$\begin{aligned} &x \geq 0, \quad y \geq 0 \\ &x \geq 0, \quad y < 0, \quad x \geq |y| \\ &x \geq 0, \quad y < 0, \quad x < |y| \\ &x < 0, \quad y \geq 0, \quad |x| \geq y \\ &x < 0, \quad y \geq 0, \quad |x| < y \\ &x < 0, \quad y < 0. \end{aligned}$$

These correspond to the six cases.

(1) This follows from Theorem 2.7 which states that whole number addition agrees with the number line model for addition.

(2) To find $a + (-b)$ when $a \geq b$, we start at a and go b units to the left, to end at the whole number $a - b$, by take-away subtraction on the number line given in Theorem 2.28. Here is an illustration of $5 + (-2) = 5 - 2 = 3$.

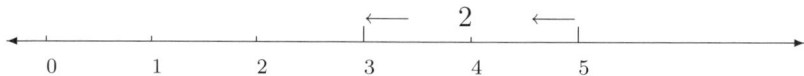

(3) To find $b + (-a)$ when $a \geq b$, we start at b and go a units to the left. After b units we are at 0 and there remain $a - b$ units to go to the left from 0. Therefore, we end at the opposite of $a - b$, which is the point $-(a - b)$. Here is an illustration of $2 + (-5) = -(5 - 2) = -3$.

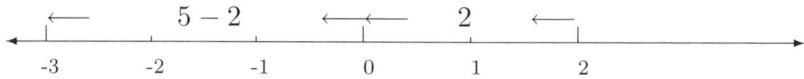

(4) To find $(-a) + b$ when $a \geq b$, we start at $-a$ and go b units to the right, to end at the point which requires $a - b$ more units to reach 0, since $-a$ is a units to the left of 0. Therefore, we have ended at $-(a - b)$. Here is an illustration of $(-5) + 2 = -(5 - 2) = -3$.

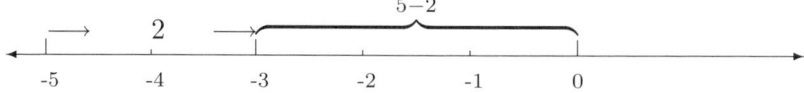

(5) To find $(-b) + a$ when $a \geq b$, we start at $-b$ and go a units to the right. We reach 0 after b units and there remain $a - b$ more units to go

7.2. Integer Addition

to the right of 0. Therefore, we end at $a - b$. Here is an illustration of $(-2) + 5 = 5 - 2 = 3$.

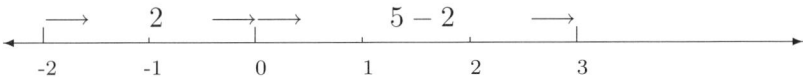

(6) To find $(-a) + (-b)$, we start at $-a$ and go b units to the left, to end at the point reached by starting at 0, going a units to the left and then b more units to the left from there, a total of $a + b$ units to the left of 0. Therefore, we end at $-(a+b)$. Here is an illustration of $(-5) + (-2) = -(5+2) = -7$.

$\square$

We must now verify that integer addition satisfies the properties of whole number addition listed in Theorem 2.4.

Theorem 7.19. *If x, y, and z are integers, then:*

(1) $x + y$ is the whole number sum if x and y are whole numbers.
(2) $x + y = y + x$ (commutative property).
(3) $x + (y + z) = (x + y) + z$ (associative property).
(4) $x + 0 = 0 + x = x$ (identity property of 0).
(5) $x + (-x) = 0$ (additive inverse is the opposite).
(6) If $x + y = 0$, then $y = -x$ and $x = -y$ (uniqueness of additive inverse).
(7) If $x < y$ and if z is any integer, then $z + x < z + y$ (order property).

Proof. (1) This follows from our definition of addition of integers together with Theorem 2.7 which proved that the number line model of addition agrees with the definition of addition of whole numbers.

(2) The commutative property follows from Theorem 7.18 and the commutative property of whole numbers. Each case must be considered separately. If a and b are whole numbers, then:

$a + b = b + a$ by the commutative property of whole numbers.

For the sum of a positive and a negative with the negative smaller, if $a \geq b$, then

$$a + (-b) = a - b, \text{ by (2) of Theorem 7.18}$$
$$= (-b) + a, \text{ by (5) of Theorem 7.18}$$

and, with the negative larger,
$$b + (-a) = -(a - b), \text{ by (3) of Theorem 7.18}$$
$$= (-a) + b, \text{ by (4) of Theorem 7.18}.$$
Finally, for the sum of negative numbers,
$$(-a) + (-b) = -(a + b), \text{ by (6) of Theorem 7.18}$$
$$= -(b + a), \text{ by commutativity of whole number addition}$$
$$= (-b) + (-a), \text{ by (6) of Theorem 7.18}$$

(3) The proof of the associative property of addition is similar to that of the commutative property, but even more tedious because of more cases. If $x = \pm a$, $y = \pm b$, and $z = \pm c$, where a, b, and c are whole numbers, then there are $2 \cdot 2 \cdot 2 = 8$ choices of sign in the expression $x + (y + z)$. When the signs of x, y, and z are all the same, we have the case of all positive
$$a + (b + c) = (a + b) + c$$
which is true by the associative property for whole numbers; and we have the case of all negative
$$(-a) + ((-b) + (-c)) = (-a) + (-(b + c)) = -(a + (b + c))$$
$$= -((a + b) + c), \text{ by associativity of whole numbers}$$
$$= (-(a + b)) + (-c) = ((-a) + (-b)) + (-c)$$
where each step uses (6) of Theorem 7.18. For each of the 6 remaining choices of sign, we must consider the 6 possible orders of a, b, and c. For example, if $a = 3$, $b = 5$, and $c = 9$, then for the case when $z = x$ and y are positive and z is negative, the 6 possible orders
$$3 + (5 + (-9)), \qquad 3 + (9 + (-5))$$
$$5 + (3 + (-9)), \qquad 5 + (9 + (-3))$$
$$9 + (3 + (-5)), \qquad 9 + (5 + (-3))$$
could each require a different argument to verify the associative property, because each could use different parts of Theorem 7.18 and of subtraction associative properties of Theorem 2.33. That amounts to at most $6 \cdot 6 = 36$ cases to be checked. Here are two sample cases, and the rest are in the written exercises.

To verify that $3 + (5 + (-9)) = (3 + 5) + (-9)$ in a way that illustrates the general case for z negative and $a \leq b \leq c$, we have
$$3 + (5 + (-9)) = 3 + (-(9 - 5)), \text{ by (3) of Theorem 7.18}$$
$$= -((9 - 5) - 3), \text{ by (3) of Theorem 7.18, since } 9 - 5 \geq 3$$
$$= -(9 - (3 + 5)), \text{ by (6) of Theorem 2.33}$$
$$= (3 + 5) + (-9), \text{ by (3) of Theorem 7.18}$$

7.2. Integer Addition

To verify a case like $5 + ((-9) + 3) = (5 + (-9)) + 3$, it is a good strategy to show that both sides equal the same thing, as follows.

$$\begin{aligned} 5 + ((-9) + 3) &= 5 + (-(9-3)), \text{ by (4) of Theorem 7.18} \\ &= -((9-3) - 5), \text{ by (3) of Theorem 7.18} \\ &= -(9 - (3+5)), \text{ by (6) of Theorem 2.33} \end{aligned}$$

and for the right hand side

$$\begin{aligned} (5 + (-9)) + 3 &= (-(9-5)) + 3, \text{ by (3) of Theorem 7.18} \\ &= -((9-5) - 3), \text{ by (4) of Theorem 7.18, since } 9 - 5 > 3 \\ &= -(9 - (5+3)), \text{ by (6) of Theorem 2.33} \end{aligned}$$

and these two results are equal because $3 + 5 = 5 + 3$. Except for the cases left to the exercises, this completes the proof of the associative property of addition of integers.

(4) For the identity property of 0, we have $x + 0 = x$, because to start at x and go 0 unit intervals is to remain at x. We have $0 + x = x$ because of the commutative property of addition, or more directly, by the definition of $0 + x$, which is the point arrived at after going $|x|$ unit intervals from 0 to the right or left according as x is positive or negative, and this is also the definition of the point x on the number line.

(5) $x + (-x)$ means start at x and go $|-x| = |x|$ units to the right or left, depending on whether $-x$ is positive or negative. If x is positive, then $-x$ is negative and going x units to the left from x ends at 0. If x is negative, then $-x$ is positive and going $|x|$ units to the right from x ends at 0. The case $x = 0$ is obvious. We conclude that in all cases, $x + (-x) = 0$.

(6) It is instructive to see two proofs of the uniqueness of the additive inverse. The first proof refers directly to the number line. The second proof is algebraic. Here is the first proof. We are given that $x + y = 0$. If x is positive, then going $|y|$ units from x and ending at 0 means we must have gone x units to the left. Therefore, $|y| = x$ and y is negative, which means that $y = -x$. If $x = 0$ and $x + y = 0$, it is evident that $y = 0$. Finally, if x is negative, then going $|y|$ units from x and ending at 0 means we must have gone $|x|$ units to the right. Therefore, $|y| = |x|$ and y is positive, from which we conclude that $y = -x$. That concludes the first proof.

Here is the second proof. If $x + y = 0$, then

$$\begin{aligned} -x &= -x + 0, \text{ identity of 0} \\ &= -x + (x + y), \text{ given } x + y = 0 \\ &= ((-x) + x) + y, \text{ associative property of addition} \\ &= 0 + y, \text{ additive inverse} \\ &= y, \text{ identity of 0} \end{aligned}$$

from which we conclude that $y = -x$. This concludes the second proof.

(7) If $x < y$, then adding z to both sides moves each number the distance $|z|$ to the right, if z is positive, to the left if z is negative, or not at all if $z = 0$. Such movement will not change the fact that x lies to the left of y on the number line. □

The opposite of an integer x is its additive inverse in the same sense as the reciprocal of a rational number is its multiplicative inverse. Compare $x + (-x) = 0$ to the statement $x \cdot (1 \div x) = 1$.

The Theorem implies that a sum of integers can be grouped and ordered in whatever way is convenient. Often, this way is first to add the positive integers and then subtract the sum of the negative integers. For example, to find the sum of the integers 87, −142, −53, 25, and 110, first sum the positive integers, and from this subtract the sum of the negative integers:

$$87 + (-142) + (-53) + 25 + 110 = (87 + 25 + 110) + ((-142) + (-53))$$
$$= (87 + 25 + 110) + (-(142 + 53)), \text{ (6) of Theorem 7.19}$$
$$= (87 + 25 + 110) - (142 + 53), \text{ (2) of Theorem 7.19}$$
$$= 222 - 195 = 27$$

Example 7.20. The Dow Jones Industrial average began the week at 9137. On Monday it gained 87 points, Tuesday it lost 53 points, Wednesday it lost 142 points, Thursday it gained 25 points, and Friday it gained 110 points. What was the net change in the Dow Jones over that week? What was the Dow Jones average at the end of the week?

Answer: The net change for the week would be the sum of the gains minus the sum of the losses. If the gains are posted as positive amounts and the losses as negative amounts, then the preceding example shows that the sum of the gains minus the sum of the losses is the sum of the integers denoting the daily changes.

The Dow Jones average at the end of the week would be its initial value of 9137, plus the net change, if the net change is a gain, but minus the net change, if the net change is a loss. Using integers to post the daily changes, the net change is the sum of the daily changes (positive if the net change is a gain, negative if the net change is a loss) and the Dow Jones average at the end of the week is the sum of 9137 and the net change:

$$9137 + (87 + (-142) + (-53) + 25 + 110) = 9137 + 27 = 9164$$

§7.2 Exercises

(1) Explain how to use Theorem 7.18 to calculate $387 + (-852)$ and do the whole number calculation required by this theorem. Explain how you determine which case of the theorem applies.

(2) Prove that $(-3) + (-5) = (-5) + (-3)$. Make your argument general enough that it illustrates how to prove the commutative property for the sum of any two negative integers. Follow the argument of the proof of (2) of Theorem 7.19.

(3) Prove that $7 + (-3) = (-3) + 7$. Make your argument general enough that it illustrates how to prove the commutative property for the sum of a positive and a negative integer in the case when the positive integer has the larger absolute value. Follow the argument of the proof of (2) of Theorem 7.19.

(4) Prove that $(-2) + ((-3) + (-5)) = ((-2) + (-3)) + (-5)$. Make your argument general enough that it illustrates how to prove the associative property for three negative integers. Follow the argument of the proof of (3) of Theorem 7.19.

(5) Prove that $(-3) + (5 + (-4)) = ((-3) + 5) + (-4)$. Make your argument general enough that it illustrates how to prove the associative property for the sum of two negative integers and one positive integer in the order negative, positive, negative and with the positive integer having the largest absolute value. Follow the argument of the proof of (3) of Theorem 7.19.

(6) The six permutations of $3, 5, 9$ in $3 + (5 + (-9))$ are

$$3 + (5 + (-9)), \quad 3 + (9 + (-5)), \quad 5 + (3 + (-9))$$
$$5 + (9 + (-3)), \quad 9 + (3 + (-5)), \quad 9 + (5 + (-3))$$

Prove the associative property for these six cases in the way of the proof of (3) of Theorem 7.19.

(7) Write the six permutations of $3, 4, 5$ in $3 + ((-5) + 9)$ and prove the associative property for each case.

(8) Same as the preceding problem for
   (a) $(-3) + ((-5) + 9)$.
   (b) $(-3) + (5 + 9)$.
   (c) $3 + ((-5) + (-9))$.
   (d) $(-3) + (5 + (-9))$.
   (e) $(-3) + ((-5) + 9)$.

(9) Follow the second proof of the uniqueness of the additive inverse to prove that $-(x+y) = (-x) + (-y)$ for any integers x and y. Namely, prove that $(x+y) + ((-x) + (-y)) = 0$.

(10) Denoting deposits with positive integers and withdrawals with negative integers, Ragna recorded her bank deposits for the month as $273, $421, $-$137, $215, and $-$211. What is her total deposit for the month? If her bank balance at the beginning of the month was $8,642, what was it at the end of the month? Explain how you do the computation. What propeties of integer addition allow you to find the answer by adding the deposits and subtracting the sum of the withdrawals?

7.3. Integer Subtraction

As usual, subtraction is the inverse of addition. It can be defined in two ways, as comparative subtraction or as take-away subtraction. Both operations yield the same value because of the commutative property of addition of integers.

Definition 7.21 (Comparative subtraction). If x and y are integers, then $x - y$ is that integer z for which $x = y + z$. In symbols, $x - y = z$ means that $x = y + z$.

For example, $2 - 5 = -3$ because $2 = 5 + (-3)$.

Definition 7.22 (Take-away subtraction). If x and y are integers, then $x - y$ is that integer z for which $x = z + y$. In symbols, $x - y = z$ means that $x = z + y$.

For example, $6 - (-8) = 14$ because $6 = 14 + (-8)$. The statement $x - y$ is read x *minus* y and the result is called the *difference*. The number x is called the *minuend* and the number y is called the *subtrahend*. We use the same notation for both kinds of subtraction, because both operations produce the same difference. Notice that the definitions make no hypothesis about $y \leq x$. We shall see that $x - y$ is defined for any two integers.

Theorem 7.23. *If x and y are whole numbers, then $x - y$ in comparative subtraction is the same as $x - y$ in take-away subtraction.*

Proof. If $z = x - y$ for comparative subtraction, then $y + z = x$. But $y + z = z + y$, by the commutative property of addition of integers, and therefore $z + y = x$ so that $z = x - y$ for take-away subtraction as well. The proof of the converse goes in the same way. □

Comparative subtraction $x - y$ is called that because it answers the question: How much larger than y is x? For whole numbers and positive

7.3. Integer Subtraction

rational numbers, we asked this question under the hypothesis that $y \leq x$, but for integers we remove this hypothesis. For example, to the question of how much larger than 5 is 3, the answer is the comparative difference $3 - 5 = -2$. Being larger by -2 means that it is actually smaller by $|-2| = 2$. The comparative adjectives larger and smaller denote directions which can be indicated by positive and negative, respectively. The sign of $x - y$ tells which number is larger, and the magnitude $|x - y|$ tells how much larger.

Oral exercise 7.24. Explain how $x - y$ answers the question of how much larger than y is x for the following cases. Use the definition and number line to calculate these differences.

(1) $10 - 6$
(2) $6 - 10$
(3) $10 - (-6)$
(4) $6 - (-10)$
(5) $(-10) - 6$
(6) $(-6) - 10$
(7) $(-10) - (-6)$
(8) $(-6) - (-10)$

According to Theorem 2.27, take-away subtraction of whole numbers $a - b$, where $a \geq b$, gives the number of elements remaining in a set of a elements from which b elements have been taken away. On the other hand, $a + b$ is the number of elements in a set of a elements if an additional b elements have been put in. The sense of *take-away* and *put-in* are opposites which can be expressed by positive and negative. To take more elements out of a set than it has is possible if we allow deficits. Associating take-away with a positive subtrahend and put-in with a negative subtrahend together with associating number of elements with a positive minuend and deficit with a negative minuend, we can interpret take-away subtraction for integers. There are actually eight cases to consider, which are the eight cases of Theorem 7.27 below. As an example, consider a case (5) situation $(-7) - 4 = -(7 + 4) = -11$. We start with a deficit of 7 and take away 4 which increases the deficit to 11.

Oral exercise 7.25. Give a take-away subtraction interpretation to the following. Theorem 7.27 below tells us how to calculate these differences. For these small numbers, use the definition and the number line to calculate them.

(1) $7 - 4$
(2) $4 - 7$

(3) $7 - (-4)$
(4) $4 - (-7)$
(5) $(-7) - 4$
(6) $(-4) - 7$
(7) $(-7) - (-4)$
(8) $(-4) - (-7)$

Theorem 7.26. *If x and y are integers, then $x - y = x + (-y)$.*

Compare this result with the result that division by a rational number is multiplication by its reciprocal. The present result says that to subtract an integer y is to add its additive inverse, that is, add its opposite.

Proof. If $z = x + (-y)$, then

$$\begin{aligned} z + y &= (x + (-y)) + y, \text{ by commutative property} \\ &= x + ((-y) + y), \text{ by associative property} \\ &= x + 0, \text{ by additive inverse property} \\ &= x, \text{ by identity of } 0 \end{aligned}$$

which proves that $z = x - y$ by the definition of take-away subtraction. $\square$

This theorem combined with Theorem 7.18 implies that all integer subtractions are done using only whole number addition and subtraction together with the possible taking of opposites. The eight possible cases are stated in the following.

Theorem 7.27. *If x and y are integers, then the difference $x - y$ falls into one of the following eight cases. For a and b, whole numbers with $a \geq b$,*

(1) $a - b$ is the whole number difference. For example, $5 - 2 = 3$.
(2) $b - a = -(a - b)$. For example, $2 - 5 = -(5 - 2) = -3$.
(3) $a - (-b) = a + b$. For example, $5 - (-2) = 5 + 2 = 7$.
(4) $b - (-a) = b + a$. For example, $2 - (-5) = 2 + 5 = 7$.
(5) $(-a) - b = -(a + b)$. For example, $(-5) - 2 = -(5 + 2) = -7$.
(6) $(-b) - a = -(b + a)$. For example, $(-2) - 5 = -(2 + 5) = -7$.
(7) $(-a) - (-b) = -(a - b)$. For example, $(-5) - (-2) = -(5 - 2) = -3$.
(8) $(-b) - (-a) = a - b$. For example, $(-2) - (-5) = 5 - 2 = 3$.

Proof. If x and y are integers, then the eight possible cases for $x - y$ can be listed systematically as in the proof of Theorem 7.18.

(1) $a - b$ is whole number subtraction, since $a \geq b$.

7.3. Integer Subtraction

(2)
$$b - a = b + (-a), \text{ by Theorem 7.26}$$
$$= -(a - b), \text{ by (3) of Theorem 7.18}$$

(3) $a - (-b) = a + b$, by Theorem 7.26.

(4) $b - (-a) = b + (-(-a))$, by Theorem 7.26, and this is $b + a$ because the opposite of the opposite of a is a.

(5)
$$(-a) - b = (-a) + (-b), \text{ by Theorem 7.26}$$
$$= -(a + b), \text{ by (6) of Theorem 7.18}$$

(6)
$$(-b) - a = (-b) + (-a), \text{ by Theorem 7.26}$$
$$= -(b + a), \text{ by (6) of Theorem 7.18}$$

(7)
$$(-a) - (-b) = (-a) + (-(-b)), \text{ by Theorem 7.26}$$
$$= (-a) + b, \text{ opposite of the opposite of } b \text{ is } b$$
$$= -(a - b), \text{ by (4) of Theorem 7.18}$$

(8)
$$(-b) - (-a) = (-b) + (-(-a)), \text{ by Theorem 7.26}$$
$$= (-b) + a, \text{ opposite of the opposite of } a \text{ is } a$$
$$= a - b, \text{ by (5) of Theorem 7.18}$$

□

As addition or subtraction of integers involves only addition and subtraction of whole numbers, no new algorithms are needed to do these calculations. For example, $894 - (-476) = 894 + 476$, by (3) of Theorem 7.27, and this sum can be found by our addition algorithm to be 1370.

Theorem 7.28. *Subtraction of integers has the following properties.*

(1) *If x is an integer, then $x - 0 = x$ (identity property of 0).*
(2) *If $x < y$ and if z is an integer, then $x - z < y - z$ (order property).*
(3) *If x, y, and z are integers, then*
  (a) $(x + y) - z = x + (y - z)$
  (b) $x - (y - z) = (x - y) + z$
  (c) $x - (y + z) = (x - y) - z$
  (d) $x - (y - z) = (x + z) - y$
  (subtraction associative properties).

Proof. The proof is the same as that of Theorem 2.33. The various inequalities in the hypotheses of that theorem are no longer required because subtraction is defined between any two integers. The proof uses only the properties of addition of integers. In fact, by Theorem 7.26, this whole theorem is contained in Theorem 7.19. For example, in the first subtraction associative property, change each subtraction to the sum of the negative and it becomes the addition associative property for the integers x, y, and $-z$. □

Theorem 7.29 (Distance). *If x and y are integers, then $|x - y|$ is the number of unit intervals between x and y. Call this number the distance between x and y.*

Proof. If $z = x - y$, then $y + z = x$ and this means start at y and go $|z|$ unit intervals to the right or left to arrive at x, thus showing that there are $|z|$ unit intervals between x and y. □

As a special case we have $|x - 0| = |x|$, which gives us the interpretation of $|x|$ as the distance between x and 0.

Example 7.30. $|7 - 3| = 7 - 3 = 4$ is the number of unit intervals between 7 and 3.

Example 7.31. $|(-7) - 3| = |-(7 + 3)| = 7 + 3$ is the number of unit intervals between -7 and 3.

Example 7.32. Interpreting the absolute value $|x - 3|$ as the distance between the integer x and 3 gives us a method for solving an equation of the form $|x - 3| = 5$. The solutions are the integers x whose distance from 3 is 5. Going a distance of 5 to the right of 3 gives us the solution $x = 3 + 5 = 8$ and going a distance of 5 to the left of 3 gives us the solution $x = 3 - 5 = -2$.

The *solution set* of this equation is the set of solutions x, which is

$$\{x \in \mathbf{Z} : |x - 3| = 5\} = \{3 - 5, 3 + 5\} = \{-2, 8\}$$

where **Z** stands for the set of all integers. An algebraic method for finding the solution set is to reason as follows. If $|x - 3| = 5$, then either $x - 3 = 5$ or $x - 3 = -5$, because ± 5 are the only two integers whose magnitude is 5. Solve $x - 3 = 5$ for x by adding 3 to both sides to get the solution

7.3. Integer Subtraction

$x = (x-3) + 3 = 5 + 3 = 8$. Solve $x - 3 = -5$ in the same way to get the solution $x = (x-3) + 3 = (-5) + 3 = -(5-3) = -2$, in agreement with the solutions found geometrically.

Example 7.33. The same geometric and algebraic methods can be used to find the solution set for an inequality such as "find all integers x such that $|x-3| < 5$." Since the only whole numbers less than 5 are 0, 1, 2, 3, and 4, this inequality can be solved by solving the five equations $|x-3| = 0$, $|x-3| = 1$, etc. But that is not an efficient method if the 5 were replaced by 500, say. Even for 5 it is not efficient.

The geometric method interprets $|x-3|$ as the distance between x and 3, which means that the solution set of the inequality is the set of all integers whose distance from 3 is less than 5. To the right of 3 the integers 4, 5, 6, and 7 are the ones of distance less than 5 from 3, while to the left of 3 the integers 2, 1, 0, and -1 are. Of course, 3 itself is a distance less than 5 from 3. Therefore, the solution set is

$$(7.1) \qquad \{x \in \mathbf{Z} : |x-3| < 5\} = \{-1, 0, 1, 2, 3, 4, 5, 6, 7\}$$

which is illustrated as follows.

These are the numbers that satisfy the two inequalities $-5 < x - 3$ and $x - 3 < 5$. The algebraic method is to solve these two inequalities, so that the solution set is

$$\{x \in \mathbf{Z} : |x-3| < 5\} = \{x \in \mathbf{Z} : x - 3 < 5 \text{ and } -5 < x - 3\}$$

It is customary to write the last two inequalities here as $-5 < x - 3 < 5$. These two inequalities have the same solution set as $|x-3| < 5$. The advantage in the two inequalities is the absence of the absolute value sign in them. Each inequality can be solved using the order property of addition or subtraction. Adding 3 to both sides of $-5 < x - 3$ gives $x > (-5) + 3 = -2$ and adding 3 to both sides of $x - 3 < 5$ gives $x < 5 + 3 = 8$. Combining these, we have

$$\{x \in \mathbf{Z} : |x-3| < 5\} = \{x \in \mathbf{Z} : -2 < x < 8\}$$

which agrees with the solution set in (7.1) obtained by the geometric method.

§7.3 Exercises

(1) Write the eight cases of Theorem 7.27 for $a = 7$ and $b = 4$. Prove each case for these values of a and b and illustrate each case on the number line.

(2) Write the eight cases of Theorem 7.27 for $a = 497$ and $b = 152$. Prove each case for these values of a and b, following the arguments given in the proof of the theorem.

(3) Alexi runs a small business. Each month she calculates the profit as the revenues minus the costs. She calls profits P, revenues R, and costs C and uses the formula $P = R - C$.
  (a) In May her revenues were \$15,780 and her costs were \$12,037. What was her profit P in May?
  (b) In June her revenues were \$14,090 and her costs were \$15,225. What was her profit P in June? Could you use her formula to calculate it? Explain.
  (c) Alexi monitors the growth of her company by calculating her profit growth, called G, which is how much larger each month's profit is than the previous month's profit. For this she uses the formula $G = P - P_l$, where P stands for this month's profit and P_l stands for last month's profit. What was her profit growth in June? Could you use her formula to calculate it? Explain.

(4) Use the geometric method of Example 7.32 to find the solution set of the equation $|x - 7| = 4$. Use the number line to illustrate your argument and to exhibit the solution set. Next, use the algebraic method of the same Example to find the solution set.

(5) Use the geometric method of Example 7.33 to find the solution set of the inequality $|x - 4| < 6$. Use the number line to illustrate your argument and to exhibit the solution set. Next, use the algebraic method of the same Example to find the solution set.

(6) Use the algebraic method of Example 7.32 to find the solution set of $|x - (-475)| = 273$.

(7) Use the algebraic method of Example 7.33 to find the solution set of $|x - (-380)| < 166$.

7.4. Multiplication of Integers

We want to extend the definition of multiplication of whole numbers to the set of integers in such a way that the properties of multiplication are

7.4. Multiplication of Integers

preserved. As a first step we see that the definition of multiplication by a natural number makes sense for any integer as multiplicand.

Definition 7.34. If a is a natural number and if y is an integer, define the product $a \cdot y$ to be y added to itself a times, which is a groups of y. Define $0 \cdot y$ to be 0.

For example, $3 \cdot (-4) = (-4)+(-4)+(-4) = -(4+4+4) = -(3 \cdot 4) = -12$ and $2 \cdot 5 = 5 + 5 = 10$. According to this definition the product of positive integers is the same as their product as whole numbers. Furthermore, $3 \cdot (-4) = -(3 \cdot 4)$ can be interpreted as the opposite of $3 \cdot |-4|$.

We want the properties of multiplication of whole numbers to carry over to multiplication of integers. We have already defined a positive times a negative integer, so the commutative property requires that a negative times a positive is also defined as this product in reverse order. For example, the commutative property requires that $(-3) \cdot 5$ equal $5 \cdot (-3) = -(5 \cdot 3)$.

Definition 7.35. If a and b are whole numbers, define $(-a) \cdot b = b \cdot (-a)$, the latter being already defined in definition 7.34.

We want to interpret this definition in such a way that we see how to define the product of two negative integers. We have

$$(-a) \cdot b = b \cdot (-a) = -(b \cdot a) = -(a \cdot b)$$

which permits the interpretation that multiplication by $-a$ is the opposite of multiplication by $|-a|$. For example, $(-7) \cdot 5 = 5 \cdot (-7) = -(5 \cdot 7) = -(7 \cdot 5)$. Therefore, multiplication by the opposite of 7 is the opposite of multiplication by 7. This is a general fact when multiplying whole numbers. Multiplication by the opposite of a whole number is the opposite of multiplication by the whole number. This formulation can be extended to multiplication of a negative integer by a negative integer.

Definition 7.36. If a and b are natural numbers, define $(-a) \cdot (-b) = -(a \cdot (-b))$.

We have already established that $a \cdot (-b) = -(a \cdot b)$, so this definition is really saying that

$$(-a) \cdot (-b) = -(-(a \cdot b)) = a \cdot b$$

because the opposite of the opposite of $a \cdot b$ is $a \cdot b$. Putting these three definitions together, we arrive at the definition of multiplication in the set of integers.

Definition 7.37 (Multiplication of integers). If a and b are whole numbers, then

(1) $a \cdot b$ is the whole number product.

(2) $a \cdot (-b) = -(a \cdot b)$.

(3) $(-a) \cdot b = -(a \cdot b)$.

(4) $(-a) \cdot (-b) = a \cdot b$.

Since any integer is a whole number or the opposite of a whole number, this definition defines the product of any two integers.

Oral exercise 7.38. (1) What is 7 times -3? 7 times -8?

(2) What is -7 times 5? -7 times 7? What is the opposite of 7 times 5? Of 7 times 7?

(3) What is -8 times -9? What is the opposite of 8 times -9?

We need to check that cases (2), (3), and (4) of the definition imply that they hold for any integers x and y.

Theorem 7.39. *If x and y are integers, then*

(7.2) $$x \cdot (-y) = -(x \cdot y)$$
(7.3) $$(-x) \cdot y = -(x \cdot y)$$
(7.4) $$(-x) \cdot (-y) = x \cdot y$$

Proof. Any integer is either a whole number or the opposite of a whole number. This means that either $x = a$ or $x = -a$, for some whole number a, and either $y = b$ or $y = -b$, for some whole number b. Each equation thus has $2 \cdot 2 = 4$ cases. If both x and y are whole numbers, then the equations are true by the above definition. The other three cases must be checked for each equation, a total of nine cases.

Case $x = a$ and $y = -b$. Using the fact that the opposite of the opposite of a number is the number again, we have $x \cdot (-y) = a \cdot (-(-b)) = a \cdot b$, while $-(x \cdot y) = -(a \cdot (-b)) = -(-(a \cdot b)) = a \cdot b$. Hence, $x \cdot (-y) = -(x \cdot y)$ in this case. The remaining eight cases are left as exercises. □

Corollary 7.40. *If x is an integer, then $(-1) \cdot x = -x$. In words, multiplication by -1 is the same as taking the opposite.*

Proof. By (7.3), we have $(-1) \cdot x = -(1 \cdot x) = -x$. □

The properties of multiplication of integers are the same as those in Theorem 2.63 for multiplication of whole numbers. The one new point is the order property for multiplication of an inequality by a negative integer.

Theorem 7.41. *Multiplication of integers has the following properties. If x, y, and z are integers, then:*

(1) *and if x and y are whole numbers, then $x \cdot y$ is the whole number product.*

7.4. Multiplication of Integers

(2) $x \cdot y = y \cdot x$ (commutative property).

(3) $x \cdot (y \cdot z) = (x \cdot y) \cdot z$ (associative property).

(4) $x \cdot 1 = 1 \cdot x = x$ (identity property of 1)

(5) $x \cdot (y + z) = x \cdot y + x \cdot z$ (distributive property of multiplication over addition)

(6) $x \cdot (y - z) = x \cdot y - x \cdot z$ (distributive property of multiplication over subtraction)

(7) Order properties:
   (a) If $x < y$ and if z is positive, then $z \cdot x < z \cdot y$.
   (b) If $x < y$ and if z is negative, then $z \cdot x > z \cdot y$.

Proof. Any integer is a whole number or the opposite of a whole number. Because multiplication is defined for each of these cases separately, the only way to prove these properties of multiplication is to check each case. For this purpose, we suppose that the three arbitrary integers are $x = \pm a$, $y = \pm b$, and $z = \pm c$, where a, b, and c are whole numbers.

(1) There is nothing to prove in this case because if $x = a$ and $y = b$, then we defined their product to be the whole number product $a \cdot b$.

(2) For the commutative property we have three cases to check.

Case $a \cdot b$. $a \cdot b = b \cdot a$ is true for whole number multiplication.

Case $a \cdot (-b)$. By (2) of Definition 7.37 we have

$$a \cdot (-b) = -(a \cdot b)$$

and by (3) of the same definition we have

$$(-b) \cdot a = -(b \cdot a)$$

and therefore the two products are equal because $a \cdot b = b \cdot a$ for whole numbers.

Case $(-a) \cdot (-b)$. By (4) of Definition 7.37 we have

$$(-a) \cdot (-b) = a \cdot b$$

and

$$(-b) \cdot (-a) = b \cdot a$$

and these are equal because $a \cdot b = b \cdot a$ for whole numbers. That completes the proof of the commutative property for integers.

(3) In a product of three integers $x \cdot (y \cdot z)$ each integer can be one of two possibilities, a whole number or negative, and therefore there are $2 \cdot 2 \cdot 2 = 8$ cases to check. Each case turns out to be a product of whole numbers or the opposite of such a product, so that the associative property holds because it is true for whole numbers. Here are the details for each case.

Case $a \cdot (b \cdot c)$. This case is true because the associative property is true for whole numbers.

Case $a \cdot (b \cdot (-c))$. Using (2) of Definition 7.37, we have
$$a \cdot (b \cdot (-c)) = a \cdot (-(b \cdot c)) = -(a \cdot (b \cdot c))$$
Using the same part of the definition again, we have
$$(a \cdot b) \cdot (-c) = -((a \cdot b) \cdot c)$$
The two results are equal because $a \cdot (b \cdot c) = (a \cdot b) \cdot c$ for whole numbers.

Case $a \cdot ((-b) \cdot c)$. Using (3) of Definition 7.37 and then (2), we have
$$a \cdot ((-b) \cdot c) = a \cdot (-(b \cdot c)) = -(a \cdot (b \cdot c))$$
Using (2) of the definition and then (3), we have
$$(a \cdot (-b)) \cdot c = (-(a \cdot b)) \cdot c = -((a \cdot b) \cdot c)$$
The two results are equal by the associative property of whole numbers.

Case $a \cdot ((-b) \cdot (-c))$. Using (4) of Definition 7.37, we have
$$a \cdot ((-b) \cdot (-c)) = a \cdot (b \cdot c)$$
Using (2) of the definition, and then (4), we have
$$(a \cdot (-b)) \cdot (-c) = (-(a \cdot b)) \cdot (-c) = (a \cdot b) \cdot c$$
The two results are equal by the associative property of whole numbers.

The four remaining cases have a replaced by $-a$ in each of the preceding four cases. These cases are left as written exercises. After you have completed them, you will have completed the proof of the associative property for integers.

After going through all eight cases, you should notice the following useful rule.

> When multiplying three integers, the sign of the product is positive or negative depending on whether there is an even or odd number of negative factors, respectively.

(4) The identity property of 1 is true when x is a whole number a. When $x = -a$, we have by (3) of Definition 7.37
$$(-a) \cdot 1 = -(a \cdot 1) = -a$$
because $a \cdot 1 = a$ for a whole number a. The commutative property of multiplication of integers then implies that $1 \cdot (-a) = (-a) \cdot 1 = -a$, as well.

(5) To verify the distributive property we must consider twelve cases. Although it may seem unreasonably tedious to go through all twelve cases, it is really the only way to convince yourself that the distributive property is true for integers. It is also the only way to gain a deep appreciation for the

7.4. Multiplication of Integers

beauty of this intricate result. We shall see that each case comes down to the distributive property for whole numbers. Here are the details.

Case $a \cdot (b + c)$. This case is true because the distributive property is true for whole numbers.

Case $a \cdot (b + (-c))$, with $b \geq c$. Using (2) of Theorem 7.18 and then the distributive property for multiplication over subtraction of whole numbers ((5) of Theorem 2.63), we have
$$a \cdot (b + (-c)) = a \cdot (b - c) = a \cdot b - a \cdot c$$
Using (2) of Definition 7.37, we get the first equality below. Then, because $b \geq c$ implies that $a \cdot b \geq a \cdot c$ by the order property of multiplication of whole numbers, we can use (2) of Theorem 7.18 again for the second equality, to get
$$a \cdot b + a \cdot (-c) = a \cdot b + (-(a \cdot c)) = a \cdot b - a \cdot c$$
The two results are equal.

Case $a \cdot (b + (-c))$, with $b < c$. We have
$$\begin{aligned} a \cdot (b + (-c)) &= a \cdot (-(c - b)), \text{ by (3) of Theorem 7.18} \\ &= -(a \cdot (c - b)), \text{ by (2) of Definition 7.37} \\ &= -(a \cdot c - a \cdot b), \text{ by (5) of Theorem 2.63} \end{aligned}$$
Since $b < c$, we know that $a \cdot b \leq a \cdot c$ by the order property of multiplication of whole numbers. Using this fact, we have
$$\begin{aligned} a \cdot b + a \cdot (-c) &= a \cdot b + (-(a \cdot c)), \text{ by (2) of Definition 7.37} \\ &= -(a \cdot c - a \cdot b), \text{ by (3) of Theorem 7.18} \end{aligned}$$
The two results are equal.

Case $a \cdot ((-b) + c)$, with $b \geq c$. We have
$$\begin{aligned} a \cdot ((-b) + c) &= a \cdot (-(b - c)), \text{ by (4) of Theorem 7.18} \\ &= -(a \cdot (b - c)), \text{ by (2) of Definition 7.37} \\ &= -(a \cdot b - a \cdot c), \text{ by (5) of Theorem 2.63} \end{aligned}$$
Since $b \geq c$, we know that $a \cdot b \geq a \cdot c$ by the order property of multiplication of whole numbers. Using this fact, we have
$$\begin{aligned} a \cdot (-b) + a \cdot c &= -(a \cdot b) + a \cdot c, \text{ by (2) of Definition 7.37} \\ &= -(a \cdot b - a \cdot c), \text{ by (4) of Theorem 7.18} \end{aligned}$$
The two results are equal.

Case $a \cdot ((-b) + c)$, with $b < c$. We have
$$\begin{aligned} a \cdot ((-b) + c) &= a \cdot (c - b), \text{ by (5) of Theorem 7.18} \\ &= a \cdot c - a \cdot b, \text{ by (5) of Theorem 2.63} \end{aligned}$$

Since $b < c$, we know that $a \cdot b \leq a \cdot c$ by the order property of multiplication of whole numbers. Using this fact, we have
$$a \cdot (-b) + a \cdot c = -(a \cdot b) + a \cdot c, \text{ by (2) of Definition 7.37}$$
$$= a \cdot c - a \cdot b, \text{ by (5) of Theorem 2.63}$$
The two results are equal.

Case $a \cdot ((-b) + (-c))$. We have
$$a \cdot ((-b) + (-c)) = a \cdot (-(b+c)), \text{ by (6) of Theorem 7.18}$$
$$= -(a \cdot (b+c)), \text{ by (2) of Definition 7.37}$$
$$= -(a \cdot b + a \cdot c), \text{ by (2) of Theorem 2.63}$$
$$= (-(a \cdot b)) + (-(a \cdot c)), \text{ by (6) of Theorem 7.18}$$
$$= a \cdot (-b) + a \cdot (-c), \text{ by (2) of Definition 7.37}$$
which proves the distributive property for this case.

Case $(-a) \cdot (b + c)$. We have
$$(-a) \cdot (b+c) = -(a \cdot (b+c)), \text{ by (3) of Definition 7.37}$$
$$= -(a \cdot b + a \cdot c), \text{ by (2) of Theorem 2.63}$$
$$= (-a) \cdot b + (-a) \cdot c, \text{ by (3) of Definition 7.37}$$
which proves the distributive property for this case.

Case $(-a) \cdot (b + (-c))$, with $b \geq c$. We have
$$(-a) \cdot (b + (-c)) = (-a) \cdot (b - c), \text{ by (2) of Theorem 7.18}$$
$$= -(a \cdot (b - c)), \text{ by (3) of Definition 7.37}$$
$$= -(a \cdot b - a \cdot c), \text{ by (5) of Theorem 2.63}$$
Since $b \geq c$, we know by the order property of multiplication of whole numbers that $a \cdot b \geq a \cdot c$. Using this, we have
$$(-a) \cdot b + (-a) \cdot (-c) = -(a \cdot b) + a \cdot c, \text{ by (3) and (4) of Definition 7.37}$$
$$= -(a \cdot b - a \cdot c), \text{ by (4) of Theorem 7.18}$$
The two results are equal.

Case $(-a) \cdot (b + (-c))$, with $b < c$. We have
$$(-a) \cdot (b + (-c)) = (-a) \cdot (-(c - b)), \text{ by (3) of Theorem 7.18}$$
$$= a \cdot (c - b), \text{ by (4) of Definition 7.37}$$
$$= a \cdot c - a \cdot b, \text{ by (5) of Theorem 2.63}$$
Since $b < c$, we have $a \cdot b \leq a \cdot c$, by the order property of multiplication of whole numbers. Using this, we have
$$(-a) \cdot b + (-a) \cdot (-c) = (-(a \cdot b)) + a \cdot c, \text{ by (3) and (4) of Definition 7.37}$$
$$= a \cdot c - a \cdot b, \text{ by (5) of Theorem 7.18}$$

7.4. Multiplication of Integers

The results are equal.

Case $(-a) \cdot ((-b) + c)$, with $b \geq c$. Since $b \geq c$ we know that $a \cdot b \geq a \cdot c$ by the order property of multiplication of whole numbers. Using this, we have

$$\begin{aligned}(-a) \cdot ((-b) + c) &= (-a) \cdot (-(b-c)), \text{ by (4) of Theorem 7.18} \\ &= a \cdot (b-c), \text{ by (4) of Definition 7.37} \\ &= a \cdot b - a \cdot c, \text{ by (5) of Theorem 2.63} \\ &= a \cdot b + (-(a \cdot c)), \text{ by (2) of Theorem 7.18} \\ &= (-a) \cdot (-b) + (-a) \cdot c, \text{ by (4) and (3) of Definition 7.37}\end{aligned}$$

Case $(-a) \cdot ((-b) + c)$, with $b < c$. We have

$$\begin{aligned}(-a) \cdot ((-b) + c) &= (-a) \cdot (c-b), \text{ by (5) of Theorem 7.18} \\ &= -(a \cdot (c-b)), \text{ by (3) of Definition 7.37} \\ &= -(a \cdot c - a \cdot b), \text{ by (5) of Theorem 2.63} \\ &= (-(a \cdot c)) + a \cdot b, \text{ by (4) of Theorem 7.18} \\ &= (-a) \cdot c + (-a) \cdot (-b), \text{ by (3) and (4) of Definition 7.37}\end{aligned}$$

which is the desired result because of the commutative property of addition of integers.

Case $(-a) \cdot ((-b) + (-c))$. We have

$$\begin{aligned}(-a) \cdot ((-b) + (-c)) &= (-a) \cdot (-(b+c)), \text{ by (6) of Theorem 7.18} \\ &= a \cdot (b+c), \text{ by (4) of Definition 7.37} \\ &= a \cdot b + a \cdot c, \text{ by (4) of Theorem 2.63} \\ &= (-a) \cdot (-b) + (-a) \cdot (-c), \text{ by (4) of Definition 7.37}\end{aligned}$$

This completes the proof of (5), the distributive property of multiplication over addition for integers.

(6) It is a relief to know that the distributive property over addition already implies it over subtraction because any subtraction of integers can be converted to a sum. We have

$$\begin{aligned}x \cdot (y - z) &= x \cdot (y + (-z)), \text{ by Theorem 7.26} \\ &= x \cdot y + x \cdot (-z), \text{ by part (5) above} \\ &= x \cdot y + (-(x \cdot z)), \text{ by (7.2)} \\ &= x \cdot y - x \cdot z, \text{ by Theorem 7.26}\end{aligned}$$

(7) If $x < y$, then subtract x from both sides and use the order property of subtraction of integers to conclude that $0 < y - x$. Conversely, starting with the latter inequality, add x to both sides to conclude that $x < y$. Therefore, $x < y$ if and only if $y - x > 0$. Now suppose that $x < y$ and that $z > 0$, which

means that z is a nonzero whole number. Then $y - x > 0$, which means it also is a nonzero whole number. The order property of multiplication of whole numbers thus implies that $z \cdot (y - x) > 0$. Using the distributive property over subtraction, we have from this $0 < z \cdot (y - x) = z \cdot y - (z \cdot x)$, which means that $z \cdot x < z \cdot y$.

On the other hand, if $x < y$ and z is negative, then $y - x > 0$ and $z = -c$, for some positive whole number c. By the preceding part, $c \cdot (y - x) > 0$, and taking the opposite we have $-(c \cdot (y - x)) < 0$. By (3) of Definition 7.37, we know that $-(c \cdot (y-x)) = (-c) \cdot (y-x)$, and this equals $(-c) \cdot y - (-c) \cdot x$ by the distributive property already proved. Knowing this is negative, and adding $(-c) \cdot x$ to both sides, we conclude that $(-c) \cdot y < (-c) \cdot x$, which is what we wanted to prove. This finally concludes the proof of Theorem 7.41. □

Oral exercise 7.42. Find

(1) $(-3) \cdot (5 + (-7))$

(2) $(-2) \cdot ((-5) + 4)$

(3) $(-2) \cdot 6 + (-2) \cdot 4$

(4) $6 + (2 \cdot (-3))$

(5) Which is larger, $(-4) \cdot 77$ or $(-4) \cdot 98$?

§7.4 Exercises

Write general proofs which illustrate one of the general cases.

(1) Prove that $3 \cdot (-5) = (-5) \cdot 3$ by following the proof in Theorem 7.41 of the appropriate case of the commutative property.

(2) Prove the associative property $3 \cdot ((-5) \cdot (-7)) = (3 \cdot (-5)) \cdot (-7)$ by following the proof in Theorem 7.41 of the associative property when the first number is positive and the second two are negative.

(3) Prove the associative property for the case $(-a) \cdot (b \cdot c)$, where a, b, and c are whole numbers.

(4) Prove the associative property for the case $(-a) \cdot (b \cdot (-c))$, where a, b, and c are whole numbers.

(5) Prove the associative property for the case $(-a) \cdot ((-b) \cdot c)$, where a, b, and c are whole numbers.

(6) Prove the associative property for the case $(-a) \cdot ((-b) \cdot (-c))$, where a, b, and c are whole numbers.

(7) Prove the distributive property for $(-2) \cdot (3+(-5)) = (-2) \cdot 3 + (-2) \cdot (-5)$ by following the proof in Theorem 7.41 of the appropriate case of the distributive property.

(8) Same as the preceding problem for
   (a) $2 \cdot (5 + (-9)) = 2 \cdot 5 + 2 \cdot (-9)$
   (b) $2 \cdot (9 + (-5)) = 2 \cdot 9 + 2 \cdot (-5)$
   (c) $(-2) \cdot (9 + (-5)) = (-2) \cdot 9 + (-2) \cdot (-5)$

(9) Do each of the following computations. Justify each step by citing the appropriate property of integer addition, subtraction, or multiplication.
   (a) $27 \cdot (84 + (-43))$
   (b) $35 \cdot (28 + (-75))$
   (c) $58 \cdot ((-96) + 14)$
   (d) $42 \cdot ((-37) + 64)$
   (e) $74 \cdot ((-28) + (-67))$
   (f) $(-89) \cdot (74 + (-37))$
   (g) $(-33) \cdot (28 + (-82))$
   (h) $(-57) \cdot ((-71) + 56)$
   (i) $(-22) \cdot ((-17) + 81)$
   (j) $(-74) \cdot ((-18) + (-91))$

7.5. Division of Integers

Division of integers is defined as the inverse operation to multiplication. Division is to multiplication as subtraction is to addition. There are two kinds of division for integers, partitive and measurement. If the divisor specifies the size of the group, it is measurement division. If it specifies how many groups, it is partitive division. These two formulations lose sense when the divisor is negative. As in the case of rational numbers, we can rephrase the partitive and measurement questions to make sense for any integer divisor.

Definition 7.43 (Partitive division). If x and y are integers, and if $y \neq 0$, then $x \div y$ is that integer z such that $y \cdot z = x$. In words, y of z is x, so z is the answer to the partitive question "y of what is x"?

For example, $18 \div (-3) = -6$, because $(-3) \cdot (-6) = 18$.

Definition 7.44 (Measurement division). If x and y are integers, and if $y \neq 0$, then $x \div y$ is that integer z such that $z \cdot y = x$. In words, z of y is x, so z is the answer to the measurement question "What of y is x?

For example $(-18) \div (-3) = 6$ because $6 \cdot (-3) = -18$.

In the notation $x \div y = z$, the number x is called the dividend, y is called the divisor, and z is called the quotient. We use the same notation for both kinds of division because they both produce the same quotient.

Theorem 7.45. *If x and y are integers, then $x \div y$ in partitive division is equal to $x \div y$ in measurement division.*

Proof. If $z = x \div y$ in partitive division, then $x = y \cdot z$. By the commutative property of multiplication of integers, $y \cdot z = z \cdot y$, and therefore $x = z \cdot y$, which means that $z = x \div y$ in measurement division. □

Definition 7.46. If x and y are integers, we define x *is divisible by* y to mean that $x \div y$ exists as an integer. That is, x divisible by y means that there is an integer z such that $x = y \cdot z$.

Theorem 7.47. *Suppose that x and y are integers with $y \neq 0$.*
(1) If x and y are whole numbers, then $x \div y$ is the whole number quotient.
(2) If one of x and y is negative and the other is a whole number, then $x \div y = -(|x| \div |y|)$.
(3) If both x and y are negative, then $x \div y = |x| \div |y|$.
(4) x is divisible by y if and only if $|x|$ is divisible by $|y|$.

Proof. (1) The definition of division of integers is the same as that for whole numbers when the integers are whole numbers. Therefore, (1) is true by definition.
(2) Either $x = a$ and $y = -b$ or $x = -a$ and $y = b$, where a and $b > 0$ are whole numbers. Then $x \div y = z$ means $x = y \cdot z$ which means that $a = (-b) \cdot z$ or $(-a) = b \cdot z$. Since $z = c$ or $z = -c$, where c is a whole number, it follows that $z = -c$. But then $a = (-b) \cdot (-c) = b \cdot c$ or $-a = b \cdot (-c) = -(b \cdot c)$ each imply that $a \div b = c$. Therefore, $z = -c = -(|x| \div |y|)$.
(3) If x and y are both negative, then $x = -a$ and $y = -b$, where a and b are natural numbers. Then $a \div b = c$ means that $b \cdot c = a$ and therefore $(-b) \cdot c = -a$ and therefore $c = (-a) \div (-b) = x \div y$. In other words, $x \div y = |x| \div |y|$.
(4) This is a consequence of (1), (2), and (3). □

Definition 7.48 (Average). The average of a set of integers is their sum divided by the number of elements in the set.

Example 7.49. If the change in the Dow Jones industrial average each day of some week were 123, −54, −114, 3, 12, then the average daily change was the average of this set of integers, which is

$$(123 + (-54) + (-114) + 3 + 12)/5 = (-30)/5 = -6$$

7.5. Division of Integers

The average daily change was −6 points. If the Dow had lost 6 points each day, the total change for the week would have been the same.

Example 7.50 (Divisibility by 11). In base ten numeration, a number is divisible by 11 if and only if the alternating sum of its digits is divisible by 11. By the alternating sum of its digits we mean the sum of the digit in the one's place plus the opposite of the digit in the ten's place plus the digit in the hundred's place plus the opposite of the digit in the thousand's place, and so on through all the digits. The proof of this rule comes from the following:

(7.5)
$$10 = 1 \cdot 11 - 1$$
$$100 = 9 \cdot 11 + 1$$
$$1000 = 91 \cdot 11 - 1$$
$$10000 = 909 \cdot 11 + 1$$

and so on. These can be derived by long division. That this pattern of remainders continues forever is most easily seen in Chapter 8 on Clock Arithmetic. To see how these formulas give the divisibility rule, consider for example the number 286. The alternating sum of its digits is $6 - 8 + 2 = 0$, which is divisible by 11, and it is easily checked that $286 = 11 \cdot 26$. Both the number and the alternating sum of its digits are divisible by 11 because

$$286 = 2 \cdot 100 + 8 \cdot 10 + 6, \text{ numeration}$$
$$= 2 \cdot (9 \cdot 11 + 1) + 8 \cdot (1 \cdot 11 - 1) + 6, \text{ by (7.5)}$$
$$= (2 \cdot 9 + 8 \cdot 1) \cdot 11 + (2 - 8 + 6), \text{ by distributive property}$$

which is a multiple of 11 plus the alternating sum of the digits. From that it follows that 286 is divisible by 11 if and only if the alternating sum of its digits is divisible by 11.

Oral exercise 7.51. Test the numbers 5286 and 746988 for divisibility by 11.

Example 7.52. Temperatures are commonly measured in either units of degrees Fahrenheit or in units of degrees Centigrade (also called Celsius). If F denotes the temperature in degrees Fahrenheit and C denotes the same temperature in degrees Centigrade, then

(7.6)
$$C = \frac{5}{9} \cdot (F - 32)$$

A temperature of 5° F is

$$\frac{5}{9} \cdot (5 - 32) = \frac{5}{9} \cdot (-27) = (5 \cdot ((-27)/9) = 5 \cdot (-3) = -15°\text{C}$$

Oral exercise 7.53. (1) Water freezes at 32° F. What temperature is that in Centigrade?

(2) At sea level, water boils at 212° F. What temperature is that in Centigrade?

(3) What is the formula for degrees F in terms of degrees C?

§7.5 Exercises

(1) Write the proof of Theorem 7.47 for the cases $12 \div (-3)$ and $(-36) \div (-4)$. Write your proof in such a way that it illustrates the general case.

(2) What is the temperature in degrees Centigrade if the temperature in degrees Fahrenheit is: (a) 95. (b) 77. (c) 41. (d) 14. (e) 0. (f) -22. (g) -40.

(3) (a) For the days of a week of January the low temperatures in degrees Fahrenheit were 6, -3, -4, 0, -5, 7, and 6, respectively. What was the average in degrees Fahrenheit of the daily low temperatures for that week? Explain how you do the computation.
(b) What was the average in degrees Centigrade of the daily low temperatures for that week? Explain why you get the same answer converting the average temperature from Fahrenheit to Centigrade as you get by converting each day's low temperature to Centigrade and finding the average of these numbers.

(4) Use the divisibility by 11 test on the numbers

$$517, \quad 3503, \quad -4180$$

(5) In base five numeration, six is 11_5 and five is 10_5. Verify the following base five version of (7.5).

(7.7)
$$10_5 = 1 \cdot 11_5 - 1$$
$$100_5 = 4 \cdot 11_5 + 1$$
$$1000_5 = 41_5 \cdot 11_5 - 1$$
$$10000_5 = 404_5 \cdot 11_5 + 1$$

What should be the next equation in (7.5) and in (7.7)?

(6) State and prove a rule in base five numeration for when a number is divisible by six.

(7) State and prove a rule in base two numeration for when a number is divisible by three.

7.6. Negative Rational Numbers

The development of the preceding five sections can be repeated, without change, for the rational numbers. Negative rational numbers are defined to the left of 0 as the opposites of the positive rational numbers. The positive and negative rational numbers, together with 0, are called the rational numbers. The arithmetic operations extend from the positive rational numbers to all rational numbers in exactly the same way the extension was done for the integers.

Opposites can be taken of irrational numbers as well. For example, $-\sqrt{2}$, $-\sqrt{3}$, and $-\pi$ are the points to the left of 0 lying a distance of $\sqrt{2}$, $\sqrt{3}$, and π, respectively, from 0.

§7.6 Exercises

(1) What is the temperature in degrees Centigrade if the temperature in degrees Fahrenheit is: (a) 90. (b) 72. (c) 20. (d) -10.

(2) What is the temperature in degrees Fahrenheit if the temperature in degrees Centigrade is: (a) -10. (b) 5. (c) 22. (d) 30. (e) 40.

Chapter 8

Clock Arithmetic

If the number line is replaced by a clock face, the four arithmetic operations can be redefined to create what is called clock arithmetic. The division algorithm and the factors of the number of hours on the clock play a large role in the properties of this arithmetic.

We shall begin with the 12 hour clock to illustrate these concepts. The 12 hour clock consists of the whole numbers from 0 through 11 equally spaced and going up in the clockwise direction. The space between two adjacent numbers is called a unit. As we shall see, 12 plays the role of 0 in this arithmetic, so the 0 goes where you would normally put 12 on a clock face. A number a on this clock is a units from 0 in the clockwise direction. To go 12 units in the clockwise direction from any number is to end up at that number again.

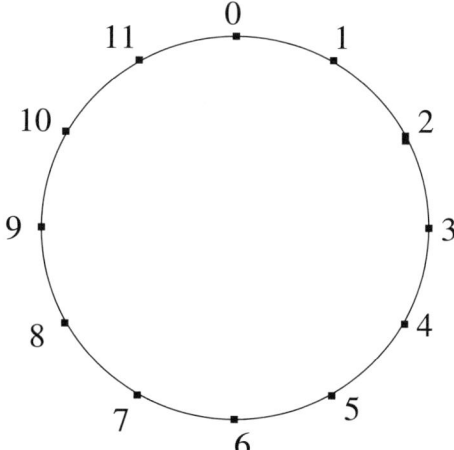

The general clock has n hours, where n can be any whole number greater than 1. We denote the n-hour clock by the set of numbers
$$\mathbf{Z}_n = \{0, 1, 2, \ldots, n-1\}$$
arranged uniformly in the clockwise direction on a clock face. For example, $\mathbf{Z}_5 = \{0, 1, 2, 3, 4\}$ is arranged like this:

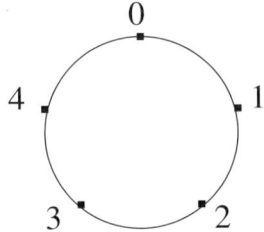

8.1. Clock Addition, Subtraction, and Multiplication

8.1.1. Clock Addition.

Definition 8.1 (Clock Addition.). If a and b are numbers on the n-hour clock, define $a+b$ to be the number on the n-hour clock arrived at by starting at a and going b units in the clockwise direction.

For example, on the 12-hour clock, $3 + 4 = 7$, $8 + 5 = 1$, and $6 + 6 = 0$. Notice that $a + 0 = a$ for any number a on this clock.

Theorem 8.2. *Addition on the n-hour clock has the following properties.*

(1) Remainder property: If a and b are any numbers on the n-hour clock, then $a + b$ on this clock is the remainder of $(a+b) \div n$, in the arithmetic of the whole numbers.

(2) Commutative property: $a + b = b + a$.

(3) Associative property: $a + (b + c) = (a + b) + c$.

(4) Identity of 0: $a + 0 = 0 + a = a$, for any number a.

Proof. We illustrate the proof on the 12-hour clock.

(1) By long division of $(a + b) \div 12$, we have $a + b = q \cdot 12 + r$, where q is the quotient and $0 \leq r < 12$ is the remainder. To go to the number a on the clock, start at 0 and go a units in the clockwise direction. Then, by the definition of clock addition, $a + b$ is found by going b more units from a in the clockwise direction. Therefore, $a + b$ is found by starting at 0 and going $a + b$ units in the clockwise direction. Going $a + b = q \cdot 12 + r$ units from 0, however, is the same as going 12 units q times, which returns one to 0, and then going r more units in the clockwise direction. Therefore, one ends at r on the clock, which means that $a + b$ is r in the 12-hour clock arithmetic.

(2) For any numbers a and b on the clock, the whole number sums $a + b = b + a$, and therefore $(b+a) \div 12$ has the same remainder as $(a+b) \div 12$. By part (1), it follows that $b + a = a + b$ on the 12-hour clock.

(3) If a, b, and c are any numbers on the 12-hour clock, then here is how we use the remainder property to calculate $a + (b + c)$ on this clock. Divide $b + c$ by 12 to get $b + c = p \cdot 12 + r$, for some remainder $0 \leq r < 12$, so that $b + c = r$ on $\mathbf{Z}_{12}$. Then divide $a + r$ by 12 to get $a + r = q \cdot 12 + s$, for some remainder $0 \leq s < 12$, so that $a + (b + c) = s$ on $\mathbf{Z}_{12}$. In whole numbers, we have
$$\begin{aligned} a + (b+c) &= a + (p \cdot 12 + r) = p \cdot 12 + a + r \\ &= p \cdot 12 + q \cdot 12 + s = (p+q) \cdot 12 + s \end{aligned}$$
which means that s is the remainder of $a + (b + c)$ divided by 12.

On the other hand, the remainder property computes $(a + b) + c$ on the 12-hour clock as follows. Divide $a + b$ by 12 to get $a + b = P \cdot 12 + R$, for some remainder $0 \leq R < 12$, so that $a + b = R$ on $\mathbf{Z}_{12}$. Then divide $R + c$ by 12 to get $R + c = Q \cdot 12 + S$, for some remainder $0 \leq S < 12$, so that $(a+b) + c = R + c = S$ on $\mathbf{Z}_{12}$. In whole numbers, we have
$$\begin{aligned} (a+b) + c &= P \cdot 12 + R + c = P \cdot 12 + Q \cdot 12 + S \\ &= (P+Q) \cdot 12 + S \end{aligned}$$
which means that S is the remainder of $(a + b) + c$ divided by 12. Since the associative property for whole number addition says that $a + (b + c) = (a + b) + c$, it follows that $s = S$. Therefore, addition on the 12-hour clock has the associative property.

(4) We have already observed that 0 has the addition identity property. □

Oral exercise 8.3. (1) On the 12-hour clock, find $3 + 5$, $8 + 3$, $9 + 6$, $(2 + 5) + 11$.

(2) In $\mathbf{Z}_{1009}$, what is $(18 + 147) + 282$?

8.1.2. Clock Subtraction.

Definition 8.4 (Clock Subtraction). If a and b are any numbers on the n-hour clock, define $a - b$ to be the number on this clock that added to b is a in clock addition. In symbols, $a - b$ is the number on this clock such that $b + (a - b) = a$ in clock addition.

For example, on the 12-hour clock $7 - 4 = 3$ because $4 + 3 = 7$, $4 - 7 = 9$ because $7 + 9 = 4$ and $6 - 6 = 0$ because $6 + 0 = 6$.

Theorem 8.5. *Subtraction on the n-hour clock has the following properties.*

(1) If a and b are numbers on the n-hour clock, then $a - b$ is the number on this clock arrived at by starting at a and going b units in the counterclockwise direction.

(2) Identity property of 0: $a - 0 = a$, for any number a on the n-hour clock.
(3) For any numbers a, b, and c on the n-hour clock:

    (a) $(a+b) - c = a + (b-c)$.
    (b) $a - (b-c) = (a-b) + c$.
    (c) $a - (b+c) = (a-b) - c$.

Proof. We illustrate the proof on the 12-hour clock.

(1) If a and b are any numbers on the 12-hour clock, then $a - b$ is that number on the clock such that $b + (a-b) = a$. Therefore, since $b + (a-b) = (a-b) + b$, we compute this by starting at $a - b$ and going b units in the clockwise direction, where we arrive at a, by definition of $a - b$. This means that if we start at a and go b units in the counterclockwise direction, we must arrive back at $a - b$, as claimed.

(2) By definition of subtraction, $a - 0$ is that number which added to 0 gives a. Since $0 + a = a$, it follows that $a - 0 = a$.

(3) (a) On the 12-hour clock, the number $(a+b) - c$ is that number which added to c gives $a + b$. We shall verify that, when $a + (b-c)$ is added to c, we get $a + b$. In fact, in clock addition

$$\begin{aligned} c + (a + (b-c)) &= (c+a) + (b-c), \text{ associative property} \\ &= (a+c) + (b-c), \text{ commutative property} \\ &= a + (c + (b-c)), \text{ associative property} \\ &= a + b, \text{ definition of } b - c \end{aligned}$$

The remaining two properties (b) and (c) are left as written exercises. Notice that the proofs of these last three properties are identical to the proofs of the same three properties in whole number arithmetic, as given in Theorem 2.33. □

Oral exercise 8.6. (1) On the 28-hour clock, what is $23 - 11$? What is $11 - 23$?

    (2) Explain why $11 - 23 = 28 - (23 - 11)$ on the 28-hour clock. This is a practical way to compute differences in clock arithmetic when the subtrahend is larger than the minuend (regarded as whole numbers).

    (3) In $\mathbf{Z}_{100}$, find $25 - 85$. *Ans.* 40

8.1.3. Clock Multiplication.

Definition 8.7 (Clock Multiplication). If a and b are any numbers on the n-hour clock, define $a \cdot b$ to be b added to itself a times.

8.1. Clock Addition, Subtraction, and Multiplication 351

For example, on the 12-hour clock, $2 \cdot 5 = 10$, because $5 + 5 = 10$ on the 12-hour clock. Another example is $3 \cdot 5 = 3$, because $5 + 5 + 5 = 3$ on the 12-hour clock. Notice that $3 \cdot 4 = 0$ and that $3 \cdot 1 = 3$ on the 12-hour clock.

Theorem 8.8. *Clock multiplication has the following properties. Let a, b, and c be any numbers on the n-hour clock.*

(1) Remainder property: $a \cdot b$ is the remainder when the whole number product of a and b is divided by n.

(2) Commutative property: $a \cdot b = b \cdot a$.

(3) Associative property: $a \cdot (b \cdot c) = (a \cdot b) \cdot c$.

(4) Identity property of 1: $a \cdot 1 = a$.

(5) Zero property: $a \cdot 0 = 0$.

(6) Distributive property over addition: $a \cdot (b + c) = a \cdot b + a \cdot c$.

(7) Distributive property over subtraction: $a \cdot (b - c) = a \cdot b - a \cdot c$.

Proof. We illustrate the proof on the 12-hour clock.

(1) By long division of $(a \cdot b) \div 12$ in whole number arithmetic, we have $a \cdot b = q \cdot 12 + r$, where q is the quotient and $0 \leq r < 12$ is the remainder. On the 12-hour clock, $a \cdot b$ is b added to itself a times. This number is the terminal point if you start at 0 and go $a \cdot b = q \cdot 12 + r$ units clockwise. Each 12 units is a complete revolution, bringing us back to the starting point, so $q \cdot 12 + r$ units is q complete revolutions and then r more units from the starting point 0. The terminal point is thus r.

(2) The commutative property follows from the commutative property of whole number multiplication together with the remainder property as follows. On the 12-hour clock, $a \cdot b$ is the remainder of the whole number product of a and b divided by 12, while $b \cdot a$ is the remainder of the whole number product of b and a divided by 12. These remainders are the same, because these whole number products are equal.

(3) To prove that $a \cdot (b \cdot c) = (a \cdot b) \cdot c$ on the 12-hour clock, we use the remainder property to calculate each product. Divide $b \cdot c$ by 12 to get $b \cdot c = p \cdot 12 + r$, for some remainder $0 \leq r < 12$, so that $b \cdot c = r$ on $\mathbf{Z}_{12}$. Divide $a \cdot r$ by 12 to get $a \cdot r = q \cdot 12 + s$, for some remainder $0 \leq s < 12$, so that $a \cdot (b \cdot c) = a \cdot r = s$ in $\mathbf{Z}_{12}$. In whole numbers,

$$\begin{aligned} a \cdot (b \cdot c) &= a \cdot (p \cdot 12 + r) = (a \cdot p) \cdot 12 + a \cdot r \\ &= (a \cdot p) \cdot 12 + q \cdot 12 + s = (a \cdot p + q) \cdot 12 + s \end{aligned}$$

which shows that s is the remainder of the whole number product $a \cdot (b \cdot c)$ divided by 12.

In the same fashion, divide $a \cdot b$ by 12 to get $a \cdot b = P \cdot 12 + R$, for some remainder $0 \leq R < 12$, so that $a \cdot b = R$ in $\mathbf{Z}_{12}$. Divide $R \cdot c$ by 12 to get

$R \cdot c = Q \cdot 12 + S$, for some remainder $0 \le S < 12$, so that $(a \cdot b) \cdot c = R \cdot c = S$ in $\mathbf{Z}_{12}$. In whole numbers,

$$\begin{aligned}(a \cdot b) \cdot c &= (P \cdot 12 + R) \cdot c = (P \cdot c \cdot 12 + R \cdot c \\ &= P \cdot c \cdot 12 + Q \cdot 12 + S = (P \cdot c + Q) \cdot 12 + S\end{aligned}$$

which shows that S is the remainder of the whole number product $(a \cdot b) \cdot c$ divided by 12. Since the associative property of whole number multiplication says that $a \cdot (b \cdot c) = (a \cdot b) \cdot c$, it follows that these two numbers have the same remainder when divided by 12; that is, $s = S$.

(4) The identity property of 1 follows from the commutative property, which says $a \cdot 1 = 1 \cdot a = a$, by definition of clock multiplication.

(5) The zero property of 0 follows from the definition of clock addition, because $a \cdot 0 = 0 + \cdots + 0$, a times, which is 0.

(6) Proof of the distributive property of multiplication over addition is similar to the proof of the associative property. It is left as a written exercise.

(7) The last property follows from the distributive property of multiplication over subtraction in the integers. Here is another proof, illustrated in a special case, using the properties already proved in clock arithmetic. By definition of subtraction, $3 \cdot (4 - 9) = 3 \cdot 4 - 3 \cdot 9$ in Z_{12}, if $3 \cdot 9 + 3 \cdot (4 - 9) = 3 \cdot 4$ in Z_{12}. We have in $\mathbf{Z}_{12}$

$$\begin{aligned}3 \cdot 9 + 3 \cdot (4 - 9) &= 3 \cdot (9 + (4 - 9)), \text{ dist. prop. over } + \\ &= 3 \cdot 4, \text{ definition of } 4 - 9\end{aligned}$$

$\square$

Definition 8.9. The *multiples* of a number a in $\mathbf{Z}_n$ are all numbers on this clock of the form $b \cdot a$, where b is any nonzero number on this clock.

In set terminology, the set of multiples of a in $\mathbf{Z}_n$ is

$$\{b \cdot a : b = 1, 2, \ldots, n - 1\}$$

Example 8.10. (1) On the 12-hour clock, the multiples of 4 are $4 = 1 \cdot 4$, $8 = 2 \cdot 4$ and $0 = 3 \cdot 4$, because any other multiple equals one of these three numbers. Notice that 0 is a multiple of 4, because $3 \cdot 4 = 0$, and that 1 is not a multiple of 4 on the 12-hour clock.

§8.1 Exercises

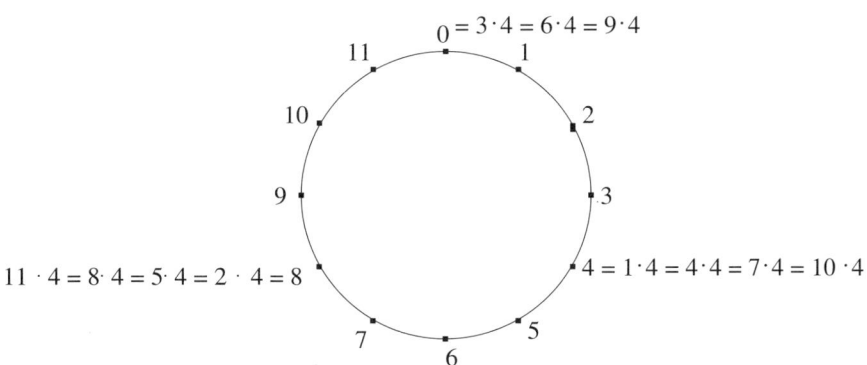

(2) On the 12-hour clock, the multiples of 5 are $5 = 1 \cdot 5$, $10 = 2 \cdot 5$, $3 = 3 \cdot 5$, $8 = 4 \cdot 5$, $1 = 5 \cdot 5$, $6 = 6 \cdot 5$, $11 = 7 \cdot 5$, $4 = 8 \cdot 5$, $9 = 9 \cdot 5$, $2 = 10 \cdot 5$, and $7 = 11 \cdot 5$. Every nonzero number on the 12-hour clock is a multiple of 5, and 0 is not a multiple of 5. In particular, 1 is a multiple of 5 on the 12-hour clock.

Oral exercise 8.11. (1) In $\mathbf{Z}_{12}$ find $5 \cdot 7$, $7 \cdot 7$, and $5 \cdot 5$.

(2) On the 12-hour clock, what are the multiples of 6? Of 7?

(3) On the 12-hour clock, which numbers have 0 as a multiple?

(4) What are the factors of 12?

(5) On the 12-hour clock, for which numbers does the set of multiples contain all nonzero numbers on the clock? Of these numbers, is zero ever a multiple?

(6) On the 12-hour clock, for which numbers does the set of multiples contain 1?

§8.1 Exercises

(1) Draw a 12-hour clock and show how to calculate $8 + 11$, $7 - 3$, $7 - 10$, $0 - 5$, $3 \cdot 7$, and $6 \cdot 10$.

(2) Draw a 13-hour clock and show the same calculations on it.

(3) In $\mathbf{Z}_{28}$ find $1 \div 11$. Use the result to find $5 \div 11$ on this clock. Use it to find $22 \div 11$ on this clock. Explain why $22 \div 11$ on this clock is the same as $22 \div 11$ in whole number arithmetic.

(4) Find a^4, which means $a \cdot a \cdot a \cdot a$, for each nonzero number a in $\mathbf{Z}_5$.

(5) For each number a in $\mathbf{Z}_{12}$ which is relatively prime to 12, find a^4.

(6) Find a^{10} for every nonzero number a in $\mathbf{Z}_{11}$.

(7) Prove that in any clock $\mathbf{Z}_n$ it is true that $(n-1)^2 = 1$. Hint: In $\mathbf{Z}_n$, you have that $n = 0$ and therefore, $n-1 = 0-1$. Find $(0-1)^2 = (0-1)\cdot(0-1)$ in $\mathbf{Z}_n$, using the distributive property over subtraction.

(8) Prove parts (3b) and (3c) of Theorem 8.5.

8.2. Clock Division

Division in $\mathbf{Z}_n$ is defined in the same way as partitive division of whole numbers is defined. As with whole numbers, some divisions do not exist. The factors of n play an essential role in determining which numbers of $\mathbf{Z}_n$ can divide every number in $\mathbf{Z}_n$.

Definition 8.12 (Clock Division). If a and b are any numbers on the n-hour clock, and $b \neq 0$, define $a \div b$ to be that number, if it exists on the n-hour clock, that multiplied by b gives a. In symbols, $a \div b$ is the number on the n-hour clock such that $a = b \cdot (a \div b)$ on the n-hour clock. If the number $a \div b$ exists on the n-hour clock, we call it the quotient of $a \div b$. If no such number exists on the n-hour clock, we say that $a \div b$ is not defined on the n-hour clock.

If $a \div b$ exists in $\mathbf{Z}_n$, then $(a \div b) \cdot b = b \cdot (a \div b) = a$, because multiplication in $\mathbf{Z}_n$ is commutative.

Example 8.13. (1) On the 12-hour clock, $5 \div 7 = 11$, because $7 \cdot 11 = 5$ on the 12-hour clock.

(2) On the 12-hour clock, $1 \div 5 = 5$, because $5 \cdot 5 = 1$.

(3) On the 12-hour clock, $5 \div 4$ is not defined, because if there were a number $5 \div 4$ on the 12-hour clock such that $4 \cdot (5 \div 4) = 5$, then 5 would be a multiple of 4. We observed above, however, that the only multiples of 4 are 4, 8, and 0. Since 5 is not a multiple of 4, we conclude that $5 \div 4$ is not defined on the 12-hour clock.

Theorem 8.14. *For a nonzero number b on the n-hour clock, the quotient $a \div b$ exists if and only if a is a multiple of b.*

Proof. If the quotient exists, then there is a number $a \div b$ on the n-hour clock such that $a = b \cdot (a \div b)$, which shows that a is a multiple of b on the n-hour clock. Conversely, if a is a multiple of b on the n-hour clock, then this means that there is a nonzero number q on this clock such that $a = b \cdot q$. Therefore, $a \div b = q$, by definition of division on the n-hour clock. □

As we shall see in the next subsection, in $\mathbf{Z}_n$, the quotient $a \div b$ exists for every number b in $\mathbf{Z}_n$ if and only if $1 \div b$ exists in $\mathbf{Z}_n$. The next theorem explains when $1 \div b$ exists in $\mathbf{Z}_n$.

8.2. Clock Division

Theorem 8.15. *For a nonzero number b on the n-hour clock, the quotient $1 \div b$ exists if and only if b and n are relatively prime, that is, gcf(b, n) = 1.*

Proof. We first prove that if $1 \div b$ exists on the n-hour clock, then b and n are relatively prime. That this quotient exists means there is a number $1 \div b$ on the n-hour clock such that $b \cdot (1 \div b) = 1$ on this clock. By the Remainder Property in Theorem 8.8, this means that the whole number division $(b \cdot (1 \div b)) \div n$ has remainder 1. If this whole number division has quotient q, then we have in whole number arithmetic, $b \cdot (1 \div b) = q \cdot n + 1$. This shows that any common divisor of b and n also divides $1 = b \cdot (1 \div b) - q \cdot n$, and therefore this common divisor must be 1.

We next prove the converse statement: If b and n are relatively prime, then $1 \div b$ exists on the n-hour clock. The proof is a careful application of Euclid's algorithm. We illustrate it on the 12-hour clock for $b = 7$. To find the gcf of 7 and 12, Euclid's algorithm goes as follows.

$$
\begin{array}{r}
1 \\
7 \overline{)\, 12} \\
\underline{7} \\
5
\end{array}
\quad \text{Divide smaller into larger}
$$

$$
\begin{array}{r}
1 \\
5 \overline{)\, 7} \\
\underline{5} \\
2
\end{array}
\quad \text{Divide remainder into preceding divisor}
$$

$$
\begin{array}{r}
2 \\
2 \overline{)\, 5} \\
\underline{4} \\
1
\end{array}
\quad \text{The gcf of 7 and 12}
$$

We rewrite this sequence of divisions as follows, in whole number arithmetic.

$$
\begin{aligned}
12 &= 1 \cdot 7 + 5 \\
7 &= 1 \cdot 5 + 2 \\
5 &= 2 \cdot 2 + 1
\end{aligned}
$$

Starting from the bottom equation and going up, solve for the remainders.

(8.1)
$$
\begin{aligned}
1 &= 5 - 2 \cdot 2 \\
2 &= 7 - 1 \cdot 5 \\
5 &= 12 - 1 \cdot 7
\end{aligned}
$$

These equations are also correct on the 12-hour clock, provided that in the third equation we replace 12 with 0. Remember that subtraction on the 12-hour clock is defined between any two numbers. On the 12-hour clock

we have
(8.2)
$$\begin{aligned}
1 &= 5 - 2 \cdot 2, & &\text{First of equations 8.1} \\
&= 5 - 2 \cdot (7 - 5), & &\text{Use second of equations 8.1, } 1 \cdot 5 = 5 \\
&= 5 - (2 \cdot 7 - 2 \cdot 5), & &\text{Distributive property over subtraction} \\
&= (5 - 2 \cdot 7) + 2 \cdot 5, & &\text{From (3b) of Theorem 8.5} \\
&= 2 \cdot 5 + (5 - 2 \cdot 7), & &\text{Commutative property of addition} \\
&= (2 \cdot 5 + 5) - 2 \cdot 7, & &\text{From (3a) of Theorem 8.5} \\
&= (2 \cdot 5 + 1 \cdot 5) - 2 \cdot 7, & &5 = 1 \cdot 5 \\
&= 3 \cdot 5 - 2 \cdot 7, & &\text{Distributive property, } 2 + 1 = 3 \\
&= 3 \cdot (0 - 1 \cdot 7) - 2 \cdot 7, & &\text{From third of equations 8.1} \\
&= 3 \cdot (0 - 7) - 2 \cdot 7, & &1 \cdot 7 = 7 \\
&= (3 \cdot 0 - 3 \cdot 7) - 2 \cdot 7, & &\text{Distributive property over subtraction} \\
&= (0 - 3 \cdot 7) - 2 \cdot 7, & &3 \cdot 0 = 0 \\
&= 0 - (3 \cdot 7 + 2 \cdot 7), & &\text{From (3c) of Theorem 8.5} \\
&= 0 - 5 \cdot 7, & &\text{Distributive property and } 3 + 2 = 5 \\
&= 0 \cdot 7 - 5 \cdot 7, & &0 = 0 \cdot 7 \\
&= (0 - 5) \cdot 7, & &\text{Distributive property over subtraction}
\end{aligned}$$

On the 12-hour clock, $0 - 5 = 7$, since $5 + 7 = 0$. Therefore, we conclude that $7 \cdot 7 = 1$ on the 12-hour clock, which means that $1 \div 7 = 7$ on this clock. □

Definition 8.16. For a number b on the n-hour clock, relatively prime to n, its *reciprocal* is the number $1 \div b$ on the n-hour clock. It is that number on this clock whose product with b is 1. If a number in $\mathbf{Z}_n$ has a reciprocal, it is called invertible.

Example 8.17. In $\mathbf{Z}_{28}$ find $1 \div 5$, the reciprocal of 5. We do this in a systematic way by using Euclid's Algorithm as in the proof of Theorem 8.15. In whole numbers, find $28 \div 5$ to get $28 = 5 \cdot 5 + 3$, then find $5 \div 3$ (previous divisor divided by remainder) to get $5 = 3 \cdot 1 + 2$, then find $3 \div 2$ to get $3 = 2 \cdot 1 + 1$, and stop because the remainder is 1.

Start with the last equation, solve for 1, use the second equation to solve for 2, and then use the first equation to solve for 3, to obtain in $\mathbf{Z}_{28}$:

$$\begin{aligned}
1 &= 3 - 2, \text{ from } 3 = 2 \cdot 1 + 1 \\
&= 3 - (5 - 3), \text{ from } 5 = 3 \cdot 1 + 2 \\
&= 2 \cdot 3 - 5, \text{ clock arithmetic properties} \\
&= 2 \cdot (0 - 5 \cdot 5) - 5, \text{ from } 0 = 5 \cdot 5 + 3 \\
&= (0 - 11) \cdot 5, \text{ clock arithmetic properties}
\end{aligned}$$

from which we conclude that $1 \div 5 = 0 - 11 = 28 - 11 = 17$ in $\mathbf{Z}_{28}$.

8.2. Clock Division

Oral exercise 8.18. On the 12-hour clock, which numbers have reciprocals? Find their reciprocals. One way to do this is to look at the multiples of the number.

Oral exercise 8.19. Explain why equations (8.2) are not valid in whole number arithmetic. How can we make sense out of something like $0 - 5$ in whole number arithmetic? Will $0 - 5$ be a whole number?

Oral exercise 8.20. On the 28-hour clock, which numbers have a reciprocal?

Oral exercise 8.21. Explain why $8 \div 4 = 2$ on the 12-hour clock, also on the 28-hour clock, in fact, on any clock with more than 8 hours. Hint: use the definition of division.

Oral exercise 8.22. For numbers a and b on the n-hour clock, explain why $a \div b$ in $\mathbf{Z}_n$ is the whole number quotient whenever it is defined.

8.2.1. Quotients from Reciprocals.

Theorem 8.23. *For a number b on the n-hour clock, relatively prime to n, the quotient $a \div b = a \cdot (1 \div b)$ for any number a on this clock.*

Proof. By Theorem 8.15 we know that $1 \div b$ exists in $\mathbf{Z}_n$ and therefore the number $q = a \cdot (1 \div b)$ exists on the n-hour clock. Then, on this clock

$$\begin{aligned} q \cdot b &= (a \cdot (1 \div b)) \cdot b \\ &= a \cdot ((1 \div b) \cdot b), \text{ Associative property of multiplication} \\ &= a \cdot 1, \text{ Definition of } 1 \div b \\ &= a, \text{ Identity property of 1} \end{aligned}$$

which means that $q = a \div b$ on this clock. $\square$

Example 8.24. (1) On the 12-hour clock, $1 \div 7 = 7$. Therefore, $2 \div 7 = 2 \cdot 7 = 2$ and $3 \div 7 = 3 \cdot 7 = 9$, on $\mathbf{Z}_{12}$.

(2) On the 28-hour clock, $1 \div 3 = 19$. Therefore, $2 \div 3 = 2 \cdot 19 = 10$ and $5 \div 3 = 5 \cdot 19 = 11$ and $12 \div 3 = 4$, on $\mathbf{Z}_{28}$.

Oral exercise 8.25. (1) On $\mathbf{Z}_{12}$ find $10 \div 7$ and $11 \div 7$.

(2) On $\mathbf{Z}_{28}$ find $10 \div 3$ and $20 \div 3$. Is $20 \div 3 = 2 \cdot (10 \div 3)$?

8.2.2. Summary.

Theorem 8.26. *The set of nonzero numbers on the n-hour clock separate into two disjoint sets. One set consists of all numbers for which 0 is a multiple. The other set consists of all numbers for which 1 is a multiple. The two sets are characterized as follows.*

(1) 1 is a multiple of $b \Leftrightarrow b$ has a reciprocal $\Leftrightarrow \gcd(b, n) = 1$.

(2) 0 is a multiple of $a \Leftrightarrow a$ has no reciprocal $\Leftrightarrow \gcd(a, n) > 1$.

Proof. (1) The first equivalence follows from Theorem 8.14. The second equivalence follows from Theorem 8.15.

(2) The second equivalence is the contrapositive of the second equivalence in (1), and therefore has already been proved. It remains to prove the first equivalence in (2). Since the second equivalence is true, the first equivalence will be true if we can prove the equivalence

$$\gcf(a,n) > 1 \Leftrightarrow a \text{ has } 0 \text{ as a multiple}$$

Let d be the $\gcf(a,n)$, so that

$$a = b \cdot d \text{ and } n = d \cdot m$$

for some whole numbers b and m.

If $d > 1$, then $b > 1$ and $n > m > 1$ and so m is a number in $\mathbf{Z}_n$ such that

$$a \cdot m = b \cdot d \cdot m = b \cdot n = 0$$

in $\mathbf{Z}_n$. Therefore, $d > 1$ implies that 0 is a multiple of a.

It remains to prove that if 0 is a multiple of a in $\mathbf{Z}_n$, then $d > 1$. But if 0 is a multiple of a in $\mathbf{Z}_n$, this means that there is a nonzero number c in $\mathbf{Z}_n$ such that $0 = a \cdot c$ in $\mathbf{Z}_n$, for some nonzero number c in $\mathbf{Z}_n$. That is, $0 < c < n$, and $a \cdot c$ is a multiple of n,

(8.3) $$a \cdot c = n \cdot m$$

for some natural number m. If $d = 1$, then a and n have no prime factor in common, which means that every prime factor of a must also be a prime factor of m. Therefore, a must be a factor of m, so $m = a \cdot q$ for some natural number q. Dividing both sides of equation (8.3) by a, we get

$$c = n \cdot q$$

which is impossible for a number $c < n$. Therefore, our assumption that $d = 1$ cannot be true, which means that $d > 1$ as was to be proved. □

Oral exercise 8.27. (1) In the 14-hour clock, list the numbers for which 1 is a multiple. List the numbers for which 0 is a multiple. In which of these two lists do all the numbers have reciprocals?

(2) Same question for the 12-hour clock.

(3) Same question for the 7-hour clock.

§8.2 Exercises

(1) On the 28-hour clock find the numbers for which 0 is a multiple. Find the numbers for which 1 is a multiple.

(2) On the 14-hour clock find the reciprocal of 5. Find $8 \div 5$ on this clock.

(3) The only numbers on the 12-hour clock which are relatively prime to 12 are 1, 5, 7, and 11. That is, there are four numbers on the 12-hour clock relatively prime to 12. Find $a^4 = a \cdot a \cdot a \cdot a$ on the 12-hour clock, for each of $a = 5, 7$, and 11. Since $a^4 = a \cdot a^3$, explain why $a^3 = 1 \div a$ for each of these numbers.

(4) How many numbers on the 14-hour clock are relatively prime to 14? Verify that any such number raised to the sixth power is 1 on the 14-hour clock. Explain how to find the reciprocals of these numbers on the 14-hour clock.

(5) Go through the proof of Theorem 8.15 for the case of 5 on the 12-hour clock. Do the same for the case of 2 on the 5-hour clock.

(6) In $\mathbf{Z}_{29}$, find $1 \div 10$. Use this to find in $\mathbf{Z}_{29}$: (a) $5 \div 10$. (b) $20 \div 10$. (c) $21 \div 10$. *Ans:* (a) 15.

(7) Find $1 \div 83$ in $\mathbf{Z}_{1008}$. Use this to find $10 \div 83$ in $\mathbf{Z}_{1008}$. Hint: Use the method of Euclid's Algorithm given in the proof of Theorem 8.15. *Ans.* 923

Chapter 9

RSA Encryption

The following Associated Press article appeared in the St. Louis Post-Dispatch on Saturday, August 28, 1999.[1]

> Amsterdam, Netherlands—A group of scientists claimed Friday to have broken an international security code used to protect millions of daily Internet transactions, exposing a potentially serious security failure in electronic commerce.
>
> Researchers working for the National Research Institute for Mathematics and Computer Science in Amsterdam said consumers and some businesses could fall victim to computer hackers if they get their hands on the right tools.
>
> But not every computer whiz has access to the equipment, worth several million dollars, and no related Internet crimes have yet been uncovered, the experts said.
>
> The scientists used a Cray 900-16 supercomputer, 300 personal computers and specially designed number crunching software to break the so-called RSA-155 code—the backbone of encryption codes designed to protect e-mail messages and credit card transactions.
>
> "Your everyday hacker won't be able to do this," said project director Herman te Reile. "You have to have extensive capacity, the money and the know-how, but we did it."

---

[1] Reprinted by permission

Te Reile said his international team of researchers, assisted by Microsoft and Sun Microsystems and professionals from Britain, Canada and Australia, took six weeks to crack the security system.

The codes are used to protect information transmitted over the Internet—such as credit card numbers, stock transactions or private e-mail messages. RSA codes—and frequently the same RSA-155 version the Amsterdam group managed to decipher—protect "many billions of dollars" worth of daily transactions at banks, stock exchanges and online retailers, the institute said.

This section presents some features of RSA encryption, a method named after the inventors, three computer scientists at MIT, Ronald L. Rivest, Adi Shamir, and Leonard Adleman, who published their method in April, 1977, in [**RSA77**]. The method is an application of clock arithmetic, in particular, of the fact that multiplication on a clock results in random permutation of the numbers on the clock.

For any positive integer n we let $\mathbf{Z}_n$ denote the n-hour clock, $\mathbf{Z}_n = \{0, 1, \ldots, n-1\}$. Remember Theorem 8.15 says that a number m in $\mathbf{Z}_n$ has a reciprocal if and only if m is relatively prime to n; that is, $\gcd(m, n) = 1$. For example, only the numbers 1, 5, 7, and 11 have a reciprocal in the 12-hour clock, $\mathbf{Z}_{12}$: $1 \div 5 = 5$, because $5 \cdot 5 = 1$, and similarly $1 \div 7 = 7$ and $1 \div 11 = 11$, in the 12-hour clock. If n is a prime number, then every nonzero number in $\mathbf{Z}_n$ has a reciprocal.

9.1. A Preliminary Example

The idea of RSA encryption is to convert letters to numbers, regard the numbers as being numbers on a clock, and then raise the numbers to a given power on this clock. The effect is a random permutation of the numbers on the clock. The decoding is accomplished by taking the corresponding root. The ultimate issue will be how difficult it is to find the decoding power, when one knows the number of hours in the clock and the coding power.

In all examples the English text is converted into a sequence of two digit numbers by the scheme

A	B	C	D	E	F	G	H	I	J	K	L	M
01	02	03	04	05	06	07	08	09	10	11	12	13
N	O	P	Q	R	S	T	U	V	W	X	Y	Z
14	15	16	17	18	19	20	21	22	23	24	25	26

9.1. A Preliminary Example

In addition, a space between words will be denoted with the number 27. Let us convert the message

(9.1) \qquad\qquad\qquad KEY UNDER MAT

into numbers. It becomes the sequence of two digit numbers

(9.2) $\qquad m = [11\ 05\ 25\ 27\ 21\ 14\ 04\ 05\ 18\ 27\ 13\ 01\ 20]$

Next we decide on what is called the *block size* for the numbers in m. The block size of m is presently $b = 2$. The block size b can be any whole number greater than 1. To put m into block size b, start from the left and gather together each successive b digits in m. For example, m put into block size $b = 3$ is

(9.3) $\qquad m = [110\ 525\ 272\ 114\ 040\ 518\ 271\ 301\ 20]$

and m put into block size $b = 7$ is

(9.4) $\qquad m = [1105252\ 7211404\ 0518271\ 30120]$

It sometimes happens, as in these examples, that the last number has fewer than b digits. Numbers in m other than the last can have effectively fewer than b digits because the first digit can be 0. A leading zero must be retained when m is reblocked to block length 2 and converted back to letters.

The next step is to choose a clock $\mathbf{Z}_n$, whose number of hours n is greater than any number in m. For our introductory examples, we want n to be a prime number, but in the real RSA coding n must be the product of two distinct primes. The size of n depends on the block size chosen. In our present example, with block size $b = 2$, the largest number in m is 27 so the smallest prime number we could use is $n = 29$. For block size $b = 3$, we need $n > 525$, and for block size $b = 7$, we need $n > 7211404$, as can be seen from equations (9.3) and (9.4), respectively.

To keep the numbers simple in our first example, we shall use block size $b = 2$ and $n = 29$. Regard the numbers in m of equation (9.2) to be numbers on the 29 hour clock, $\mathbf{Z}_{29}$. We select $c = 3$ as the *coding power*. This means that to encode the message m we raise every number in m to the third power, using multiplication in this clock. That is, $m^3 = (m \cdot m) \cdot m$, where the two multiplications are in this clock and each such multiplication can be carried out using the remainder property for clock multiplication.

Let M denote the resulting sequence of numbers. That is,

(9.5) $\begin{aligned} M &= [11^3\ 05^3\ 25^3\ 27^3\ 21^3\ 14^3\ 04^3\ 05^3\ 18^3\ 27^3\ 13^3\ 01^3\ 20^3] \\ &= [26\ 09\ 23\ 21\ 10\ 18\ 06\ 09\ 03\ 21\ 22\ 01\ 25] \end{aligned}$

For example, $11^3 = 11^2 \cdot 11$ and $11^2 = 121 = 5$ in $\mathbf{Z}_{29}$, so $11^2 \cdot 11 = 5 \cdot 11 = 55 = 26$ in $\mathbf{Z}_{29}$. The sequence of numbers M in $\mathbf{Z}_{29}$ constitute the RSA encryption of the message in equation (9.1).

The encoding procedure used the three numbers: $n = 29$, the number of hours in the clock; $b = 2$, the block size; and $c = 3$, the coding power.

Definition 9.1. The set of numbers n, b, c is called the *key* of the RSA code.

In the remaining sections of this chapter we shall address the following questions.

(1) How does one decode an encoded message?

(2) For a given clock $\mathbf{Z}_n$, which coding powers c allow decoding?

(3) Given the key, what additional knowledge is needed to determine how to decode?

(4) Does the number of hours, n, have to be a prime number?

(5) In a given clock $\mathbf{Z}_n$, is there a shortcut to raising numbers to a large power? For example, computation of 26^{19} in $\mathbf{Z}_{29}$ requires 18 multiplications in this clock. Can it be done with fewer multiplications?

Oral exercise 9.2. (1) Convert m to block size $b = 5$.

(2) Describe the process of using the RSA code with key $31, 2, 7$ to encode the message in equation (9.1).

9.2. Decoding

Definition 9.3. For an RSA code with key n, b, c, its *decoding power* d is a whole number with the property that $(m^c)^d = m$, for any number m in $\mathbf{Z}_n$.

In other words, the power d reverses the effect of raising a number to the c power. The power d is like taking the cth root in the given clock. For this reason, we also call d the power taking the cth root in $\mathbf{Z}_n$.

In the example of the preceding section, for which $n = 29$, $b = 2$, and $c = 3$, the decoding power is $d = 19$. This means that $(m^3)^{19} = m^{3 \cdot 19} = m$, for every number m in $\mathbf{Z}_{29}$. How do we know this? We could try to verify this fact for each number in $\mathbf{Z}_{29}$. For example, 11^{57} requires 56 products in this clock. Let us consider how $d = 19$ was found and how we can know, even without carrying out these calculations for all numbers in $\mathbf{Z}_{29}$, that it has the desired property.

The answer involves an application of a theorem discovered by Pierre Fermat in 1640. This is the same Fermat whose famous Last Theorem was finally proved in 1994 by Andrew Wiles at Princeton University. A wonderful brief biography of Fermat can be found at the web site

www-groups.dcs.st-and.ac.uk/~history/Mathematicians/Fermat.html

9.2. Decoding

Theorem 9.4 (Fermat's Theorem). *If p is a prime number, then $m^{p-1} = 1$ for every nonzero number m in $\mathbf{Z}_p$.*

The proof of this theorem is elementary, not beyond the level of this book, but it does go beyond the scope of the book. We will accept its truth on the basis of some verifications, and on the basis of the proof which appears in every book on elementary number theory. Observe that the conclusion of the theorem is not true if p is not a prime number. For example, in $\mathbf{Z}_6$ we have $2^5 = 2$, $3^5 = 3$, $4^5 = 2^2 = 4$, and $5^5 = 5$. The correct theorem for clocks $\mathbf{Z}_n$, when n is not a prime number, is called Euler's Theorem, which will be stated in a later section, as it is needed for RSA encryption.

9.2.1. Fermat's Theorem in $\mathbf{Z}_{11}$. We verify Fermat's Theorem in $\mathbf{Z}_{11}$ by computing the tenth power of every nonzero number in this clock.

$$(9.6) \quad \begin{aligned} 2^{10} &= (2^5)^2 = 10^2 = 1, \text{ since } 2^5 = 10 \\ 3^{10} &= (3^5)^2 = 1^2 = 1, \text{ since } 3^5 = 3^{2 \cdot 2 + 1} = 9^2 \cdot 3 = 4 \cdot 3 = 1 \\ 4^{10} &= (2^2)^{10} = (2^{10})^2 = 1^2 = 1 \\ 5^{10} &= (5^2)^5 = 3^5 = 1, \text{ since } 5^2 = 3 \\ 6^{10} &= 2^{10} \cdot 3^{10} = 1 \cdot 1 = 1 \\ 7^{10} &= (7^2)^5 = 5^5 = (5^2)^2 \cdot 5 = 3^2 \cdot 5 = 9 \cdot 5 = 1 \\ 8^{10} &= (2^3)^{10} = (2^{10})^3 = 1^3 = 1 \\ 9^{10} &= (3^2)^{10} = (3^{10})^2 = 1^2 = 1 \\ 10^{10} &= (2 \cdot 5)^{10} = 2^{10} \cdot 5^{10} = 1 \cdot 1 = 1 \end{aligned}$$

Oral exercise 9.5. (1) Verify Fermat's Theorem in $\mathbf{Z}_5$, $\mathbf{Z}_3$, and $\mathbf{Z}_7$.

(2) Find a nonzero number m in $\mathbf{Z}_{10}$ for which $m^9 \neq 1$. Hint: What is 9^2 in this clock? From that, what can you say about 9^9? How do you explain why the conclusion of Fermat's Theorem isn't true for $\mathbf{Z}_{10}$?

9.2.2. Allowable Coding Powers.

Theorem 9.6. *Consider a clock $\mathbf{Z}_n$, where n is a prime number. If the whole number $c < n$ is relatively prime to $n - 1$, then c has the decoding power $d = 1 \div c$ in $\mathbf{Z}_{n-1}$.*

Proof. Begin by observing what this theorem says about $\mathbf{Z}_{29}$ for the case of $c = 3$, which is relatively prime to $29 - 1 = 28$, as can be seen from the prime factorization $28 = 2^2 \cdot 7$. According to the theorem, $c = 3$ has the decoding power $d = 1 \div 3$, the reciprocal of 3 in $\mathbf{Z}_{28}$, which exists because $\gcd(3, 28) = 1$, by Theorem 8.15. Then $3 \cdot d = 1$ in $\mathbf{Z}_{28}$, that is, $3 \cdot d = 28 \cdot q + 1$ for some whole number q. Therefore, for any nonzero number m in $\mathbf{Z}_{29}$,

$$(9.7) \quad (m^3)^d = m^{3 \cdot d} = m^{28 \cdot q + 1} = (m^{28})^q \cdot m^1 = 1^q \cdot m = m$$

because $m^{28} = 1$ by Fermat's Theorem.

Here is the general proof. If $c < n$ is relatively prime to $n - 1$, then c has a reciprocal, $d = 1 \div c$, on the clock $\mathbf{Z}_{n-1}$. That means that $c \cdot d = 1$ on this clock. By the remainder property for computing products on $\mathbf{Z}_{n-1}$, this means that $c \cdot d = (n-1) \cdot q + 1$, for some whole number q. If m is any nonzero number in $\mathbf{Z}_n$, then

(9.8) $\quad (m^c)^d = m^{c \cdot d} = m^{(n-1) \cdot q + 1} = (m^{n-1})^q \cdot m^1 = 1^q \cdot m = m$

since, by Fermat's Theorem, $m^{n-1} = 1$ for any nonzero number in $\mathbf{Z}_n$. $\square$

9.2.3. Example of $\mathbf{Z}_{29}$. Since 29 is a prime number, Fermat's theorem says that $m^{28} = 1$ for every nonzero number in $\mathbf{Z}_{29}$. For which powers $c < 29$ is there a decoding power d? According to Theorem 9.6, the answer is those numbers c which are relatively prime to 28. The numbers c in $\mathbf{Z}_{28}$ relatively prime to 28 and greater than 1 are 3, 5, 9, 11, 13, 15, 17, 19, 23, 25, and 27.

Given one of these coding powers c, its decoding power is $d = 1 \div c$ in $\mathbf{Z}_{28}$, according to Theorem 9.6. For example, if $c = 3$, then $d = 1 \div 3$ in $\mathbf{Z}_{28}$.

The proof of Theorem 8.15 showed us how to use Euclid's Algorithm to find $1 \div 3$ in $\mathbf{Z}_{28}$. In whole numbers, the long division $28 \div 3$ gives $28 = 9 \cdot 3 + 1$. In $\mathbf{Z}_{28}$, this last equation becomes $0 = 9 \cdot 3 + 1$, from which we get $(0 - 9) \cdot 3 = 1$ in $\mathbf{Z}_{28}$. Therefore,

$$d = 1 \div 3 = 0 - 9 = 28 - 9 = 19$$

is the decoding power for $c = 3$. Indeed, for any number m in $\mathbf{Z}_{29}$, using rules of exponents, we have

$$(m^3)^{19} = m^{3 \cdot 19} = m^{28 \cdot 2 + 1} = (m^{28})^2 \cdot m^1 = 1 \cdot m = m$$

because $m^{28} = 1$ for every nonzero number m in $\mathbf{Z}_{29}$, by Fermat's Theorem. In the clock $\mathbf{Z}_{29}$, raising numbers to the nineteenth power is analogous to taking the cube root, because it *undoes* the effect of taking the third power.

Oral exercise 9.7. (1) In $\mathbf{Z}_{29}$, what power d corresponds to taking fifth roots? Ninth roots? Eleventh roots? Nineteenth roots? *Ans.* 17, 25, 23, and 3, respectively.

(2) Explain why the decoding power d corresponding to an allowable coding power c in $\mathbf{Z}_{29}$ must be itself an allowable coding power in this clock.

(3) In $\mathbf{Z}_{11}$, which power corresponds to taking the seventh root?

(4) In $\mathbf{Z}_{11}$, which power corresponds to taking the ninth root?

(5) In $\mathbf{Z}_5$, which powers (other than 1) have a corresponding root? What is the power giving the root?

9.2. Decoding

(6) In $\mathbf{Z}_{19}$, which powers (other than 1) have a corresponding root? *Ans.* Numbers relatively prime to $18 = 2 \cdot 3^2$.

(7) In $\mathbf{Z}_{19}$, which power d gives fifth roots? *Ans.* $d = 1 \div 5$ in $\mathbf{Z}_{18}$. Use the method of the proof of Theorem 8.15 to find d.

9.2.4. Decoding the Key n, b, c **when** n **is Prime.** When n is prime, the allowable coding powers are the numbers c in $\mathbf{Z}_{n-1}$ which are relatively prime to $n - 1$. The decoding power of such a c is $d = 1 \div c$ in $\mathbf{Z}_{n-1}$.

Example 9.8. The encoded message M in equation (9.5) was encoded with the key $n = 29$, $b = 2$, $c = 3$. Therefore, the decoding power for $c = 3$ is $d = 1 \div 3 = 19$ in $\mathbf{Z}_{28}$. Decode M by raising each number in it to the nineteenth power in $\mathbf{Z}_{29}$. For example, the first number in M is 26 and the calculation of 26^{19} requires 18 multiplications in this clock. This appears to be too tedious to bother with, even when we see how it goes in principle. The first few multiplications in the 29 hour clock are $26^2 = 676 = 9$, $26^3 = 9 \cdot 26 = 234 = 2$, $26^4 = 2 \cdot 26 = 52 = 23$, etc.

In Section 9.4 below we'll see how to reduce this calculation to only six multiplications in $\mathbf{Z}_{29}$. Unbelievable as it sounds, this reduction is related to the base 2 numeral for 19.

Example 9.9. Decode the message $M = [309\ 675\ 303\ 659]$, which has been encoded with the RSA key $n = 1009$, $c = 605$, $b = 3$. *Solution.* The decoding power for this key is $d = 1 \div 605$ in $\mathbf{Z}_{1008}$, which we find using the proof of Theorem 8.15. Dividing 1008 by 605 and so on until we have remainder 1, we have

$$\begin{aligned} 1008 &= 605 \cdot 1 + 403 \\ 605 &= 403 \cdot 1 + 202 \\ 403 &= 202 \cdot 1 + 201 \\ 202 &= 201 \cdot 1 + 1 \end{aligned}$$

Regarding these four equations as being in the clock $\mathbf{Z}_{1008}$, we start with the last equation and work up to find

$$\begin{aligned} 1 &= 202 - 201, \text{ fourth eqn.} \\ &= 202 - (403 - 202) = 2 \cdot 202 - 403, \text{ third eqn.} \\ &= 2 \cdot (605 - 403) - 403 = 2 \cdot 605 - 3 \cdot 403, \text{ second eqn.} \\ &= 2 \cdot 605 - 3 \cdot (0 - 605) = 5 \cdot 605, \text{ first eqn.} \end{aligned}$$

from which we conclude that $d = 1 \div 605 = 5$ in $\mathbf{Z}_{1008}$.

If M is a number in $\mathbf{Z}_{1009}$, then $M^5 = (((M \cdot M) \cdot M) \cdot M) \cdot M$ requires four multiplications. For example, using the remainder property for computing products in $\mathbf{Z}_{1009}$, we find 309^5 in $\mathbf{Z}_{1009}$ as follows, where the whole number product is designated with the equal sign and the remainder after dividing by 1009, the product in this clock, is designated with the $\equiv$ sign.

$$309 \cdot 309 = 95481 \equiv 635$$
$$635 \cdot 309 = 196215 \equiv 469$$
$$469 \cdot 309 = 144921 \equiv 634$$
$$634 \cdot 309 = 195906 \equiv 160$$

These calculations can be done easily on the Explorer Plus calculator, which will compute the remainders. Raising each of the four numbers in M to the fifth power in $\mathbf{Z}_{1009}$ we obtain $m = [160\ 501\ 30\ 5]$. In reblocking this to block size $b = 2$, we must write the third number, 30, as 030, but no change is made to the last number in m. We obtain $m = [16\ 05\ 01\ 03\ 05]$, which converted to letters is the message PEACE.

§9.2 Exercises

(1) Verify Fermat's Theorem in $\mathbf{Z}_5$. That is, verify that in $\mathbf{Z}_5$ every nonzero number raised to the fourth power is 1, namely, $2^4 = 1$, $3^4 = 1$, $4^4 = 1$, and, of course, $1^4 = 1$.

(2) For which powers c is there a corresponding root in $\mathbf{Z}_5$? Find the corresponding power d in each case. Ans. $c = 3$, $d = 1 \div 3$ in $\mathbf{Z}_4$.

(3) Verify that $m^6 = 1$ for every nonzero number m in $\mathbf{Z}_7$.

(4) Which powers c have a decoding power d in $\mathbf{Z}_7$? Find the d for each c. Ans. $c = 5$, $d = 1 \div 5 = 5$ in $\mathbf{Z}_6$.

(5) Find a nonzero number m in $\mathbf{Z}_{100}$ for which $m^{99} \neq 1$. Why isn't the conclusion of Fermat's Theorem true for $\mathbf{Z}_{100}$?

(6) Prove that in any clock $\mathbf{Z}_n$, it is true that $(n-1)^2 = 1$. Hint: In $\mathbf{Z}_n$ you have $n = 0$ and therefore $n - 1 = 0 - 1$. Square this last expression in $\mathbf{Z}_n$ using the distributive property over subtraction.

(7) Find the decoding power d for the key $n = 29$, $b = 2$, $c = 15$.

(8) Find the decoding power d for the key $n = 1009$, $b = 3$, $c = 11$.

(9) Decode the message $M = [611\ 155\ 411]$, which has been encoded with the RSA key $n = 1009$, $c = 605$, $b = 3$.

9.3. Nonprime Number of Hours

We have defined the key of an RSA code to be a triple of numbers (n, c, b). The natural number n gives the number of hours on the clock in which we work, the encoding power is a nonzero element c in $\mathbf{Z}_n$, and the block size is a natural number $b > 1$. In our examples, we took n to be a prime number, in which case c had to be relatively prime to $n - 1$ and the block size cannot

9.3. Nonprime Number of Hours

exceed the number of digits in n. In this case, knowing n and c we can quite easily calculate the decoding power d, because we know that $d = 1 \div c$ in $\mathbf{Z}_{n-1}$.

In order for an agent to send me an encoded message, I would have to tell her n, the encoding power c, and the block size b. If n were a prime number, then the agent, or anyone else with knowledge of the key, could figure out the decoding power d. To keep the decoding power secret in this case, I would have to use a secret courier to send n, c, and b to my agent. Public key encryption seeks to eliminate the use of secret courier by simply making the key public knowledge. For example, I could post it daily on my web page. The point is that d cannot be easily found from knowledge of the key when n is not a prime number. Let us see how RSA encryption works when n is not prime.

9.3.1. Example with $n = 1147$. The publicly known key is $n = 1147$, $c = 7$, and $b = 2$. We receive the encoded message

$$(9.9) \qquad M = [128\ 583\ 1100\ 675\ 1\ 763\ 326]$$

which is intercepted by an enemy agent whose knowledge of RSA encryption is limited to the preceding sections of this chapter. He goes to work to find what he thinks is the decoding power $d = 1 \div 7$ in $\mathbf{Z}_{1146}$, which exists because $\gcd(7, 1146) = 1$. Using Euclid's Algorithm method of the proof of Theorem 8.15, he finds that $d = 655$. To his consternation, however, he calculates that $128^{655} = 311$ in $\mathbf{Z}_{1147}$, which he knows cannot be right because the block size being 2 requires that this be a two digit number. What has gone wrong in this agent's analysis? His problem comes from the fact that the number n is not prime.

A systematic trial of divisibility of 1147 by primes, starting with 2, soon finds that 31 divides 1147 and that $1147 = 31 \cdot 37$ is the prime factorization of 1147. Because 1147 is not prime, the hypotheses of Fermat's Theorem do not hold for 1147 and thus we should not expect the conclusion of Fermat's Theorem to hold. Namely, we cannot expect that the decoding power d for $c = 7$ is given by the reciprocal of 7 in $\mathbf{Z}_{1146}$, because that result is based on Fermat's Theorem. What is needed is a generalization of Fermat's Theorem found about 100 years later by Leonhard Euler. For a biography of Euler, see the web site

`www-groups.dcs.st-and.ac.uk/~history/Mathematicians/Euler.html`

The statement of his theorem requires the idea of his totient function.

Definition 9.10. For a positive integer n, Euler's *totient function*, $\varphi(n)$ (read "Fye of n"), is the number of positive integers less than n which are relatively prime to n.

For a prime number p, the value of $\varphi(p)$ is $p-1$, because every integer from 1 through $p-1$ is relatively prime to p. On the other hand, $\varphi(4) = 2$, since the numbers less than 4 and relatively prime to 4 are 1 and 3. Also, $\varphi(6) = 2$, since the numbers less than 6 and relatively prime to 6 are 1 and 5.

Oral exercise 9.11. (1) Find $\varphi(10)$.

(2) Find $\varphi(35)$.

The following theorem states only two special cases of a more general result. These cases are all we need for our discussion of RSA encryption.

Theorem 9.12. *If p is a prime number, then $\varphi(p) = p - 1$. If p and q are distinct prime numbers, then $\varphi(p \cdot q) = (p-1) \cdot (q-1)$.*

Proof. The proof for the case of a prime number was given above. The proof for the case of $p \cdot q$, for distinct primes p and q can be understood from the proof of the case $5 \cdot 7 = 35$. A number less than 35 is not relatively prime to 35 precisely when either 5 or 7 is one of its prime factors. Therefore, the numbers less than 35 which are not relatively prime to 35 are $5 \cdot 1, 5 \cdot 2, 5 \cdot 3, 5 \cdot 4, 5 \cdot 5, 5 \cdot 6$ and $1 \cdot 7, 2 \cdot 7, 3 \cdot 7, 4 \cdot 7$, for a total of $6 + 4 = (7-1) + (5-1)$ numbers less than 35 which are not relatively prime to 35. The rest of the numbers less than 35 must be relatively prime to 35, and there are $(35-1)-(6+4) = (35-1)-((7-1)+(5-1)) = 35-7-5+1 = (7-1)\cdot(5-1)$ of these. □

Theorem 9.13 (Euler's Theorem). *If m is a number in $\mathbf{Z}_n$ relatively prime to n, then $m^\varphi = 1$ in $\mathbf{Z}_n$, where $\varphi = \varphi(n)$.*

Proof. The proof of this theorem is beyond the scope of this course, although not much beyond the level of this course. It can be found in any elementary number theory book. Observe that if n is a prime number, then $\varphi = n - 1$, and Euler's Theorem becomes Fermat's Theorem. □

Example 9.14. For $n = 35 = 5 \cdot 7$ we have $\varphi = (5-1) \cdot (7-1) = 24$. Thus $m^{24} = 1$ for any invertible number m in $\mathbf{Z}_{35}$. For example, verify that $8^{24} = (8^4)^6 = 1^6 = 1$ and $12^{24} = (12^{12})^2 = 1^2 = 1$, etc.

Back to the initial example above, an enemy agent with knowledge of Euler's Theorem would first factor $1147 = 31 \cdot 37$ to find that $\varphi = 30 \cdot 36 = 1080 = 2^3 \cdot 3^3 \cdot 5$. The encoding power $c = 7$ is relatively prime to φ and the agent can proceed as in the earlier examples to find the decoding power $d = 1 \div 7$ in $\mathbf{Z}_\varphi = \mathbf{Z}_{1080}$. Using Euclid's Algorithm as in proof of Theorem 8.15, he finds that $d = 463$. The agent decodes the message M by calculating M^{463} in $\mathbf{Z}_{1147}$ for each number in M. How he actually carries out such a calculation is discussed in the next section.

9.3.2. Summary. In order to find the decoding power d of an RSA encryption code with key n, b, c, one must know the value of Euler's totient function $\varphi(n)$. The only way to find $\varphi(n)$ is from the prime factorization of n. Security of a public key lies in the difficulty in finding the prime factorization of n. Even with very modern computers, this factoring problem increases dramatically as the number of digits in n increases.

In 1977 Rivest, Shamir, and Adelman estimated the time needed to factor a 50 digit number was about 3.9 hours, for a 75-digit number about 104 days, for a 100-digit number about 74 years, and for a 200-digit number over a billion years. The Post-Dispatch article quoted at the beginning of this chapter did not specify the size of n, but we may assume that what te Reile and his team were doing was to factor n and this effort required over six weeks. The encryptors have two ways to counter increasing speed in factoring large numbers. They can increase the number of digits in n and they can change the key frequently, say daily.

Simple computer software, such as the Arithmeticulator, can generate very large primes and calculate their product. For example, the numbers

$$p = 3,672,073,456,729,609,278,080,930,548,357,876,121,308,303$$

and

$$q = 11,846,818,513,593,217,487,851,681,091,819,893$$

are prime numbers given in the manual of the Arithmeticulator. The number $n = p \cdot q$ will have 80 digits. For this choice of n the block size could be any number less than 35 digits, since any number of less than 35 digits will be relatively prime to n. The Arithmeticulator will exhibit integers of up to 140 digits. It will carry out the RSA encryption for such numbers in a fraction of a second. Decryption without the key requires factorization of n, a process that we know will take weeks, if not years, with the very latest computers and software.

§9.3 Exercises

(1) Find the prime factorization of 2773 and compute Euler's totient function $\varphi(2773)$. Ans. $2773 = 47 \cdot 59$, $\varphi = 46 \cdot 58 = 2668 = 2^2 \cdot 23 \cdot 29$.

(2) For the RSA encryption key $n = 2773$, $b = 2$, c, can $c = 3$? Can $c = 23$? Explain.

(3) For the RSA key $n = 2773$, $b = 2$, $c = 5$, find the decoding power d.

(4) Use the RSA key $n = 2773$, $b = 2$, $c = 3$ to encode the message FATE.

(5) Convert the message FATE into a sequence of numbers and then block these with $b = 3$. Are the resulting numbers invertible in $\mathbf{Z}_{2773}$? Explain why the RSA key $n = 2773$, $b = 3$, $c = 5$ could be used to encode the message FATE.

(6) For the RSA key $n = 1022117$, $b = 3$, $c = 5$, find the decoding power d.
Ans. 816077.

(7) Verify that if c is the encoding power in $\mathbf{Z}_n$ and d is the decoding power, then one can reverse their roles and use d as the encoding power and c as the decoding power.

9.4. Raising Numbers to Large Powers in $\mathbf{Z}_p$

Any positive integer power can be calculated by only squaring numbers or multiplying by the original number. Computer scientists call the procedure *exponentiation by repeated squaring and multiplication*. Consider the example of 26^{19}.

$$\begin{aligned}
26^{19} &= 26^{9 \cdot 2 + 1} = (26^9)^2 \cdot 26 \\
&= (26^{4 \cdot 2 + 1})^2 \cdot 26 = ((26^4)^2 \cdot 26)^2 \cdot 26 \\
&= ((26^{2 \cdot 2})^2 \cdot 26)^2 \cdot 26 = (((26^2)^2)^2 \cdot 26)^2 \cdot 26
\end{aligned}$$

To calculate 26^{19} in $\mathbf{Z}_{29}$ one follows this formula from the inside out. In detail, it goes like this. By "mod 29" we mean the remainder after division by 29, this remainder being the value of the calculation in $\mathbf{Z}_{29}$.

$$\begin{aligned}
26^2 &= 676 \equiv 9 \bmod 29 \\
(26^2)^2 &= 9^2 = 81 \equiv 23 \bmod 29 \\
((26^2)^2)^2 &= 23^2 = 529 \equiv 7 \bmod 29 \\
((26^2)^2)^2 \cdot 26 &= 7 \cdot 26 = 182 \equiv 8 \bmod 29 \\
(((26^2)^2)^2 \cdot 26)^2 &= 8^2 = 64 \equiv 6 \bmod 29 \\
(((26^2)^2)^2 \cdot 26)^2 \cdot 26 &= 6 \cdot 26 = 156 \equiv 11 \bmod 29
\end{aligned}$$

Therefore, $26^{19} = 11$ in $\mathbf{Z}_{29}$. The calculation was made with six multiplications rather than with the 18 multiplications of the definition of raising to the nineteenth power.

The scheme followed here is using the base 2 numeral for the exponent 19. To see this, recall the following method for finding this base 2 numeral.

$$\begin{aligned}
19 &= 9 \cdot 2 + 1, \text{ 1 goes in the one's place} \\
9 &= 4 \cdot 2 + 1, \text{ 1 goes in the two's place} \\
4 &= 2 \cdot 2 + 0, \text{ 0 goes in the four's place} \\
2 &= 1 \cdot 2 + 0, \text{ 0 goes in the eight's place} \\
1 &= 0 \cdot 2 + 1, \text{ 1 goes in the sixteen's place}
\end{aligned}$$

The base 2 numeral for 19 is given, left to right, by reading the remainders from bottom to top: $19 = 10011_2$, where the subscript 2 indicates a base 2

9.4. Raising Numbers to Large Powers in $\mathbf{Z}_p$

numeral. The zeros and ones in this numeral, read left to right, tell us how to raise a number to the nineteenth power. For example, to calculate 26^{19}, the first 1 says start with the number, 26 in this case. After the first 1 on the left, each 0 means square what you have and each 1 means square and then multiply by 26 (or whatever number we started with). Following this rule, we carry out the following steps for each digit of 10011_2.

$$
\begin{array}{rcl}
1 & : & 26 \\
0 & : & 26^2 \\
0 & : & (26^2)^2 \\
1 & : & ((26^2)^2)^2 \cdot 26 \\
1 & : & (((26^2)^2)^2 \cdot 26)^2 \cdot 26 = (26^8 \cdot 26)^2 \cdot 26 = 26^{(8+1) \cdot 2 + 1} = 26^{19}
\end{array}
$$

Compare this with the preceding scheme to see that both are the same.

9.4.1. Decoding the Key $n = 1147$, $c = 7$, $b = 2$. Let us carry out the decoding of the message $M = [128\ 583\ 1100\ 675\ 1\ 763\ 326]$ given in equation (9.9). In the preceding section we used the fact that $\varphi(1147) = 1080$ in order to determine that the decoding power for this key is $d = 655$, whose base two numeral is 1110011111_2. Calculating in $\mathbf{Z}_{1147}$, we find 128^{655} by the above described method, which has reduced the number of multiplications from 644 to 14.

$$
\begin{array}{rcl}
1 & : & 128 \\
1 & : & 128^2 \cdot 128 = 326 \cdot 128 = 436 \\
1 & : & 436^2 \cdot 128 = 841 \cdot 128 = 977 \\
0 & : & 977^2 = 225 \\
0 & : & 225^2 = 157 \\
1 & : & 157^2 \cdot 128 = 562 \cdot 128 = 822 \\
1 & : & 822^2 \cdot 128 = 101 \cdot 128 = 311 \\
1 & : & 311^2 \cdot 128 = 373 \cdot 128 = 717 \\
1 & : & 717^2 \cdot 128 = 233 \cdot 128 = 2
\end{array}
$$

Patiently doing the same calculations for the remaining numbers in M, we find that the decoded message is

$$m = [2\ 25\ 27\ 12\ 1\ 14\ 4]$$

which translated to letters becomes the English phrase BY LAND.

Oral exercise 9.15. Decode the message

$$M = [27\ 01\ 17\ 17\ 21\ 19\ 11\ 22\ 09]$$

which has been encoded with the RSA key $n = 29$, $b = 2$, $c = 3$. We have already found that the decoding power is $d = 19$, whose base two numeral is

10011_2. Use the above method of exponentiation by repeated squaring and multiplication. For example, find 27^{19} by the following steps:

$$\begin{array}{rcl} 1 & : & 27 \\ 0 & : & 27^2 = a, \text{ remainder of } 27^2 \div 29 \\ 0 & : & a^2 = b, \text{ remainder of } a^2 \div 29 \\ 1 & : & b^2 \cdot 27 = c, \text{ remainder of } (b^2 \cdot 27) \div 29 \\ 1 & : & c^2 \cdot 27 = d, \text{ remainder of } (c^2 \cdot 27) \div 29 \end{array}$$

This method reduced the number of multiplications from 18 to 6 in $\mathbf{Z}_{29}$. Use the Explorer Plus calculator to continue in the same way to raise the remaining 8 numbers in M to the nineteenth power. Then convert the resulting sequence $m = [d, \ldots]$ to letters in order to read the message.

Ans. CALL HOME

§9.4 Exercises

(1) With paper and pencil only, carry out the method of exponentiation by repeated squaring and multiplication to compute 9^{16} in $\mathbf{Z}_{29}$.

(2) Use a calculator (such as the Explorer Plus or TI 34 II) to carry out the method of exponentiation by repeated squaring and multiplication to compute 15^{27} in $\mathbf{Z}_{29}$.

(3) Decode the message $M = [1003]$, which was encoded with the RSA key $n = 10094$, $c = 11$, $b = 4$.

9.5. Signatures

If an important encoded message arrives from my agent, how can I be sure that she, rather than an enemy agent, sent it? This is the issue of putting a signature on an encrypted message. The agent can put her signature on a message by posting her own public key $\{N, C, b\}$ for an RSA encryption, to which only she has the knowledge needed to find the decoding power D. Suppose my public key is $\{n, c, b\}$ with decoding power d. Knowledge of c and C is public, but only she knows D and only I know d.

To send me an encoded message with her signature, in the case when $n < N$, she first encodes it with c of my key in $\mathbf{Z}_n$, and then encodes the result with D of her key in $\mathbf{Z}_N$. (If $N < n$, then she reverses the order of these encryptions). I decode her signature with C in $\mathbf{Z}_N$ and then decode that with my d in $\mathbf{Z}_n$. (Reverse order if $N < n$). If the result is a readable message, then I know that applying her C to the received message must have decoded something coded with her D, something known only to her. Only she could have sent such a message.

9.5. Signatures

The signing process works because the role of c and d can always be reversed. To see this, recall that c and d are characterized by the property that for any invertible number m in $\mathbf{Z}_n$ we have $m^{c \cdot d} = m$ in $\mathbf{Z}_n$, which happens if $c \cdot d = 1$ in $\mathbf{Z}_\varphi$. The order of c and d can be reversed because multiplication is commutative in $\mathbf{Z}_\varphi$ and in the integers.

This process can fail if some m^c in $\mathbf{Z}_n$ is a number not relatively prime to N. This problem has several solutions. One possibility is for my agents and me each to have a public key encryption and a public key signature code, with the signature n always smaller than the prime factors used in the encryption n.

Example 9.16 (A signed encryption). I have the RSA public key $n = 1147$, $c = 7$, $b = 2$ and my agent has the RSA public key $N = 2773$, $C = 17$, $b = 2$. She sends me the message

$$MS = [1584 \ 2072 \ 2417 \ 456 \ 1143 \ 77 \ 749 \ 2207 \ 2258]$$

which she has encrypted with my key, using $c = 7$ in $\mathbf{Z}_{1147}$, then signed with her key, using D (which I don't know) in $\mathbf{Z}_{2773}$. She encodes and signs in that order because $n < N$.

To decode this message, I first use her coding power $C = 17 = 10001_2$ in $\mathbf{Z}_{2773}$ to get

$$M = [583 \ 1108 \ 321 \ 1100 \ 875 \ 762 \ 1126 \ 763 \ 788]$$

a sequence of numbers in $\mathbf{Z}_{2773}$ which my agent originally got from $\mathbf{Z}_{1147}$. Then I use my decoding power, which I know is $d = 463 = 111001111_2$, because I know how to factor $n = 1147$, in order to decode M to get

$$m = [25 \ 15 \ 21 \ 27 \ 19 \ 20 \ 09 \ 14 \ 11]$$

Converted back to letters, m is the message YOU STINK, no doubt showing that she secretly likes me.

9.5.1. Some Luck Involved. In the preceding example, my agent began with the message sequence m. Regarding the numbers in m as numbers in $\mathbf{Z}_{1147}$, she raised each one to the $c = 7$ power and obtained the sequence of numbers M. She then regarded the numbers in M as numbers in $\mathbf{Z}_{2773}$ when she raised each to the power D in order to sign the message. The whole process will work only if each number in M has a reciprocal in $\mathbf{Z}_{2773}$, that is, each number in M must be relatively prime to 2773. Since the numbers in M are not less than the prime factors of 2773 (which you have probably already found to be 47 and 59), we cannot be certain that they are relatively prime to 2773. Namely, it could happen that 47 or 59 divides one of the numbers in M. How lucky was my agent that this didn't happen?

The question amounts to asking, what fraction of all the numbers in $\mathbf{Z}_{2773}$ have no reciprocal? This is easy for us to answer, once we know the prime factorization $2773 = 47 \cdot 59$, because Euler's totient function $\varphi(2773)$ is the number of numbers less than 2773 which are relatively prime to 2773. Therefore, the number of numbers less than 2773 which are not relatively prime to 2773 is $2773 - \varphi(2773) = 47 \cdot 59 - 46 \cdot 58$. As a fraction of the 2773 numbers in $\mathbf{Z}_{2773}$, this is
$$\frac{47 \cdot 59 - 46 \cdot 58}{47 \cdot 59} = 1 - \frac{46}{47} \cdot \frac{58}{59} = \frac{105}{2773}$$
which is approximately .0379. We conclude that most of the numbers in $\mathbf{Z}_{2773}$ have a reciprocal. Only 105 of the 2773 numbers have no reciprocal. The larger the two prime factors of N become, the smaller this fraction becomes.

§9.5 Exercises

(1) Your public key is $n = 1147$, $c = 7$, with block size 2 and your agent's public key is $N = 2773$, $C = 17$ with block size 2. Your agent sends you the following signed, encrypted message: $MS =$ [2751 2752 456 1715 1584 456 2766 1 588 2752 2207 77 749 2207 2752]. What is the message?

Bibliography

[AKS02] Manindra Agrawal, Neeraj Kayal, and Nitin Saxena. PRIMES is in P. Department of Computer Science & Engineering, Indian Institute of Technology, Kanpur, India, August 2002.

[Ask99] Richard Askey. Knowing and teaching elementary mathematics. *American Educator*, 23:6–13, 1999. The Professional Journal of the American Federation of Teachers.

[Bak73] Augusta Baker, editor. *Nursery Rhymes and Songs*. Parents' Magazine Enterprises, New York, New York, 1973.

[Bec02] Sybilla Beckmann. *Mathematics for Elementary Teachers*. Addison-Wesley, 2002.

[Dav48] Charles Davies. *Arithmetic*. A. S. Barnes, New York, New York, 1848.

[Euc56] Euclid. *The Thirteen Books of Euclid's Elements*. Dover, New York, NY, second edition, 1956. Introduction and commentary by Thomas L. Heath, Three volumes.

[GS93] I. M. Gelfand and A. Shen. *Algebra*. Birkhäuser, Boston, 1993.

[How99] Roger Howe. Knowing and teaching elementary mathematics. *Notices of the American Mathematical Society*, 46:881–887, 1999.

[KR70] John L. Kelley and Donald Richert. *Elementary Mathematics for Teachers*. Holden-Day, San Francisco, CA, 1970.

[Ma99] Liping Ma. *Knowing and Teaching Elementary Mathematics*. Lawrence Erlbaum, Mahway, NJ, 1999.

[otMS01] Conference Board of the Mathematical Sciences. *The Mathematical Education of Teachers*, volume 11 of *Issues in mathematics education*. American Mathematical Society, 2001.

[PB00] Thomas H. Parker and Scott Baldridge. Elementary mathematics for teachers. Book manuscript, May 2000.

[Rau77] Albert N. Raub. *The Complete Arithmetic*. Werner Company, Chicago, 1877.

[RSA77] Ronald L. Rivest, Adi Shamir, and Leonard Adleman. A method for obtaining digital signatures and public key cryptosystems. Technical Memo 82, MIT Laboratory for Computer Science, 1977.

[Tea99]   Primary Mathematics Project Team. *Primary Mathematics*, volume 1A–5B. Federal Publications, Singapore, third edition, 1999. Project Director Kho Tek Hong, Curriculum Planning and Development Division, Ministry of Education, Singapore.

[Wu]   H. Wu. What is so difficult about the preparation of mathematics teachers? to appear. Available at www.math.berkeley.edu/~wu/.

[Wu99]   H. Wu. Basic skills versus conceptual understanding, a bogus dichotomy. *American Educator*, 23:14–19, 50–52, 1999. The Professional Journal of the American Federation of Teachers.

Index

absolute value, 316
  distance, 330
addition
  algorithm, 114
    other bases, 116
    scratch, 115
  algorithm with straws, 115
  decimal, 266
  fraction, 206
  integers, 318
    calculation, 319
  number line, 42
  properties, 40
    integers, 321
    rational numbers, 207
  rational numbers, 205
  whole numbers, 39
Adleman, 362
Agrawal, 159
AKS algorithm, 159
algorithm
  AKS, 159
  Euclid's, 169
angle, 32
Archimedes, 97, 294
area, 27
  circle, 237, 253
  rectangle, 60, 94
  reference square, 27
  triangle, 94
array
  rectangular, 60
associated rectangle, 94
average, 146, 225
  integers, 342
  speed, 146

Avogadro's number, 313

base five
  counting to five, 19
base two
  counting to two, 19
block size, 363
borrowing, 119, 123

cancellation, 185, 222
capacity, 30
  English units, 30
carrying, 113
Cartesian product, 61
casting out
  3's and 9's, 163
change making, 48
circle, 34
  area, 237
circumference, 34
  measure, 97
clock
  n-hour, 348
  addition, 348
    properties, 348
  division, 354
    existence, 354
    reciprocal, 357
  multiple, 352
  multiplication, 350
    properties, 351
  reciprocal, 356
  subtraction, 349
    properties, 349
coding power, 363
  allowable, 365

common factor, 168
common multiple, 173
complement, 9
composite, 155
compound fraction, 236
compound interest, 304
contrapositive, 158
counting
  outcomes, 71, 132
  subsets, 131
cross multiply, 198

decimal, 257
  addition, 266
  fraction, 261, 262
    arithmetic, 265
  long division, 270
  multiplication, 268
  nonrepeating, 284
  $\sqrt{2}$, 287
  order, 259
  repeating, 273
  round off, 259
    error, 260
  subtraction, 267
decoding power, 364
decomposing, 119, 123
denominator, 91
  common, 198
diameter, 34
difference, 48
dimidial fraction, 264
disection method, 287
disjoint, 9
distance between integers, 330
dividend, 76
divides, 154
divisibility, 160
  tests, 161
    base five, 163
divisibility rules
  by 11, 343
divisible, 77
  integers, 342
division
  algorithm
    estimating quotients, 140, 141
    measurement, 145
    other bases, 143
    standard, 138
  decimal, 270
  fractions, 230
  integers
    calculate, 342
  long, 85
  measurement
    integers, 341

problems, 78
rational numbers, 229
whole numbers, 76
partitive
  integers, 341
  problems, 78
  rational numbers, 229
  whole numbers, 76
properties, 80, 232
divisor, 76

empty set, 2
∅, 2
encryption
  signed, 375
equilateral triangle, 33
equivalent sets, 5
Eratosthenes
  sieve, 155
Euclid's Algorithm, 169
Euler, 369
Euler's Theorem, 370
even number, 156
expansion
  $(a+b) \cdot (c+d)$, 67
  $(a+b)^2$, 67
  $(a-b) \cdot (a+b)$, 130
exponent
  laws, 311
  negative, 311
exponents, 309
  large, 372
  laws, 309

factor, 154
  common, 168
    greatest, 168
  finding all, 184
factorial, 72
factorization, 154
Fermat, 364
  theorem, 365
  web site, 364
finite, 6
  definition, 21
fraction
  compound, 236
  dividing the whole, 90, 190
  equal value, 199
    lowest terms, 200
  improper, 192
  long division, 193
  lowest terms, 197
  mixed, 192
  of a fraction, 93
  of line segment, 96
  order, 202, 203

Index

proper, 192
reduce, 197
same value, 197
value, 190
    lowest terms, 201
Fundamental Theorem of Arithmetic, 178

gcf, 168
    by prime factors, 180
    difference of multiples, 171
    several numbers, 171
greatest common factor, 168
Gregorian calendar, 36

half square, 94
history of math url, 97
hour
    definition, 35
hypotenuse, 33

improper fraction, 192
infinite, 6
interest
    compound, 304
$\cap$, 8
invert and multiply, 230
irrational, 287
    $\sqrt{2}$, 286
isosceles triangle, 33

key
    (n, c, b), 368
key of RSA code, 364

Lambert, Johann, 295
laws of exponents, 309, 311
lcm, 173
    from gcf, 175
    prime factors, 181
    simplified method, 182
least common multiple, 173
length, 25
lengths
    English units, 26
    metric, 26
less than, 6
    fraction, 202, 203
    integers, 317
light-year, 312
line
    definition, 25
line segment
    definition, 25
line segment diagram, 97
long division, 85
    decimal, 263, 270
    fraction, 193, 206

longitude, 35, 149
lowest terms, 197

mean, 225
measurement, 24
measurement division
    whole numbers, 76
meridian, 35
minuend, 48
minus
    subtraction, 48
mixed fraction, 192
money, 31
    decimal, 271
month
    nursery rhyme, 36
multiple, 154
    clock, 352
    common, 173
    least, 173
multiplication
    algorithm, 128
        other bases, 129
    area, 60
    by -1, 334
    by 10, 126
    by 9, 127
    Cartesian product, 62
    decimal, 268
        by 10, 269
    fraction, 219, 221
    integers, 332
        sign, 334
    mixed fractions, 225
    properties, 62, 223
        integers, 334
    rational number, 222

$n(A)$, 6
natural numbers, 14
negative number, 315
Newton's method, 289
noon, 35
number
    composite, 155
    even, 156
    negative, 315
    odd, 156
    prime, 155
    rational, 190
        negative, 345
number line
    definition, 14
    rational number, 190
    $\sqrt{a}$, 290
number of elements of, 6
numeration

any base, 105
counting to ten, 16
existence, uniqueness, 109
formal definition, 102
long division, 87
role of
  exponents, 310
role of multiplication, 66
sums, 43
numerator, 91

odd number, 156
oil
  consumed, 314
  reserves, 314
one-to-one correspondence, 4
opposite, 316
order
  decimals, 259
  fraction, 202, 203, 235
  integers, 317
  numeration, 108
outcome, 71

partition, 227
partitive division
  whole numbers, 76
percentage, 300
  less, 303
  method
    partitive division, 307
    unitary, 307
  more, 303
  part unknown, 305
  unknown, 301
  whole unknown, 307
perimeter, 27
permutation, 72
  number of, 72
$\varphi(n)$, 369
π, 35
  Archimedes's estimate, 97
  definition, 252
  irrational, 295
  places computed, 295
  web site, 97
Post-Dispatch, 361
predecessor, 13
price, 247
prime, 155
  dividing product, 178
  infinite number, 164
  relatively, 172, 355
prime factorization, 157
  existence, 157
  uniqueness, 179
prime factors, 179

to find gcf, 180
to find lcm, 181
proper fraction, 192
properties
  addition, 40
  set operations, 10
  subtraction, 51
proportion, 242, 245
protractor, 33
Pythagorean Theorem, 68, 291, 297

quintal fraction, 263
quotient, 76

radius, 34
rate, 70
  catch up, 84
  combining, 84
  of approach, 70
  of separation, 70
ratio, 242
  changing, 248
rational number, 190
  repeating decimal, 275
ray, 32
reciprocal, 231
reduce fraction, 197
relatively prime, 172
remainder property
  clock addition, 348
repeating decimal
  rational number, 277
rest mass, 312
right angle, 33
right triangle, 33
Rivest, 362
round off, 259, 280
  error, 260, 281
RSA
  Rivest, Shamir, Adleman, 362
ruler and compass, 291

same number of elements, 5
scientific notation, 312
scratch addition, 115
set, 1
set operations properties, 10
Shamir, 362
Sieve of Eratosthenes, 155
significant digits, 312
similar triangles, 251
simplified method
  find lcm, 182
  find lcm and gcf, 183
solution set, 330
  inequality, 331
square

Index

half, 94
square root, 287
$\sqrt{2}$, 286
straight angle, 32
subset, 3
   proper, 3
subtraction
   algorithm, 121
      other bases, 123
   comparative, 47
      integers, 326
      rational numbers, 212
   decimal, 267
   integers
      additive inverse, 328
      calculation, 328
   mixed fractions, 216
   number line, 214
   properties, 51
      integers, 329
      rational numbers, 214
   take-away, 47
      integers, 326
      rational numbers, 212
subtrahend, 48
successor, 13
sum, 39
supplement, 32

temperature
   Fahrenheit and Centigrade, 343
text to numbers, 362
time, 35
   longitude, 149
totient function, 369
   $\varphi(p)$
      p prime, 370
triangle
   area, 94
   associated rectangle, 94
   equilateral, 33
   inequality, 46
   isosceles, 33
   right, 33
   ruler and compass, 47
   similar, 251
   sum of angles, 44
Turnbull web site, 97

union, 8
∪, 8
unit cube, 28
unit line segment, 14
unitary method, 80

value of a fraction, 190
vertex, 32

vertical pair, 32
volume, 28
   displacement, 31
   rectangular solid, 61

weight, 29
   metric units, 30
whole angle, 33
whole numbers, 13
Wiles, Andrew, 364
word problem
   advice, 147

Z_n, 348